The Tramways of Portugal

Edited by
Brian King and Carl Isgar

LRTA
Since 1937

Published by the Light Rail Transit Association
8 Berwick Place
Welwyn Garden City
AL7 4TU

www.lrta.org

Copyright © Light Rail Transit Association 2022

Designed by:
Steve Herbert-Mattick
sherbertdesign@gmail.com

Printed and bound in the UK by:
Page Bros Group Ltd
Mile Cross Lane
Norwich
NR6 6SA
www.pagebros.co.uk

ISBN 978-0-948106-55-2

Front cover:
Car 3 in yellow livery seen in Sintra
Vila in the 1950s nicely captures the
charm and character of the period.
B. Lennox collection

Rear cover:
Top: C008 is seen arriving at the
Ferry terminus, Cacilhas on 13 May
2009 with the rather fine sailing
vessel Fragata Dom Fernando II
Glória dating from 1845 prominent
in the background. *Michael Russell*

Bottom: On 21 June 2008 Metro do
Porto car 2 is seen at Senhora da
Hora on service C to ISMAI. *L Folkard*

Contents

Introduction

Welcome to the fifth edition of the Tramways of Portugal. I was not alone in thinking that there would be little more to tell when the fourth edition was published in 1995. I am very happy to have been proved wrong. Portuguese tramways have long held a fascination for most tramway enthusiasts, particularly British and American, not least because of the influence that both countries have exercised over operations and vehicles. Moreover, as much of the rolling stock and equipment survived for so long into the second half of the twentieth century, it was possible to see fascinating examples of the early products of pioneering British and American tramcar manufacturers in regular service long after they had disappeared elsewhere. After extensive closures in Lisbon the situation stabilised, with a rejuvenated fleet, and recent developments are positive. Across the river in Almada a new system has opened. Porto survived to celebrate a century of electric traction despite gloomy predictions to the contrary, and the long-promised heritage tramway blossomed, affording the opportunity to ride on classic Brill-pattern semi-convertible tramcars in everyday service. The new Metro has won worldwide acclaim. Sintra has reopened the challenging upper section and now terminates within a short distance of the town centre.

When the first edition was published in 1964 Portugal was a 'closed' dictatorship, isolated from the rest of Europe beyond Spain, which was also a 'closed' dictatorship. Travel by air was only for the wealthy, the alternative being a lengthy journey by land or sea, and consequently little was known about the country itself, let alone the tramways. Portugal was almost invisible, being overshadowed by Spain. This isolation, together with financial constraints, meant that the tramways, like much about the country, survived in a time warp. Low key tourism began in the late 1960s; the country was opened up after the bloodless 1974 Revolution, joined the EU in 1986, and is now an integral part of Europe. This has resulted in a significant increase in our historical knowledge with archive material now made available in Lisboa, Porto and Sintra, thanks to our Portuguese friends. Progress in photographic and reproductive techniques has facilitated a vast improvement in the quality and quantity of the illustrations, many of them historic and newly discovered. This new fifth edition of the Tramways of Portugal encapsulates the systems and operations covered in the previous publications but has been expanded to include the new installations in Lisboa and Porto. It also provides wider coverage of the trolleybuses, thus providing a comprehensive overview of electric street traction in Portugal. Whereas previous editions have been of a relatively small format, it was decided from the outset that with the volume of material available, this new publication would not be a handbook, but a full-blown guide in order to do justice to the subject and provide a comprehensive overview of what continues to be a fascinating country for the transport enthusiast. All of this has resulted in a considerable increase in pages from 34 in the first edition and 92 in the fourth edition to some 328 in this edition.

Mule trams were introduced in Portugal commencing with Porto in 1872, then Lisboa in 1873, Coimbra in 1874 and Braga in 1877. Electric trams followed commencing with Porto in 1895, the first in the Iberian Peninsula, Lisboa in 1901, Sintra in 1904, Coimbra in 1911 and Braga in 1914. Braga closed in 1963; it was replaced by a trolleybus system, but this closed in September 1979. Coimbra closed in January 1980 but the city still operates trolleybuses, it has a two-line system, which commenced operations in 1947, and is now the only remaining trolleybus system in the country, following the closure of Porto in 1997.

There is an old Portuguese adage which states that "Porto works and Braga prays, Coimbra studies and Lisboa plays". This is still a relevant reflection on the nature of the tramway systems. Lisboa has become a compact and characterful mainstream European capital. Porto has transformed itself in recent years, and besides being a busy commercial and industrial centre has become a significant tourist destination. Sintra retains its many charms while Braga and Coimbra remain relatively undiscovered.

But even now you can still step back into history by boarding a tram. Lisboa's narrow twisting switchback streets have made route 28 an international tourist "must do". Porto's classic semi-convertible trams serve the imposing central area and the riverside, which is very popular at weekends, resulting in very busy trams. Sintra continues to exude a unique rural charm as it wends its way through delightful surroundings between town and ocean. Add to this a most pleasant semi-tropical climate, friendly people, a relaxed approach to life sustained by good and inexpensive food and wine, and we still have much to be thankful for.

Brian King, Poole, Dorset, May 2022

Acknowledgements

Some of the information in this this book is derived from research carried out over many years by the late John Price and the co-editor of this work, Brian King. However it has been expanded significantly by more recent research carried out by a team of transport enthusiasts from the UK, Portugal and The Netherlands who have between them worked tirelessly for some five years and have compiled most of the chapters. The contributions have been edited by Carl Isgar and Brian King who have attempted to produce a consistent style.

We are therefore extremely grateful to our team of contributors, without their efforts this new work would not have been possible. In alphabetical order, they are Owen Brison, Pedro Costa, João Firmino, Ernst Kers, Bob Lennox, Pedro Mendes, Pedro Milheiro, the late Andy Steel, Garth Tilt and Salamão Vieira. Unfortunately Andy Steel passed away following a heart attack in November 2019. We are also grateful to Roger Smith, the LRTA's cartographer, for preparing clear, accurate and detailed maps of all the systems without which this book would be the poorer.

The LRTA has been fortunate to benefit from the cooperation of a number of photographers who have made their collections available, these include Tony Belton, Les Folkard, Kath Lomas-King, Peter Haseldine, Alan Murray, the late John Meredith, David Pearson, Mike Russell, Hugh Taylor, Luis Vieira and our many friends from Portugal. In addition we have been provided with an extensive selection of images from two significant collections. Firstly the TMS library at Crich, through the generous cooperation of Laura Waters, Curator, Collections and Library, has provided scans from the collections of Pam Eaton, N N Forbes, J H Price, J J W Richards and others. Secondly through the good offices of Martin Jenkins, the Online Transport Archive (OTA) has made available an extensive selection of colour slides mostly taken by W C Janssen and W R Stillman, some taken as early as 1954. These have been scanned and are all tram views, taken in Lisboa, Porto, Sintra, Coimbra and Braga.

Ponte do
Lima ○ 11
Viana do ○ 11,13
Castelo

Braga ●

Mirandela ○ 18

Povoa de ○ 1
Varzim

Lixa ○ 3
Penafiel ○ 3

Vila
Real ○ 2
Regua ○ 2

Porto ●

Douro

Entre-os-Rios ○ 3

Douro

España

7 ○ Braçal
7 ○ Foz do Rio Mau

8C
São Jacinto ○
Aveiro ○
8A

Viseu ○ 15

Areão ○ 5
5 ○ Mira

Mondego

Covilhã ○ 16

Figueira ○ 4
da Foz

● Coimbra

Portugal

14 ○ Leiria

12
Nazaré ○
São Martinho ○ 14
do Porto

Torres Novas
Alcanena ○ ○ 10
10

Tajo

Legend

1 2 3 etc. for further details, see
"Other Tramways"

Elvas ○ 8B

Sintra ○
Oeiras ○
17,9

Lisboa

Almada

Oceano
Atlântico

0   50   100
kilometres

Oceano

Atlântico

España

© *R.A.Smith, January 2019. No. 2186, v1.2.*

Gallery 1
Semi Convertible

Trams in Portugal have a distinctive character, due in no small part to the semi-tropical climate. In the early years of tramways operators in sunnier climes resorted to having two fleets; open for summer and enclosed for winter. This was obviously expensive in terms of purchase, poor utilisation and depot accommodation, so some operators tried one set of trucks and two sets of bodies, but this was not an ideal arrangement either. Enterprising car builders offered convertible trams with removable sides, but conversion meant time out of service, while the sides had to manhandled on and off and still storage space was needed. Thus the semi-convertible came about, in which the conductor, or even passengers, could lower the windows into a side pocket, to give much of the benefit of a crossbench car. The pockets encroached on internal space, thus limiting seating capacity, and were widely regarded as a hygiene (and fire) risk because of discarded rubbish and cigarette ends, to say nothing

Porto Bogie car 249 was delivered by Brill in 1904 and was the first semi-convertible tram in Portugal. It is seen in 1964 in Matosinhos demonstrating windows opened to various heights. *Photo-print*

An interior view of one of the Porto semi-convertible two axle cars showing the distinctive tunnel shaped ceiling coving that accommodates the window sections when in the raised open position. *R. Lomas*

of spitting, which was then a common hazard.

In 1900, the J.G.Brill Company of Philadelphia, U.S.A. unveiled the "Brill Patent Grooveless Post Semi-Convertible" design. Instead of dropping down grooved window pillars into side pockets, the windows were now guided upward at the optimum angle by profiled cappings screwed to the pillars (enabling easy pillar or window replacement if needed), and stowed in pockets in the roof cavity, giving rise to a distinctive tunnel shaped ceiling coving. Now there was less encroachment on interior seating width, and no chance of unwanted rubbish causing trouble. When the car was moving, open bulkhead doors ensured adequate

cooling breezes, and when stationary warm air rose and exited the car through open clerestory windows. Provision of catches to provide intermediate window positions, individual sun blinds, armrests, footrests and rattan seat coverings ensured optimum passenger comfort.

The design was a great success, and was readily adopted in Portugal, where four out of the five electric systems purchased, built, or rebuilt cars using the system from 1904 to 1946. There were over two hundred such cars, more than a quarter of the national fleet at maximum extent, and it is still possible to ride on examples in daily service in Porto.

LISBOA

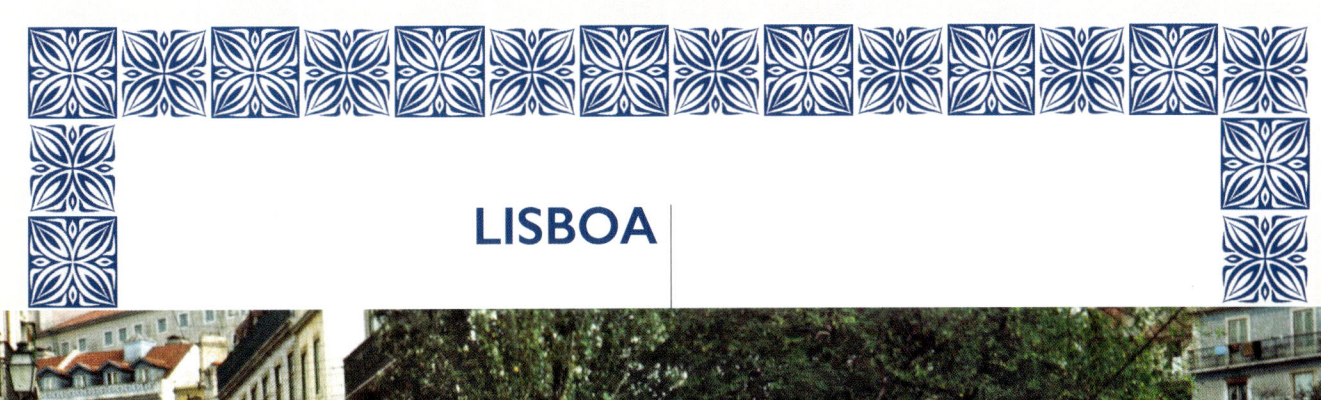

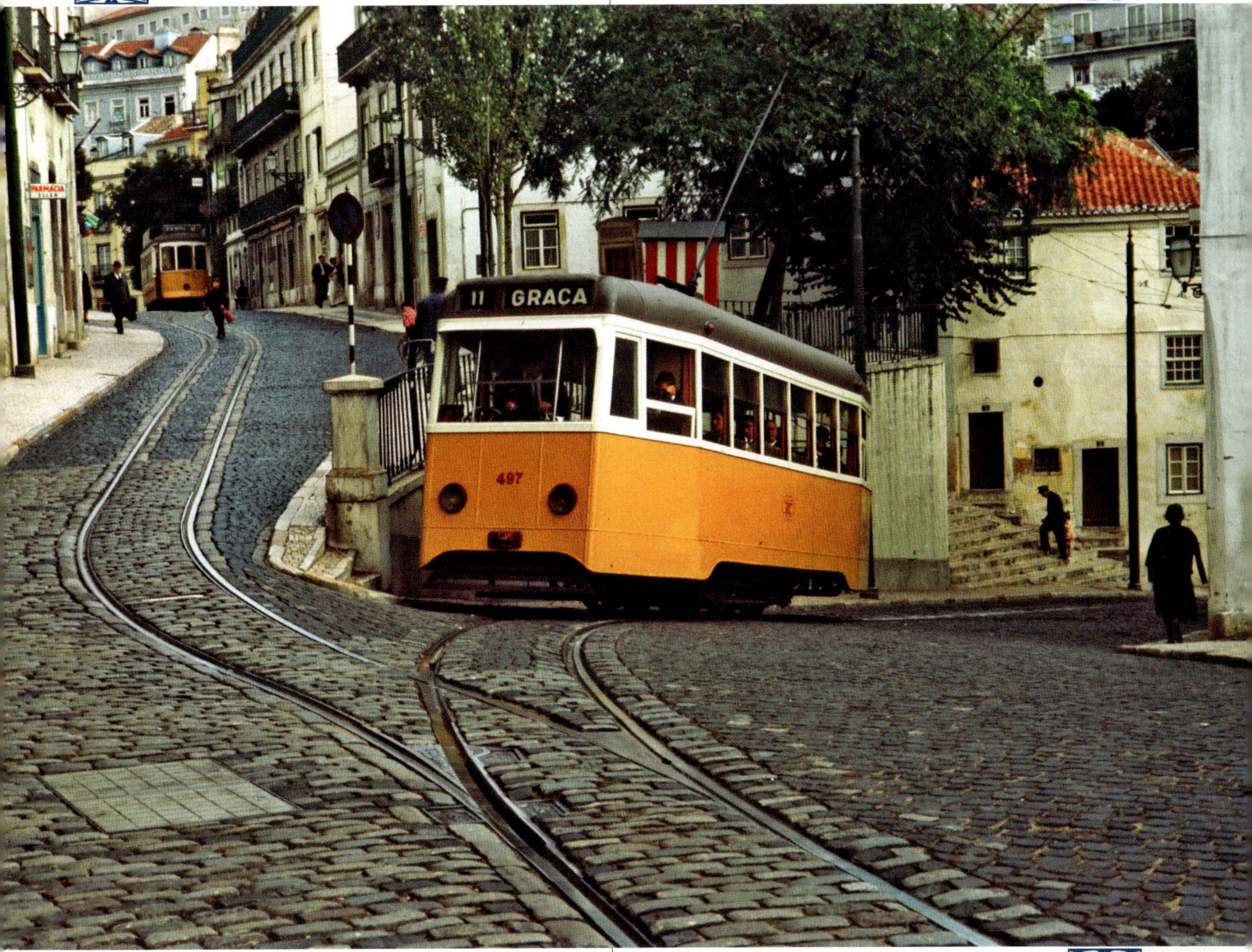

Tramways

By Ernst Kers, Owen Brison & Pedro Mendes

497 seen at Sao Tome in November 1965.

W R Stillman/Online Transport Archive

Above: Lisboa mule car 166 is seen at Praça do Comércio in this view taken prior to the electrification of this route in 1901. *Illustraçao Portugesa*

LISBOA, the Capital of Portugal, is superbly situated on the north bank of the Rio Tejo. This proximity to the river means that even in mid-summer Lisboa is refreshed by a gentle breeze, the more so as the City is supposedly built on seven hills (although the visitor may feel that the actual number seems more like seventy). Baedeker wrote that Lisboa "vies in beauty of position with Naples and that despite the absence of a mountain background, it possesses a beauty of its own with the wide expanse of the Tejo and the luxuriant vegetation of its public gardens". Lisboa remains a most attractive city today.

Public transport in Lisboa and the surrounding area is provided by a number of undertakings. The surprisingly compact urban area is served by the publicly owned Companhia Carris de Ferro de Lisboa (CCFL, popularly called Carris), operating trams and buses, and the underground railway, Metropolitano de Lisboa. Suburban railway services to Sintra, Cascais and Azambuja as well as on the South side of the Tejo between Barreiro and Setúbal are operated by Comboios de Portugal (CP), the state owned railway operator. The cross-river suburban railway however is operated by the private Fertagus company. A number of private operators provide rural bus services. There are also ferry services provided by Transtejo to dormitory towns south of the river, and a number of smaller coach and bus undertakings. The tramway system was always contained within the city boundary, except for the portion between Algés and Cruz Quebrada. In the Greater Lisboa area are two more tram systems, the metre gauge line of Sintra and the 1435 mm gauge Metro Transportes do Sul light rail system on the south bank of the Tejo, which are both described in separate chapters.

Mainly mules

Before the arrival of the tramways, road transport was provided by horse or mule drawn buses of many different sizes and types, both open and closed, operated by many different companies. The Companhia de Carruagens Omnibus was formed in 1835

and operated large omnibuses. In February 1865 most of the omnibuses and horses of this company were lost in a huge fire in their depot at Rua do Cruxifixo. Other companies took the opportunity to fill the gap. One of them was the Companhia de Carruagens Lisbonenses which had their depot at Largo de São Roque and operated Char-a-bancs (small open buses), to Dafundo, Ajuda, Calhariz, Benfica, Carnide, Carriche, Ameixoeira, Charneca, Portela and Olivais. What all the horse buses had in common was lack of comfort for their passengers. The steamboats which departed every half hour from Cais do Sodré to Belém were more comfortable.

The Câmara Municipal de Lisboa (CML, Municipality of Lisboa) had received the first request for a concession for a horse tramline on 14 May 1870, which was followed by a similar request on 17 June 1870 by the brothers Francisco and Luciano Cordeiro de Sousa and requests from ten more applicants until 3 October 1872. CML had already in October 1870 approved in principle the application of the Cordeiro brothers, but this had to be authorised by the court. Finally on 27 October 1872 the municipality granted a concession to the Cordeiro brothers to build and run tramways. This they transferred to the newly formed Companhia Carris de Ferro de Lisboa (Carris) on 14 February 1873, a company registered in Rio do Janiero. It became a Portuguese company on 31 May 1876.

The lines that Carris had to build and operate according to the concession were:

1. Santa Apolónia - Ruas do Jardim do Tabaco and Alfândega - Terreiro do Paço (Praça do Comercio) - Largo do Pelourinho (Praça do Município) - Rua do Arsenal - Praça dos Remolares (Cais do Sodré) - Aterro (Av.24 de Julho) - Rua das Janelas Verdes - Pampulha - Alcântara (provisional because the Aterro / landfill between Santos and Alcântara wasn't yet available). Branches had to be created from Terreiro do Paço to the Passeio Publica (Restauradores),

via Rua dos Fanqueiros to Praça da Figueira and to all the public quays of the proposed new docks. Those docks had not yet been built and those last branches were never built.

2. Largo do Corpo Santo - Rua de São Paulo - Rua da Boavista - Largo de Conde Barão - Rua de São Bento - Rato with branches from Rato to Principe Real and to Estrela via Santa Isabel.

3. Rossio - Rua Nova da Palma - Largo do Intendente - Igreja dos Anjos as far as the city boundary.

Other prospective routes were too steep for mule cars, and were eventually provided with cable tramways, funiculars and vertical lifts operated by separate companies, as will be explained later.

Track laying began on 14 April 1873, the tracks being 1435 mm (standard) gauge. The first 30 tramcars, 22 closed and eight open, were ordered from John Stephenson & Co, New York. Starbuck had also delivered a closed and an open car. CML gave permission to build a provisional shed for the depot and stables at the Santos garden. Test running started in September. There were concerns that the mules wouldn't be able to pull the tramcars up the Santos ramp to Rua das Janelas Verdes, but it appeared to be no problem when using two pairs of mules instead of just one pair. On 10 November CML gave permission to start the services. On Monday 17 November 1873 the first line was officially opened with a parade of all 32 available tramcars, between Santos and the Santa Apolónia railway station. After returning to Santos this was celebrated according to Portuguese tradition with a huge meal attended by 160 guests. The next day normal public services began.

On 14 January 1874 the line was extended from Santos via Pampulha to Alcântara and on 6 February to Belém. On 11 February services commenced between Praça do Comércio and Intendente with a branch to Restauradores. On 11 August 1874 the Belém line was extended to Algés. In 1874 the number of tramcars was increased to 54, 14 open and 28 closed cars from Stephenson and 1 open and 11 closed cars from Starbuck, and the number of horses and mules from about 135 to over 400. The Santos depot and stable complex had become too small and in 1875 a large complex was built at Santo Amaro.

On 1 January 1876 the line to Largo de Conde Barão was opened. The landfill between Santos and Alcântara made it possible to open a line via the Marginal on 30 September 1877. This line was on 6 October extended to Calvário where it connected with the existing line via Pampulha. From now on the trams between Santos and Belém alternately used the Pampulha or the new Aterro (landfill) route. On 30 November 1877 the Conde Barão line was extended to Santos.

Although the Santo Amaro premises were for the moment large enough to accommodate the tramcars and the livestock, there was a need for a second depot closer to the city centre to replace the original provisional depot at the Santos garden. In December 1877 Carris bought a site between the ramp of Santos and the river side close to the bottom of the ramp. Here a new depot with stables was realised. This Santos depot was never used by the electric trams.

In February 1878 a Winterthur steam locomotive was tested, followed a few months later by locomotives of Henschel and Merryweather. The Winterthur and Henschel locomotives were tested before in Porto. Although in Porto these tests resulted in the acquisition of Henschel locomotives, Carris decided for the moment not to continue with steam traction. The Henschel

locomotive returned to Porto where it remained in service for about 37 years. It is thought that both the Winterthur and Merryweather locomotives went elsewhere.

On 1 November 1879 the line Conde Barão - Rato - Principe Real was opened. This line was combined with the route Principe Real - Rua do Alecrim - Cais do Sodré on 1 April 1882. Carris now extended its network by using cars suitable for use both on rails and on the road. The first of these cars was acquired in 1882 from a factory in Würzburg, the reason why Carris called them Würzburg cars. That didn't prevent Carris from building more of the type themselves. By 1888 Carris had 34 Würzburg cars, 17 open and 17 closed. These Würzburg cars were used off-rails on the routes from Rato to Estrela from 1 November 1882, to Arco do Cego and Largo de Andaluz from 1 November 1883, the latter service extended to Jardim Zoologico on 1 July 1884, and finally to Poço do Bispo on 5 November 1885. On 9 November 1890 these services to Estrela and Jardim Zoologico were suspended. In 1882 Carris acquired a site at Arco do Cego intended as depot for new tramlines. As no rails had reached Arco do Cego until 1890, the site was probably first used by the Würzburg cars.

In September 1887 experiments started with a battery tram, delivered by the Belgian company Julien. A second battery tram designed by the Lisboa resident electrical engineer Maximiliano Herrmann was available for tests from November. These experiments, both with and without passengers, but which seem to have been modestly successful, lasted until September 1890. In 1889 Carris acquired two steam locomotives from Merryweather for the service between Cais do Sodré and Algés. They were meant for the heavy traffic on Sundays and holidays during the bathing season. However due to strikes in England they only arrived in October 1889. These steam tram services ended in 1892.

The success of the trams encouraged other applicants to request their own concessions for tramlines. At first CML appeared to be positive to these requests. But in 1888, Carris was granted a 99-year exclusive tramway concession for the city, at the end of which the track and rolling stock would pass to the municipality.

According this new concession Carris had to operate a network comprising the existing lines and a number of new lines.

The existing lines in 1888 were:

4. Rua da Bica do Sapato (Eastern end) - Santa Apolónia - Ruas do Jardim do Tabaco and Alfândega - Praça do Comercio - Rua do Arsenal - Corpo Santo - Cais do Sodré - Rua 24 de Julho - Rua Fradesso da Silveira - Calvário - Belém - Algés.

5. Corpo Santo - Rua de São Paulo - Conde Barão - Santos - Rua das Janelas Verdes - Pampulha - Alcântara - Calvário with connection to the first line. At Santos a line connected to the first line.

6. Line in Rua Vieira da Silva connecting the first two lines with the Alcântara railway station. At that time Alcântara was the terminus of the Sintra railway opened in 1887. The Rossio station and tunnel opened in 1891.

7. Praça do Município - Rua de São Julião (opposite direction via Praça do Comercio) - Rua Áurea - Rossio - Restauradores - Avenida da Liberdade until Rua das Pretas.

8. Rossio - Rua da Betesga - Praça da Figueira - Rua Nova da Palma - until the North end of the Largo do Intendente.

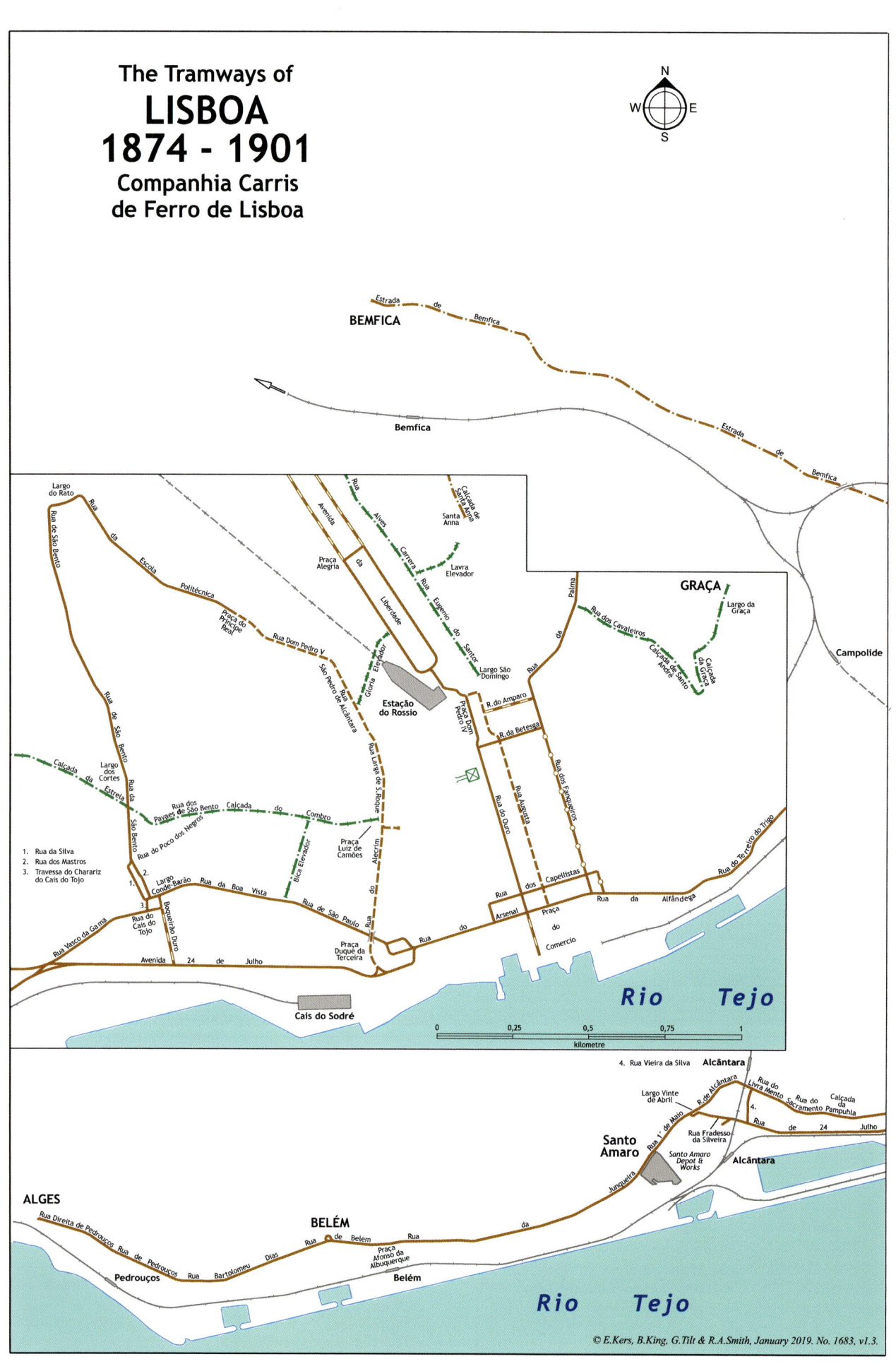

The Tramways of
LISBOA
1874 - 1901
Companhia Carris
de Ferro de Lisboa

BEMFICA

Bemfica

GRAÇA

Campolide

Largo do Rato

Santa Anna

Largo da Graça

Praça Alegria

Lavra Elevador

Largo São Domingo

Estação do Rossio

R. do Amparo

R. da Betesga

1. Rua da Silva
2. Rua dos Mastros
3. Travessa do Charariz do Cais do Tojo

Largo dos Cortes

Praça Luiz de Camões

Bica Elevador

Praça do Comercio

Praça Duqué da Terceira

Cais do Sodré

0   0,25   0,5   0,75   1
kilometre

Rio Tejo

4. Rua Vieira da Silva   Alcântara

Largo Vinte de Abril

Santo Amaro

Santo Amaro Depot & Works

Alcântara

ALGES

BELÉM

Praça Afonso da Albuquerque

Pedrouços

Belém

Rio Tejo

© E.Kers, B.King, G.Tilt & R.A.Smith, January 2019. No. 1683, v1.3.

LUMIAR

Campo
Grande

Braço
de
Prata

Areeiro

Entre
Campos

AREEIRO

POÇO DO
BISPO

Largo do
Dr. Afonso
Pena

Marvila

Rego

Chelas

Estrada de Benfica

SÃO
SEBASTIÃO

Arco de Cego
Depot

ARCO DO
CEGO

Avenida   Duque   de   Ávila

Rua Pascoal
de Melo

Largo de
Arroyos

Xabregas

Rua Alexandre Herculano

Largo
S. Barbara

Campo
Mártires
da Pátria

Largo do
Intendente

Rua de
Santa
Apolónia

Largo da
Estrela

Calçada da Estrela

Largo
Muzeu
Artilheria

Cáis dos Soldados

Rua San
Francisco
de Paula

Rua
Janelas Verdes

Rua   de   24   Julho

Rua   de   24   Julho

Rio   Tejo

Legend

	mule	cable
tramways constructed 1874 - 1880		
tramways constructed 1874 - 1880, removed by 1901		
tramways constructed 1881 - 1888		
tramways constructed 1889 - 1894		
tramways constructed 1889 - 1894, removed by 1901		
tramways constructed 1894 - 1901		
vertical passenger lift (Rua S.Justa)	⊠	
state railway (CP) and station		
Estoril Railway and station		

© E.Kers, B.King, G.Tilt & R.A.Smith, January 2019. No. 1684, v1.3.

9. Conde Barão - Rua de São Bento - Rato - Principe Real - Rua do Alecrim - Corpo Santo with a branch line to Rua Garrett. Carris considered the option to make a funicular in Rua do Alecrim, but soon electrification was recognised as a better alternative

New lines to be constructed were:
10. Rato - Rua do Sol ao Rato - Rua São Bernardo - Rua da Bela-Vista - Rua da Lapa
11. Largo de São Sebastião - Campolide
12. Largo do Intendente - Rua dos Anjos - Arco do Cego
13. Arco do Cego - Lumiar
14. Arco do Cego - São Sebastião - Benfica
15. Largo de Santa Barbara - Estrada de Sacavém - Areeiro
16. Largo Dona Estefania - Gomes Freire - Campo dos Martires da Patria - Travessa do Torel. The Companhia dos Ascensores Mechanicos de Lisboa was supposed to construct a tramline between the top of the Lavra funicular (opened April 1884) and the Travessa do Torel in connection with this Carris line.
17. Line in the future Avenida Almirate Reis replacing the line in Rua dos Anjos.
18. Extending the line from Rua Bica do Sapato to Poço do Bispo.
19. Line through Rua do Instituto Industrial.
20. Line on the future Avenida Marginal do Tejo.

Not all these new lines were built, or constructed according to the original plans. As the 1888 concession was for trams with animal traction, other companies could continue to operate cable trams and funiculars, or gain concessions for them, in the hilly areas of the city.

By 1889 the number of tramcars had risen to 154 for 1435 mm gauge, of which 56 were normal open, 64 normal closed and 17 each open and closed of the Würzburg type.

In 1876 Carris had asked the authorities to take action against competitors who mostly used char-a-bancs with similar wheel distance as the trams on the tracks of Carris. The municipality however said that the rails were part of the public road and everybody could use them. Matters however worsened when in 1882 the Ripert company was founded, using cars with wheels especially designed for running both on rails and road. This system was invented by Antoine Ripert from Marseille (France). Soon these types of cars were used by other competitors too. The court didn't consider themselves authorised to decide about the issue and so Carris had to fight its competitors with other means like lowering fares. A technical solution considered in 1889 was decreasing the gauge to 900 mm, which it was thought, in vain, would make it impossible to produce a practicable road vehicle capable of using the rails. Since 1890 all new tramlines were constructed with this narrow gauge: 14 August 1890 Intendente - Largo Santa Barbara - Arco do Cego - Lumiar, 1 November 1890 Torel - Arco do Cego - Jardim Zoologico, 12 February 1891 Jardim Zoologico - Benfica and on 17 December 1891 Largo Santa Barbara - Areeiro. At the end of 1891 a third rail was installed between the existing tracks from Intendente to Rossio to make through running of the Areeiro, Lumiar and Benfica trams possible. In 1890 Carris acquired its first 14 closed and eight open tramcars for 900 mm gauge. At the end of 1890 Carris had 1210 mules and horses.

To show how severe the competition was, Carris published in their annual report the number of trams and competing vehicles in Rua Junqueira on 4 January 1889. Carris provided 185 trips in both directions, Ripert and the other rivals 145 up to Belém and 143 in the other direction. It is worth remarking that while the Carris provided services until after 1 o'clock at night, the competitors ceased operating just after 9 o'clock in the evening.

Another method to fight the competitors was to buy them out. In March 1892 Carris succeeded in doing this with fourteen rival companies. Some of them had only one or two cars, but others up to 25. With these buy-outs Carris became owner of their assets, which included in total 96 cars and 484 mules and horses. During the rest of the year more rival companies were bought out, which brought another 60 cars into the possession of Carris. As some of them were withdrawn, the total fleet comprised 309 cars at the end of the year. On 19 April 1894 Carris succeeded in doing the same with the Ripert company, and with that gained another 68 cars and 429 mules and horses. As more cars had been withdrawn, the fleet at the end of 1894 totalled 316 cars, of which 36 were abandoned in 1895. In July and August 1896 four more competitors were bought out. However Eduardo Jorge with the Carros de Choras didn't give up and in 1896 a new rival company, the Luzitania entered the scene

Electrification and extension

In 1892 Alfredo da Silva, a shareholder and later director of the company, visited on behalf of Carris several European cities to study options for modernisation. Based on his reports Carris decided in 1894 to develop plans for electric traction. The track gauge of the whole network was to be reduced to 900 mm, as already used on the lines in the northern area of the city. On 27 February 1896 Carris sent a proposal to CML to electrify the network. Of course the electrification plans meant a huge change for the existing 1888 concession and also the technical aspects and the effects for the city had to be studied by commissions of the authorities. Protests by the rivals followed. The Nova Companhia dos Ascensores Mechanicos de Lisboa, operating funiculars and cable tramlines, realised that electric trams were, in contrast to mule trams or even steam trams, able to penetrate into "their" territory, the Lisboa hills. On 6 May 1897 they sent a letter to the King to protest against the plans of Carris. The direct Carris rival Luzitania, which at that moment operated 30 mule cars, protested loudly, but all in vain. The government gave its authorisation to the plans in May 1897 and the municipality agreed one month later. Carris could continue with the electrification. It was necessary to operate a mixed-gauge system for a time, and new cars were ordered from Brill in America which could serve initially as mule cars and then as trailers after electrification. At the end of 1898 the fleet totalled 329 cars. Eduardo Jorge moved the operations of his Carros da Chora to the rural environments of the city and later became one of the large bus operators in that area.

A new company, Lisbon Electric Tramways Ltd (LET), was incorporated in London in July 1899 to lease the system and provide the necessary capital, which was arranged through the merchant bank Wernher, Beit & Co, with whom LET shared offices at London Wall Buildings. Wernher, Beit & Co had connections with the South African gold and diamond interests. Another new company, the Portuguese Construction Co. Ltd. but also called the Lisbon Construction Co. or just the Construction, was incorporated in November 1899 to realise the electrification.

As everywhere the electrification meant building a complete

An early view of Santo Amaro depot with Brill crossbench four wheel cars being assembled. *Museu da Carris*

new network with new rails suited for the heavier electric trams and installing overhead wires, as well as building and equipping a power plant at Santos, acquiring almost 200 electric tramcars, replacing the old buildings of the Santo Amaro and Arco do Cego depots by new car-barns, realising workshops where the electric trams could be maintained and preparing all the staff. The Santo Amaro site was also extended to the south by buying from the Portuguese railways the adjacent land. Many major roads and streets were largely blocked for months because of the works, causing many protests. But works progressed quite steadily and on 31 August 1901 the first electric line was opened between Cais do Sodré and Algés. A year later the mule trams had disappeared, although some Würzburg cars were used for a few more years on off-rail routes. As Lisboa was a fast developing city, new lines were opened. Also the line from the Baixa via Rua Limoeiro and Escolas Gerais to Graça was opened in 1906. One line that Carris wanted was from Rossio via Rua do Carmo and Rua Garrett to Largo das Duas Igrejas (now Largo do Chiado). This would give a direct connection between the Baixa and the Bairro Alto. However the narrow, steep and very busy Rua do Carmo was considered by many not to be suited for trams. The

question raised many discussions and petitions pro and contra, but the line was never built.

This connection was eventually made when the former Estrela cable line was electrified and extended to Rossio.

New lines that were built were:
7 December 1904 Santos - Estrela
1 March 1905 Baixa - Limoeiro - São Tomé - Arco de Santo André
2 June 1905 Estrela - Domingos Sequeira - Ferreira Borges - Campo de Ourique - São João dos Bemcasos - Amoreiras - Rato
15 July 1906 reopening Torel - Palhavã
17 July 1906 São Tomé - Escolas Gerais - Graça
7 November 1906 Carmo - São Roque
In the meantime the cable trams and funiculars of the NCAML became obsolete. The takeover of this company by Carris was announced by A Gazeta dos Caminhos de Ferro on 1 January 1909. In fact it was taken over by Carris on 31 December 1912 but drawings prove that Carris was already preparing the electrification of the NCAML cable trams and funiculars in 1910. Because the NCAML concession was more favourable than

411 is seen in 1949 at Praça dos Restauradores. *Courtesy of the National Tramway Museum, photographer N.N Forbes*

the Carris concession, the NCAML was only dissolved in 1926. Between 1914 and 1926 trams and tickets on the Estrela line carried the name of the NCAML and were limited to this route.

Carris 1910s to 1950s

The last decades of the 19th and first decades of the 20th century were a period of much political and social unrest in Portugal. A revolution in 1910 ended the monarchy and installed a republic. However the social and political situation remained unstable. The participation in WWI made things worse and the capital in particular was the scene of demonstrations, strikes and even sabotage. Carris was also severely affected by these. Trams were halted or didn't leave the depots during the multiple strikes. Old competitors like Eduardo Jorge with his Chora cars, but also opportunistic others appeared temporarily again in the Lisboa streets with their old mule cars during several of these strikes, but also when technical failures caused total power shutdowns.

A first bus-line was opened on 24 November 1912: Sete Rios - Carnide. More bus-lines followed during 1913, some to destinations outside Lisboa. Some of them closed the same year, the others were closed in 1915.

On 25 June 1917, just after midnight, a fire started in the print shop of Santo Amaro. The fire spread to the carpenter's shop, the paint shop and oil storage next to it. Two cars, 473 and 474 were burnt out and were rebodied.

During WWI prices, especially of coal, had risen. Carris wanted to increase the fares but CML refused to approve this. In the

Cross bench car 311 approaching the end of the line at Carnide on 13 June 1954. *W C Janssen/Online Transport Archive*

meantime the staff wanted increased wages because the costs of living went up too. Multiple strikes followed during the years 1918 until 1922, some of only a day but others lasted for weeks. Carris lost lots of money and finally the government took action by overruling CML and installing an arbitrary committee to decide about the fares. Carris could raise the fares and concede the demands of the staff.

In the 1920's several new routes were opened, for example to

Carnide (1924) and Ajuda (1927), and an outer belt line in the 1930s, though the eastern part beyond Alto de São João was not completed until 1958. The short extension from Cruz Quebrada to Estadio, often used for driver-training, was added in 1945, but was closed in 1991 when the stadium moved elsewhere to make way for a new road.

New trams were acquired and Carris started to build its own new trams. In the early 1930's the typical small Lisboa tramcar evolved. Late in the 1930's Carris started to show the route numbers on the trams as information to the public. Until then route numbers were meant for internal use only. Public information on the trams had been limited to two blinds in the destination box: one showing the destination and the other an intermediate location.

In 1934 LET bought a site at Amoreiras for a new depot. Construction works didn't start until 1937. In 1940 the site was extended with the acquisition of a number of adjacent plots. Amoreiras also became the bus depot and works.

Workmen's fares were introduced on all routes from 1 August 1935, and remained unchanged for 40 years, until the introduction of the Passe Social. Ordinary fares remained unchanged for 36 years from 1926 to 1962. The reputation of Carris as an employer generally ensured that there was a waiting-list for jobs, and benefits included summer and winter uniforms, medical services, grocery discount shops, and 25 barbers whose services were free (shaving twice a week, hair cutting twice a month) to the platform staff. Workshop staff who preferred home-cooked lunches could have their pre-cooked meals placed on any car of route 19, which terminated at the works at Santo Amaro where the lunch containers would be

transferred to the works kitchen and kept warm until required. At all depots and the Geradora power station there were canteens and washrooms.

In 1940 the Exposição do Mundo Português ran from 23 June until 2 December. This exposition was to promote and legitimise the "Estado Novo" by connecting it with a mythical past. In other words, the fascistic regime of Salazar wanted to show itself in a favourable light in an ideal but fake world. The exposition took place in Belém between the Mosteiro de Jerónimos and the Tejo. Of course the trams had a large part in transporting the crowds, but they could not follow their normal route in front of the Mosteiro. Instead the trams were diverted between Largo

Brill clerestory bogie 330 arrives at Carnide and negotiates the very tight corner at this point in the 1960s. *G B Claydon*

Brill clerestory bogie 332 approaching the Carnide terminus in 1967, showing one of the tight clearances encountered in Lisbon. *J. Jordan*

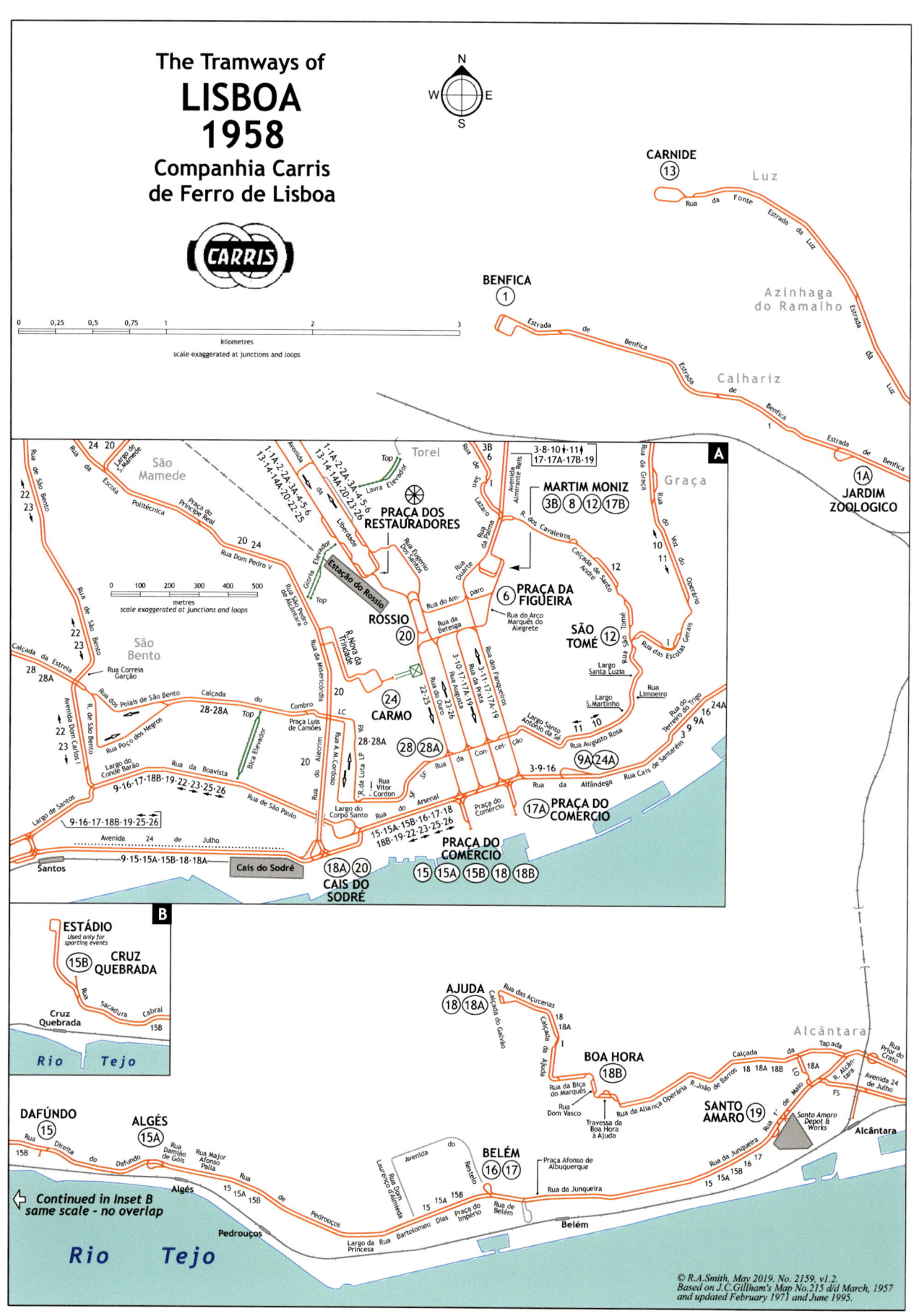

The Tramways of
LISBOA
1958
Companhia Carris
de Ferro de Lisboa

CARRIS

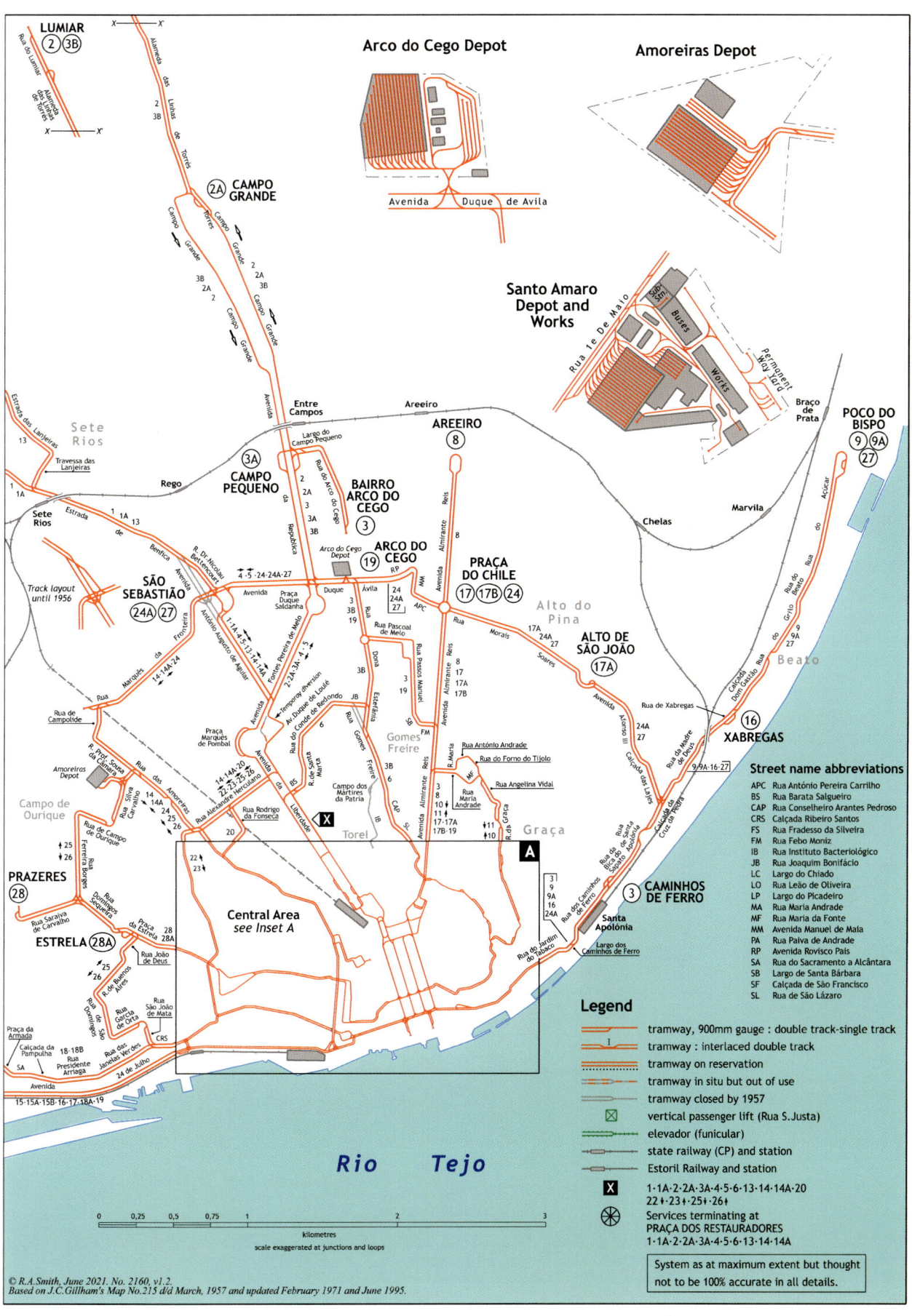

Tracklaying in Lisboa's narrow streets often means closing the entire road. This scene is at Rua Conceição in October 1959, with car 424, when reconstruction was taking place to provide double track to enable line 28 to be diverted to reach Graça. *Courtesy of the National Tramway Museum, photographer J.H. Price*

705 is in Rua das Escolas Gerais attracting the attention of some local dogs with the signalman standing in a doorway. *B. King*

Jerónimos and Rua Bartolomeu Dias via Rua dos Jerónimos and Encosta da Ajuda (Av. Restelo).

In 1951 Carris started to use concrete foundations when track renewal or new track installations were carried out. This was at the request of CML to maintain the paving in good condition. At the end of 1955 already over 30 km of tracks had concrete foundation.

Carris 1960s to 1970s

The first modern buses were six single-deck AECs bodied by Weymann obtained in 1940 for special events, notably the Belém exposition, and used from 1944 to serve the airport. These and most of their pre-1968 successors were of the half-cab type, with the layout reversed in relation to those in Britain. Most of the 143 buses purchased in 1946-8 were single deck AECs, but 362 double-deckers of the same make were put into service between 1950 and 1967, and some of the single-deckers were rebodied as double-deck. The double-deckers delivered up to 1960 had Weymann bodies built in Britain, but after a change in the law favouring local manufacture, subsequent deliveries had bodies of similar design by the Portuguese Union of Transport Operators for Import and Commerce (UTIC). Later purchases included 55 rear-engined Daimler Fleetlines and 26 high-capacity single-deckers. No British buses were acquired after 1974.

The bus livery was green and cream (or latterly all-over green) until nationalisation, and subsequently orange and mushroom. By 1972, 560 buses were operating on 56 routes from garages at Amoreiras and Cabo Ruivo, and further acquisitions had to be kept in the permanent way yard at Santo Amaro. Until the 1970's the bus lines were additional to the tram network. The buses were not yet replacing the trams. The first significant reduction of the tram services was caused by the opening of the Metropolitano de Lisboa in December 1959, with trams being removed in 1960 from Avenida Fontes Pereira de Melo and from the Avenida da Liberdade. A new track was laid across Avenida da Liberdade from Rua Alexandre Herculano towards Gomes Freire, its opening in August 1960 marking the end of tramway operation to Restauradores. Trams were withdrawn from Rossio, except for route 28, and a proportion of journeys to and from Benfica and Carnide were terminated at Sete Rios, passengers to and from central Lisboa being offered transfer tickets valid on the Metropolitano. Tram feeder services also operated between Entre Campos and Lumiar.

247, a Brill 21E CCFL standard, with bus 217 at Rua Correia Garção / Calçada de Estrela, São Bento on route 23 on 9.June 1973. *A. G. Murray*

Lisboa 805 and 804 are seen in the sunshine at Sete Rios metro interchange on 4 May 1964. *L. Folkard*

Further changes were brought about by the opening of the Tejo road bridge in August 1965. The access road from the bridge joins the riverside road at Alcantara, and this brought about the closure of two short sections of tramway in favour of alternative routes. Further metro extensions occasioned the closure of routes 2/2A to Lumiar on 15 May 1971, and route 8 to Areeiro on 18 December 1972. In 1973 motorway construction work at Palhavã occasioned the closure of the Benfica and Carnide routes, despite the construction in 1960 of the tram/metro interchange station at Sete Rios.

Late in 1972 Carris announced that trams would be phased out completely over a period of five years. The previously immaculate fleet began to look shabby, and Carris appealed for authorisation to increase fares, which was refused. Instead, the municipality offered to buy 67% of the company's shares, and this was accepted (the remaining shares changed hands in 1980). On 1 January 1974 Lisbon Electric Tramways Ltd changed its name to LET (Holdings) Ltd and was left to develop surplus Carris property, including Amoreiras bus garage, which was on a valuable site and would be redeveloped as a shopping centre. The bus works was moved to new premises at Miraflores north of Algés.

Changes since 1974

On 25 April 1974 the Government was overthrown, and as a direct result Portuguese involvement in the overseas colonies was reduced to the extent that one million Portuguese returned home, many of them to Lisboa. Few of them had private transport, and perforce turned to public transport. Every available bus and tram was pressed into service to satisfy the upsurge in demand, many of the withdrawn vehicles being put back into service almost irrespective of condition. Orders were quickly placed for 400 air-suspension single-deck buses on Volvo and MAN chassis, which were delivered during 1975-9.

Government policy was to encourage public transport, and a national policy of ticket prepurchase was adopted for urban areas. Carris followed this directive by introducing a monthly subsidised season ticket (Passe Social) on 1 January 1976; this scheme was extended to include the Metropolitano, Transtejo ferries and Rodoviária Nacional buses in the Lisboa area on 1 January 1977, and further extended to include CP suburban rail services on 1 May 1977. Pre-payment of fares was encouraged by establishing ticket-sales and information kiosks at principal

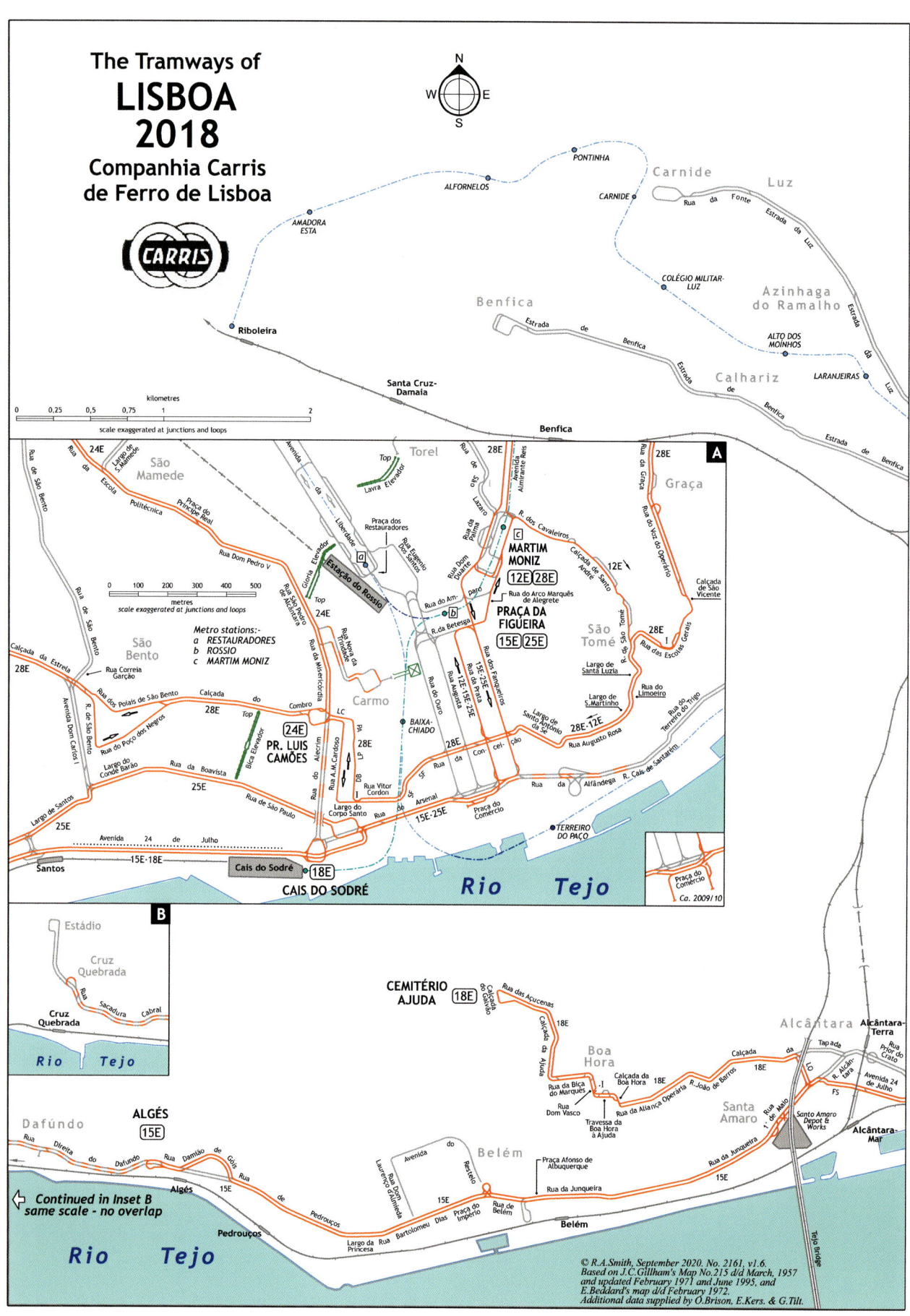

The Tramways of
LISBOA
2018

Companhia Carris
de Ferro de Lisboa

CARRIS

N W E S

kilometres

0   0.25   0.5   0.75   1   2

scale exaggerated at junctions and loops

metres

0   100   200   300   400   500

scale exaggerated at junctions and loops

Metro stations:-
a RESTAURADORES
b ROSSIO
c MARTIM MONIZ

Pontinha
Alfornelos
Amadora Esta
Carnide
Colégio Militar-Luz
Alto dos Moinhos
Laranjeiras
Riboleira
Santa Cruz-Damaia
Benfica

Carnide   Luz
Azinhaga do Ramalho
Calhariz

A

Torel   Graça
São Mamede
24E   28E   28E
Lavra Elevador
Top

MARTIM MONIZ
12E   28E

PRAÇA DA FIGUEIRA
15E   25E

São Tomé
12E
28E

Estação do Rossio

São Bento
28E

PR. LUIS CAMÕES
24E

BAIXA-CHIADO
28E   28E-12E

25E

15E·25E

TERREIRO DO PAÇO

Rio   Tejo

Praça do Comércio
Ca. 2009/10

Santos

CAIS DO SODRÉ
18E

B

Estádio
Cruz Quebrada
Cruz Quebrada

CEMITÉRIO AJUDA   18E
Boa Hora
18E

Alcântara   Alcântara-Terra
Alcântara-Mar

Santa Amaro
Santo Amaro Depot & Works

Dafúndo   ALGÉS   15E
Belém

Algés   15E
Pedrouços   15E
Belém   15E

← Continued in Inset B
same scale - no overlap

Rio   Tejo

© R.A.Smith, September 2020. No. 2161, v1.6.
Based on J.C.Gillham's Map No.215 d/d March, 1957
and updated February 1971 and June 1995, and
E.Beddard's map d/d February 1972.
Additional data supplied by O.Brison, E.Kers. & G.Tilt.

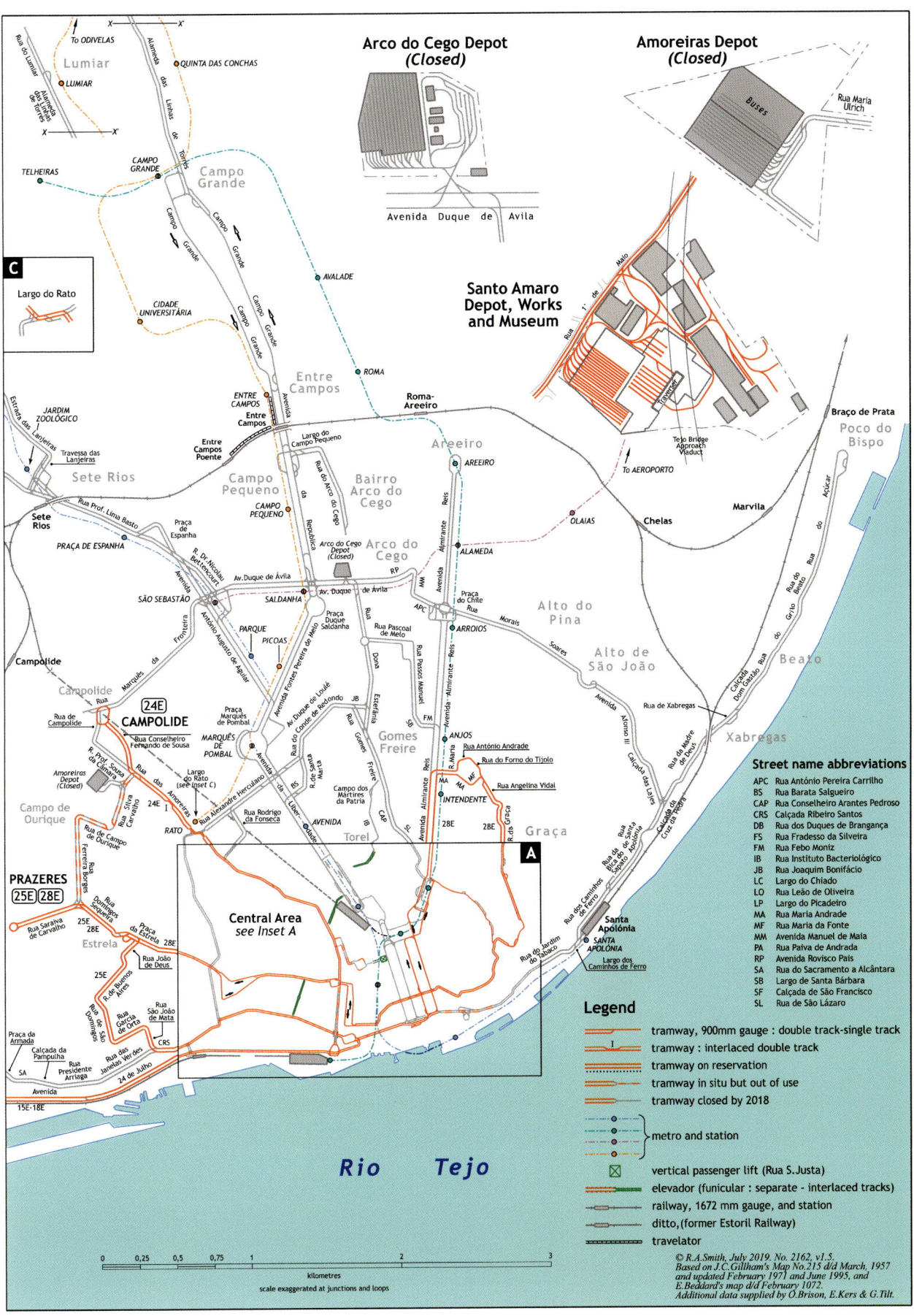

Car 467, a rebodied former St Louis car is seen at Largo do Rato during track reconstruction works, sometime in the late 1960s.

Brian G Dutton/Online Transport Archive

Above: A rare sight of car 905 at Santa Apolónia on 3 May 1964. These cars were normally confined to route 15. *L Folkard*

locations and equipping buses with ticket-cancellers for use by passengers, while the cost of Tarifa Unico flat-fare tickets purchased from the bus drivers was fixed at a higher level than other tickets.

The tramway system continued as it had been at the time of the municipal takeover, no further contraction being implemented for several years. In 1977 it was decided to resume general overhaul and retain the existing fleet until 1985. Workshop facilities were expanded, and two overhauls were completed per week; subsequent overhauls included the changes required for one-man operation. Meanwhile, in 1976 a commission from the Zürich undertaking (VBZ) and the Lausanne Transport Institute (EPFL) was invited to report on the tramways. In 1978 the tramway system carried 90 million passengers, representing about 19.5% of the total Lisboa ridership.

The report "Lisboa Tramways- Rehabilitation or Substitution" was published in August 1978. The report recommended the retention and rehabilitation of the tramways in general and of the riverside and hilly routes in particular. As a result Carris staff visited various European manufacturers to discuss car types and possible licencing arrangements, retaining the 900-mm gauge. Carris prepared a draft specification for a six axle articulated design with chopper control as a basis for discussions with manufacturers. New cars would enter service between Cruz Quebrada and Praça de Figueira, thus providing surface light rail transit over the route of the projected riverside Metro, which was abandoned on grounds of cost. In the event, this project was also delayed by about ten years for financial reasons.

In May 1982, Carris announced its intention to buy new two-axle cars for the hilly-routes (12, 18, 25 and 28). An earlier plan

Cars 405 and 278 are seen in Rua das Amoreiras on 3 May 1964. *L. Folkard*

Lisboa 702 climbing up to the junction at São Tomé on 3 May 1964. *L. Folkard*

"Caixote" 284 negotiates open trackwork on Rua de Pedroucos, near Algés, in 1967. *J. Jordan*

204 and 210 are seen at Benfica terminus in 1967. *J. Jordan*

"Caixote" 428 stands at Lumiar terminus in 1967. *J. Jordan*

to keep route 20 was dropped, and the other surviving routes would be replaced by buses as and when their cars and track wore out. After experiments with car 277, Carris decided that the best option was to re-truck and re-equip 45 existing small trams. Most of those selected were the remaining cars of the series 221 to 282, and this process is described in the section on the remodelled cars.

In the 1980's and 1990's the major part of the trams were replaced by bus lines. Five tramlines were kept after line 17 was suspended in 1997: 12, 15, 18, 25 and 28. In a few cases closing a tramline was announced as temporary, but re-opening never happened until in the Spring of 2018 part of line 24 reopened. In some places tracks and overhead lines are still in place.

Lisboa today

Having outlined Lisboa's tramway history to date, we must now set down the principal features of the routes that operate today. As previously mentioned, the city is built on a series of hills rising from the river. The most tortuous tramways are those in the Graça area, which is perched some 80 metres above the Baixa, as the city centre is popularly known. Lisboa is nowadays visited by millions of tourists and the old small trams are part of the sights.

Route 12 is worthy of note, despite being the shortest route. For many years it was about 1 km (Martim Moniz - São Tomé) and operated by only one car. This route was single track but for three, now disconnected, passing loops, and climbs steeply for almost its entire length with a maximum 14.5% gradient in the upper part of Calçada de Sto. André. Line 12 was changed to a clockwise circular line in November 1997 going uphill on

its original route. Approaching São Tomé the car breasts the summit on a very narrow reverse curve at which the road camber changes abruptly, calling for extremely skilled driving as the tram heels over from one side to the other at walking pace. São Tomé is also noteworthy for the track formation at the junction with the Graça route (28), incorporating single track (in the Spring of 2018 changed to double track) and pointwork on sharply diverging gradients. Downhill from São Tomé to the Baixa the route of line 28 is followed and via Rua da Prata the trams go to Praça da Figueira and then to nearby Martim Moniz again. Changing line 12 from a shuttle line into a circular line made it possible to use the single-ended Remodelados and withdraw the last double-ended trams from the regular service, and also resulted in increased ridership, not least because it now serves the Portas do Sol & Sta. Luzia area, comes very close to the Castelo de São Jorge and passes the Sé, all major tourist hotspots in the city.

Line 15 is the only remaining flat route and the only one on which the articulated trams can be used. It runs between Praça da Figueira and Algés, but there are often short workings outwards to Belém and inwards to Cais do Sodré. Praça da Figueira is a major bottle-neck in the current network as trams of the lines 12, 15 and 25 and both the red and green tourist services have to share the sole track. It's not uncommon for a queue of trams to be waiting in Rua da Prata for their turn. Belém is a major tourist destination and that makes line 15 trams very crowded for most of the day.

Line 18 is the route of the Ajuda area, the hills north of Belém. Once a very busy line, it's now reduced to a line with minor

Above: 210 on single track necessitated by trackworks alongside Santa Apolonia station, controlled by Carris man in uniform with lollipop in 1969. *P. Haseldine*

Above, right: A pleasant view taken in 1969 looking down some steps at São Tomé to the rear of "Caixote" 316 as it meanders towards Graça. *P. Haseldine.*

Right: Not much room for pedestrians; 613 is seen in the narrow Rua das Escolas Gerais on route 28 in 1969. *P. Haseldine*

Below: 547 + trailer, and a group of Carris staff in uniform at Praca Areeiro in 1969. This terminus closed soon afterwards following extension of the metro. *P. Haseldine*

Car 906 in Praça do Comércio on 30 March 1974. *Michael Russell*

traffic volumes, suffering competition from parallel bus-lines. The line is only operated on weekdays and Saturday mornings, but not at all during August. In 2013 the line was planned to be closed, but this wasn't carried out. However service was further reduced and limited to Cais do Sodré - Ajuda only. Tourists are not seen on this line, but for tram enthusiasts it has quite some interesting points, e.g. the Boa Hora track layout and Calçada da Ajuda with a 10.5% gradient.

To almost everybody's surprise, line 24 was reopened in the Spring of 2018, 28 years after temporary closure, although only between Praça Luís de Camões and Campolide. Rails, switches and overhead lines had been retained on this stretch during all these years, so reinstating was technically quite easy. It's expected that the line will be extended again from Praça Luís de Camões to Cais do Sodré. The tracks in Rua do Alecrim are still in place but must be reconnected at Praça Luís de Camões. In Rua do Alecrim the overhead lines must be reinstated in front of a new building that replaced old ones during recent times. To ease demolition of the old and construction of the new building, the overhead wires were removed. In the rest of Rua do Alecrim they are still in place. At the Cais do Sodré end tracks were reconnected with the change of the layout in the area that occurred in 2017. Rua das Amoreras with the interlaced tracks when crossing under the Aqueduto das Aguas Livres is one of the most noteworthy spots on the network.

The current line 25 is what remains from the once large circular line 25/26. It operates between Praça da Figueira (until 2016 from Rua da Álfândega) and Prazeres, but only during weekdays. The section between Estrela and Prazeres is shared with line 28.

The use of this line by tourists is still limited, but it runs through interesting areas and also has its own 10.5% gradient: Rua de São Domingos. If line 24 is extended from Praça Luís de Camões to Cais do Sodré, it will be possible again to see trams of line 25 in Rua de São Paulo under the viaduct in Rua do Alecrim and trams of line 24 on that same viaduct, although luck will be needed to have them both at the same moment on a photo. At Corpo Santo and Estrela are turning loops which are frequently used for short workings to get trams on schedule again.

Route 28 is a strong contender for the title of "The World's Most Exciting Tram Ride". It boasts many steep slopes of which a total of almost 500 m has gradients of 12% to 14.5%, another 500 m about 10% and other stretches with a total of about one kilometre have gradients of at least 8%.

In Alfama (Escolas Gerais) is an extremely narrow section with an interlaced curve, followed by a single-track street so narrow that it is possible to reach from the car and touch buildings on either side. Until the installation in 1990 of colour-light signals, traffic control was by three hand signalmen who stationed themselves at strategic points, usually outside a minuscule bar, to regulate the passage of the trams (and perforce all other traffic) by signalling to tram drivers, and each other, with red and green indicator bats. At night, oil lamps with coloured lenses were used.

Rua da Graça is a busy shopping street just wide enough (by Lisboa standards) to have double track. The trams here run very close to the pavements which themselves are hardly over 60 cm wide.

Line 28 also connects several tourist hotspots like Portas do Sol, Sta. Luzia and the nearby Castelo de São Jorge, the Sé, the

Baixa with Rua Augusta and Chiado and the Bairro Alto. These places are continuously crowded with tourists. Largo da Graça with the nearby Miradouro and the Basilica da Estrela are also along the route and on the "must do" list of many tourists. This and the spectacular route itself means that the trams of line 28 are completely overcrowded most of the time. During a large part of the day there is often a queue of hundreds of people at the Martim Moniz terminus waiting for a tram.

Turning loops that can be used for short workings are at Graça, Luís de Camões and Estrela, the latter shared with line 25 and the red tourist service.

Operating reliably and safely in Lisboa conditions with its many

steep slopes demands a high standard of maintenance and Carris can carry out virtually every aspect of tramcar engineering at Santo Amaro, from body building to armature winding. Even a routine Carris overhaul can involve the replacement of structural components. The site at Santo Amaro is also the headquarters of Carris, and is dominated by the Tejo road bridge, three of the supports of which march across the premises. In the shadow of the bridge lies the 20-road depot and the tram works, the body shop being reached by a traverser, too short to accommodate articulated cars. Considerable alterations to the track layout were carried out in 1993-4 to accommodate the future articulated cars; the ordinary service cars having been transferred in 1986 to Arco do Cego. The latter depot was closed in 1996 and Santo Amaro is now the only remaining depot for the trams.

Permanent way

The permanent way department at Santo Amaro contrives to provide and maintain a high standard of workmanship despite the challenges posed by the nature of many of the routes. The unusually narrow gauge allows generous curves to be laid in restricted spaces: indeed many of Lisboa's routes would probably not have been possible had the standard gauge been retained. The rail was formerly spiked to wooden sleepers on ballast foundations, but current practice is to clip to anchors made from upturned rail and complete it with tie-bars on a bed of concrete. In company days, rail was bought from South Durham Steel and Iron Co to sections BS8 and BS86, but recent work has been German, to NP4 section. In 1948, when new rail was almost unobtainable, the rail from the former Gosforth Park

804 poses at the Estadio terminus on 31 March 1974. *Michael Russell*

801 has just left Algés on 31 March 1974 en-route to Cruz Quebrada. Note the overhead feeder wires. *Michael Russell*

An atmospheric night shot of "Caixote" 491 at Carmo on 3 April 1974.

Michael Russell

Light Railway at Newcastle was bought and used. Points formerly came from Hadfields, but are now increasingly double-tongued pointwork of German manufacture.

In 1981 a start was made at Santo Amaro on manufacturing new crossings for route 15 in anticipation of the new rolling stock. Cars run through crossings on their wheel flanges, the groove being built up to allow a smooth passage; joints are Thermit-welded for the same reason. Renewal work is carried out at night, but it is frequently necessary to close the narrower streets while work is in progress. In Lisboa, rail has sometimes to be bent vertically as well as horizontally to accomplish some of the more tortuous changes of gradient and direction. The use of slipper brakes formerly promoted rapid rail wear on certain inclines. At many locations the alignment of the track is critical, so small are the clearances between trams and buildings.

Overhead

Until the 1990's, Lisboa overhead represented traditional American practice, with current-collection by fixed-head trolley pole, from round-section wire held in clinch ears with cap-and-cone insulators (capecone in Portuguese). The tubular traction poles, painted dark green, are of British type and some still have bracket arms. Trolley wire height is 6 metres, with lightly-tensioned span wires, to permit the operation of double-deck buses. In 1982 a section of the riverside route at Dafundo was renewed in grooved wire (fio ranhurado) with flexible stirrup-wire suspension and nonfouling ears to test single arm pantographs fitted to cars 333 and 339. Motor tower wagons are stationed at strategic points on the system, in radio contact with the control centre at Santo Amaro and the linesmen.

The whole network is still suitable to be worked with trolley-poles, but a large part can now also be used with pantographs. On the lines 15, 18 and 25 pantograph operation is the standard. The lines 12 and 28 are operated with trolley-poles because in some places the tracks are sufficiently close to the buildings that people leaning from a window or a small balcony on the first floor could be in danger of being hit by a pantograph. The turning loops of Graça and Camões however are suitable to be used with pantograph to make it possible that line 28 trams can turn here (needing to go backwards) from the wrong direction in case the route through the Baixa is temporary blocked. Line 24 is also operated with trolley-poles as the overhead wires were never modified for working with pantographs.

741 *(below)* and 701 (above) at the at the bridge carrying Rua do Alecrim (route 20) over Rua Sao Paulo, Cais do Sodré. *Michael Russell*

Power supply

For the power supply of the electric trams a generating plant often called the Geradora was built at Santos at the riverside. This made it possible to deliver the coal by ship. On the day of opening of the first electric tramline, 31 August 1901, the power plant had boilers from Babcock & Wilcox of Glasgow, compound steam machines from McIntosh, Seymour & Co of New York and generators and other necessary electric equipment from General Electric. During the years that followed the capacity was gradually enlarged with extra and also more modern boilers, machines and electric installations.

In 1923/24 the first of the oldest steam engine/generator combinations was replaced by a new turbine from Brush. From 1925 on more old installations, were gradually replaced by new, more modern ones. The last pre-WWI installations were decommissioned just before WWII.

The most common fuel used in the Geradora was coal acquired from the United Kingdom. During the World Wars the British coal was in short supply. In 1917 two boilers were modified to burn wood. Although the coal shortage had ended, two other boilers were modified to burn oil in 1920. Using oil next to coal was continued until closure of the Geradora.

In 1928 the first experiments were done to use coal from Portuguese mines. However this coal was of much poorer quality and the Santos boilers were not made to burn this successfully. The Portuguese coal had to be mixed with larger amounts of British coal to keep the boilers on pressure. Coal and oil shortages during WW2 required coal from Portuguese mines to be used along with olive waste and husks, crushed olive stones, cork bark and lignite. This was continued after the war because the Estado Novo regime of Salazar wanted the country to be self-supporting as much as possible.

In later years substations which were connected to the Geradora by high tension (6.6 kV) lines were installed: Arco do Cego in 1916, Glória, on the site of the former engine house of the funicular in 1925, Santo Amaro in 1929 and Amoreiras in 1939. In 1942 a mobile substation entered service. It was used at Dafundo because of the new line to the Estádio Nacional, but also at Santo Amaro to increase the capacity. It didn't get its power from Santos, but from the Companhias Reunidas Gás e Electricidade (CRGE or Reunidas). Soon Reunidas started to supply extra electrical power to Carris.

Underground feeder cables were used to distribute the power over the network except beyond Alges. The growing network and the installing of the substations caused frequent changes of the feeder network.

By November 1951 most power came from the hydro-electric plant of Zêzere of the Companhias Reunidas. At peak hours a quarter still came from the Geradora.

In 1952 it was decided to shut down the Santos power plant and change completely to the national grid. To realise this more connecting cables had to be installed, the most important one a 12.8 km long high tension cable between Moscavide and Santos. The Geradora was closed with a small ceremony on 24 November 1955.

Lisboa Rolling Stock Until the 1990s

Mule cars

Tramway operation began in Lisboa in November 1873 with a fleet of 30 mule cars built to the standard European 1435-mm gauge by the John Stephenson Company of New York and two built by Starbuck. 23 were small enclosed cars and nine were of open-sided cross-bench design. By mid-1874 the fleet had increased to 54 cars (and 421 mules). For a few years no new cars were acquired. The annual report of 1877 gives an overview of the fleet on 31 December of that year. There is no specification about the type of tram, but this can be determined when comparing with the numbers in the 1891 overview which follows the 1877 overview below.

The 1877 overview:
22 American trams with 22 places (1873 closed trams from Stephenson)
6 American trams with 12 places (1874 closed trams probably from Stephenson)
10 English trams with 16 places (1874 closed trams probably from Starbuck)
1 English tram with 18 places (1873 closed tram from Starbuck)
8 American trams with 32 places (1873 open trams from Stephenson)
6 American trams with 32 places (1874 open trams probably from Stephenson)
1 English tram with 32 places (1873 open tram from Starbuck)

Apparently in 1877 Carris made a distinction between "americanas" and "inglezas" based on the country of their supplier. An overview in the annual report of 1891 for the cars existing on 31 December of that year shows "Americanas" to

be used for all trams notwithstanding the country they came from, as well as "De viação mixta" (of mixed road) for cars which could be used on both rails and road. This type was also called "Würzburg" because the first of these cars was acquired from NOELL in Würzburg, although most were built by Carris workshops. The Würzburg system was similar to the Ripert system used by rival companies, the difference mainly the first supplier. In the 1891 report an overview with the year they entered service was given:

- 1873 nine open trams with 32 places (8 Stephenson, 1 Starbuck), 22 closed trams with 20 places (Stephenson) and one closed tram with 18 places (Starbuck). The capacity of cars with longitudinal benches common in closed trams could differ somewhat depending on the space supposed to be occupied by one passenger. This explains that trams specified in 1877 with having 22 places, had in 1891 only 20.
- 1874 six open trams with 32 places (probably Stephenson), 10 closed trams with 16 places (probably Starbuck) and six closed trams with 12 places (probably Stephenson).
- 1875 to 1877 no new trams.
- 1878 six open trams with 32 places and one closed tram with eight places and space for luggage. All seven trams built by Carris workshops.
- 1879 eight open trams with 20 places and one closed tram with 12 places. (all Carris workshops)
- 1880 eight open trams with 24 places (bought from unknown supplier), two open trams with 20 places (Carris workshops) and two closed trams with 12 places (Stephenson).
- 1881 six open trams with 24 places and three closed trams with 12 places. (all Carris workshops)
- 1882 two open trams with 24 places, five open trams with 20 places, four closed trams with 12 places (all Carris workshops) and one closed Würzburg car (or viação mixta) with 10 places (NOELL).
- 1883 two closed trams with 16 places, four closed trams with 12 places and four closed Würzburg cars with ten places. (all Carris workshops)
- 1884 three closed trams with 12 places and five open Würzburg cars with 16 places. (all Carris workshops)
- 1885 two open trams with 24 places, one closed tram with 18 places and five open Würzburg cars with 16 places. (all Carris workshops)
- 1886 four closed trams with 10 places (Stephenson), four open Würzburg cars with 16 places (Carris workshops) and two closed Würzburg cars with 10 places (local supplier).
- 1887 one open Würzburg cars with 16 places.
- 1888 two open Würzburg cars with 16 places and two closed Würzburg cars with 12 places
- 1889 two open trams with 24 places and one closed Würzburg car with 12 places.
- 1890 two open trams with 28 places, six open trams with 24 places and 14 closed trams with 18 places. These 22 trams were for the new 900 mm gauge lines and acquired from (an) unknown supplier(s).
- 1891 no new trams (or Würzburg cars)

This meant that by 1891 there were 120 trams for 1435 mm gauge, 34 Würzburg cars and 22 trams for 900 mm gauge, a

A view of some extensive track reconstruction at Dafundo in preparation for the introduction of the low-floor articulated cars, which in the event did not run beyond Algés, October 1982. *B. King*

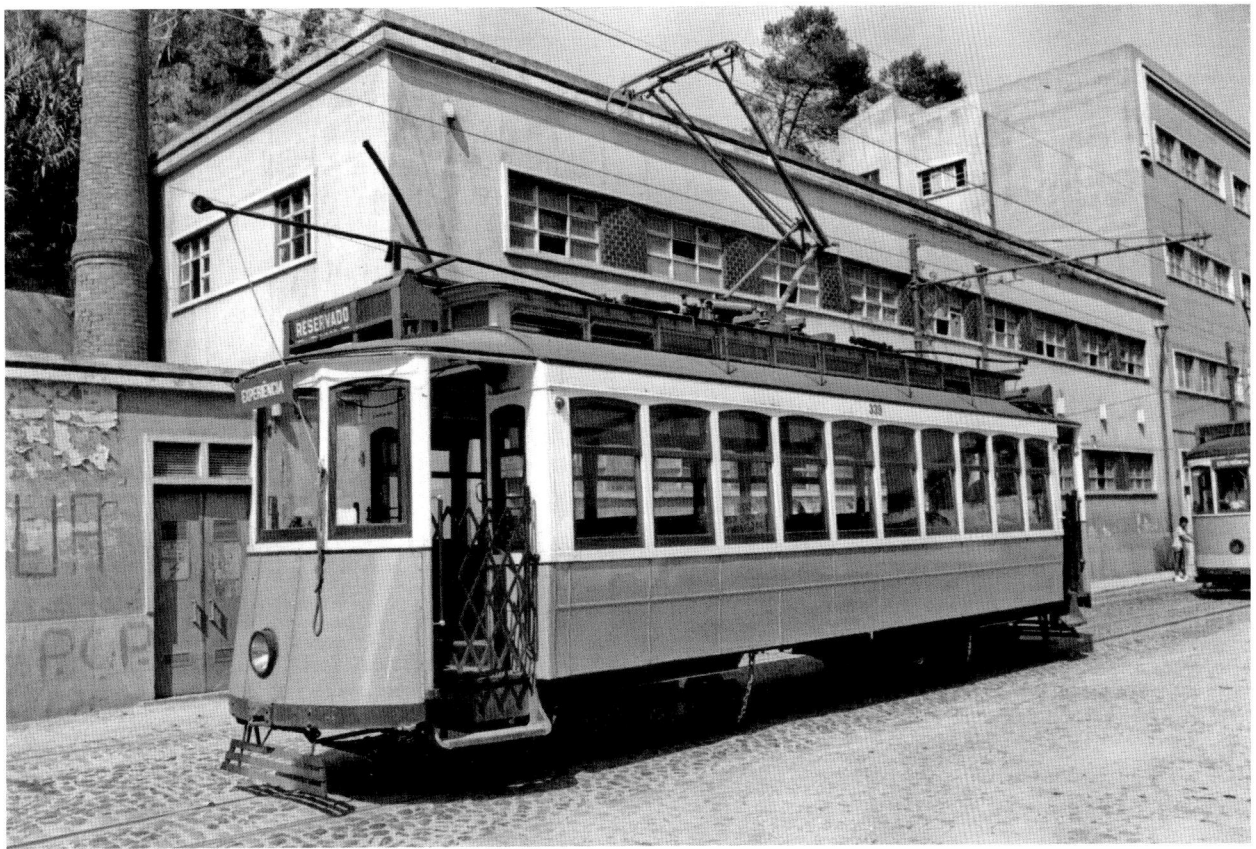

339 is seen fitted with a pantograph to check the alignment of the reconstructed overhead at Dafundo, 7 August 1982. *O Brison*

On 12 July 1982, 724 has arrived at Rua de San Lazaro to assist an unidentified "Caixote" car that appears to have a broken axle. *L. Folkard*

An interior view of 342 taken on 15 July 1982, note the windows that slide into the roof space, and the tunnel shaped ceiling containing the window pockets. When cars carried conductors, passengers boarded at the rear, left at the front, and the last passenger off was expected to close the bulkhead doors. *L. Folkard*

Low floor car 508 passes the Jeronimos Abbey at Belem on 31 August 2001. *Michael Russell*

Tourist car 7 is seen negotiating the reverse curve at the summit of Calçada da San Francisco on 2 September 2001. *Michael Russell*

total of 176 vehicles. In 1892 Carris started to buy out rivals and thereby acquiring the rolling stock of those companies. Most of these cars were probably of the Ripert type or another type that could be used on both rails and road. Carris gained in 1892 and 1894 more than 220 cars that had belonged to many different competitors. It was a diverse collection of cars in many sizes and shapes, often of poor quality. Carris withdrew many immediately or soon after they got them. By the end of 1895 Carris owned a total of 280 vehicles but at the end of 1898 it was 329 vehicles, that number probably including the 15 open trams ordered in 1897 from Brill. Carris workshops built a replica in 1951 for use in a film, but apparently it no longer exists. A further replica was built for the museum in circa 2000.

Mules and horses

While in many European countries horses were normally used to haul trams, in Portugal the preferred animals were mules. In Lisboa the majority of the animals were mules, but Carris also used some horses. Mules and horses took the major share of the costs for the tramway companies. In the period 1873-1890 Carris bought a total of 2,564 mules and horses and sold 1,028. Another 324 had died, which means that at the end of 1890 there were 1,210 mules and horses. When Carris bought out rival companies in 1892 to 1894, they got also a large number of animals, but these were often of less quality.

Electric cars

Viewed historically, Lisboa's electric trams fall into four periods: American, Standard, Caixote (Box cars) and Modern. From

577 ascends Calçada da Ajuda on 4 May 2004. *Michael Russell*

1900 to 1914 complete trams were imported from America in dismantled form and assembled locally. Between 1925 and 1943 Carris built cars to its own design, using trucks and equipment purchased in Britain. From 1947 to 1963 new cars were built mostly to a simplified open-plan design, single-ended and using reconditioned trucks and equipment from older cars, but with aluminium panels and fibreglass domes.

Class types
At first Carris classified the tram types with "Pequeno aberto" (Small open), "Pequeno fechado" (Small closed), Grande aberto (Large open), "Grande 4 motores" (Large 4 motors), "Grande 2 motores" (Large 2 motors) and "Carros do Povo" (2nd class). The indication "Grande" for large cars was later replaced by "Salão" (Saloon)

All new cars of the 1910's and 1920's and most of the 1930's were closed and had two axles which put them all in the category "Pequeno fechado".

In the 1930's the "Pequeno aberto" cars disappeared and the term "Pequeno fechado" was replaced by "Ligeiro" (Lightweight). All two axle cars were part of this "Ligeiro" class. If useful in the context indications as "São Luís", "Brill", "Radiax" or "Estrela" could be used, but for the typical Lisboa tramcar that evolved in the late 1920's and built in large numbers in the 1930's, Carris never adopted its own class name. They were just "Ligeiros". The same happened with the small tramcars built from the late 1940's until early 1960's. Those were classified "Ligeiros" too.

Historians however like to classify the trams by model. For that reason the late J.H. Price introduced in the first edition of "The

Tramways of Portugal" published in 1964 the term "Standard" for the 2-axle trams built in the 1930´s as they were the most numerous of all types and known as the typical Lisboa tramcar. "Modern" was used in the same book for the post WWII cars. These terms were never internally used within the organisation of Carris itself.

In the late 1990's the post WWII cars were all withdrawn and could not be considered modern; instead these cars are now called "Caixote" (Box) by tram historians and enthusiasts. The advantage of these non-Carris terms is that they not only distinguish between the different types of "Ligeiros", but also can be used for the bogie cars with similar design of similar design.

The modern cars are those re-trucked and re-equipped from 1994 onwards (termed Remodelados), and the low-floor articulated cars which entered service from April 1995. No tramcars have ever been bought at second-hand; this being ruled out by the unusual track gauge and the terms of the concession, though some equipment was purchased from Sheffield and Dundee.

First generation electric trams
Owing to the unusually narrow gauge of 900 mm, a special motor had to be evolved to fit into the trucks. The wheels were dished in convex form (as opposed to the normal concave form), and a special narrow motor was produced by the General Electric Company of America. It was designated GE59, being a development of the GE58, which is normally quoted with a one-hour rating of 37 hp, though Lisboa always quoted the continuous rating (25 hp). Some of the original motors were in

Museum car 330 on private hire approaching Belem in 2009. *P. Haseldine*

use until the 1990's, converted to roller-bearing armatures. The controllers were GE K10 unless stated otherwise.

203-282: These were Lisboa's first electric cars. They were ordered from the J. G. Brill Co of Philadelphia in September 1899 and delivered in 1901. The original order was for 120 cars, but the last 40 were cancelled and replaced by bogie cars 283-322. The Brill 21E trucks of 1.98 metres wheelbase were fitted with hand-operated slipper brakes using oak shoes on both the wheels and the rails, to enable the cars to descend steep gradients. These open-sided cars were popular with passengers.
Disposals: Except for five cars all were withdrawn and the bodies scrapped 1932-37; trucks and equipment were overhauled and added to the float available for use with the new Standard cars built in the same period. Five trams were retained in service until 1948 with an "A" added to their number to avoid confusion with the existing Standard cars with the same numbers: 228A, 235A, 247A, 251A and 267A.

400-474: The bodies of these cars were ordered from the St Louis Car Company in November 1899, and had drop windows. The trucks were ordered in September 1900 from Brill. They entered service in 1901 and some were put to work on the hilly routes, for which purpose they were fitted with the same types of hand operated slipper brakes as the open cars 203-282. Originally they had K-10D controllers. From 1911 the first fifteen were fitted with Dick, Kerr DB1 Form K3 controllers and magnetic track brakes, arranged by the EMB Company to Maley's patent. 473 & 474 burnt out in 1917 but were rebuilt in 1918, becoming similar to the cars of series 476-499. Body strain on the steep hills led to rebuilding of the other 73 cars between 1927 and 1931, during which they received windscreens. The drop windows were changed to lifting windows, which gave the cars extra width internally by removing the window pockets. Seating arrangement was changed from 2+1 to 2+2 raising the seating capacity from 20 to 24. They were the last cars to remain in service without air brakes. From 1914 to 1926 sixteen cars bore the title of a subsidiary company (NCAML), as explained in the section on cable tramways.
Disposals: 415 and 468 in 1935 and 455 and 467 in 1940. Their numbers were re-used for new Standard trams. 462 became a Caixote in 1952 and 473 in 1953. 22 cars were withdrawn and rebuilt as Caixote in 1961-1963. Nine cars (423, 426, 431, 441, 454, 456, 458, 460 & 471) were scrapped in 1963 without

replacement. The other 38 were withdrawn from regular service between 1964 and 1973. 437 and 435 became tourist cars 1 & 2 in 1965 in a special red livery. They are now only used as shuttles between the museum locations on the Santo Amaro site. 444 became an information office and is now part of the museum collection restored with open platforms, drop windows and the original 2+1 seat arrangement; ten cars sold to USA; remainder scrapped.

283-322: These cars were used on the long suburban routes. As delivered, the trolleys were mounted at the roof-ends and that not in use had to be swung to the centre. After a collision 305 was rebuilt in 1936 receiving a saloon body of the Standard type. All others were eventually fitted with windscreens. In 1955 car 283 was painted grey and used for driver-instruction until 1961.
Disposals: Except for 283 all were scrapped 1952-1955. The body of 283 went in 1963 to Alvito children's park, returned to Carris 1981 and was rebuilt as a working museum car in 1985.

323-342: These saloon bogie cars were the first Lisboa cars of the patented Brill "grooveless post" semi-convertible design which proved very suitable in Lisboa's sunny climate. The window sashes slid upwards into the coving of the clerestory roof. 328, 329, 338 and 340 got a Standard body in 1932 and 327 in 1937. Nine cars (324/5/6, 331/4/5/6/7 and 341) reconditioned with plain arch roofs in 1959-61. 323, 330, 332, 333, 339 and 342 retained clerestory roofs. 333/9 were used for pantograph trials in 1982-5.
Disposals: 325,331/5/8 scrapped before 1995 after accidents. 330 restored as museum car in 1993-4; all others withdrawn in 1995/6 and sold. 323/6 are owned by a Lisboa enthusiast.

343-362: The bodies of these cars were built by the Brill subsidiary John Stephenson & Co of New York and delivered in 1907. These cars were outwardly similar to 323-342, but had four-motor GE equipment and Brill 27GE1 equal-wheel bogies ordered in September 1906 from Brill, and were intended for the longer suburban routes. They were the first Lisboa cars to have air brakes from new, which worked on the wheels. Hand track brakes appear to have been fitted from new but appear to have been removed by 1931. 345, 355 and 356 received a Standard body in 1932 and 353 in 1933. Thirteen cars, (343/4/7/9, 350/1/2/4/7/8/9 and 360/1 were reconditioned with arch roofs in 1959-61. 362 retained the original roof with the clerestory removed, and 346/8 retained clerestory roofs until withdrawn. By December 1995 five cars (343. 347, 350, 351 and 353) were fitted with pantographs.
Disposals: 344 and 354 scrapped after accidents before 1995. 348 restored for Carris museum. All others withdrawn 1995/6. 355 exhibition car (no motors), later used as information/ticket office on Praça do Comércio for the tourist service, now in storage. 356/7 scrapped, all others sold. 361 went to the Black Country Museum (UK) but is now owned by a Lisboa enthusiast.
363-367: These cars were for use on the river side route, running the low-fare Carros de Povo service. The windows had originally no glass but only vertically-striped canvas shades. The large platforms were at the same height as the saloon floor and they had two steps at each entrance. The sparsely-lit interiors were fitted with longitudinal slatted seats. This was the least comfortable type that existed. This was in line with who they

were meant for: the people who had too little money to pay normal fares. Carris didn't want to abstract revenue from the normal service trams. Everything must have been done to deter people who could afford to pay normal fares from using the Carros do Povo. After the cheap-fare service ceased they were fitted with transverse seating in 2+1 configuration for the five central bays, leaving 16 seats on the longitudinal benches. Carris now also glazed the windows. Still with their high platform floors and large standing capacity they must have been less comfortable than other cars and might have been used only for the carros operários.

Disposals: Withdrawn and scrapped in 1948.

475: This car was built in 1909 by the workshops, probably using spare body parts, on a Brill 21E truck. No photos are known of this car, but drawings show a typical Brill semi-convertible

549 is in Rua Alianca Operário near Largo do Ria Seco on 12 May 2009.
Michael Russell

car with six windows, like 343-362. Because of its width it was nicknamed "O Cruzador" (The Cruiser). or "O Couraçado" (The Battleship). The car was rebodied in 1952 becoming a Caixote.

476-499, 503-507: After car 475 there was a reversion to the older type of body with drop windows. J. G. Brill supplied these 5.18-metre-body closed cars in 1912 to 1914. They were similar to 401-474, but had more rounded ends and came with windscreens. The series of 485-499 was used on the newly electrified Estrela line. 503-507 had steel underframes instead of the usual timber.

Disposals: 483 received a Standard body in 1935. Twenty-four (476-482, 484-499 and 506) received Caixote bodies: 476-480, 482, 492 & 506 in 1951-1953 and 481, 484-491 & 493-499 in 1961-1963. The other four (503-505) were withdrawn and scrapped by 1971, and 507 by 1973.

500-502: These 26-seat J. G. Brill cars of 1913, but entering service in March 1914, were mounted on Brill Radiax E1 trucks. The bodies with drop windows were a longer version of those of the other Brill cars ordered at about the same time. The order was not repeated, and all three cars were withdrawn in 1968. Radiax (radial axle) trucks were developed by Brill in 1910 as an attempt to have a 2-axle truck with a longer wheel base than common trucks and suitable for longer cars without the necessity to use bogies. 500-502 were considered hard on the track and tended to be used on straight routes such as the 16.

Lifeguards and liveries

All of the pre-1915 American-built cars were supplied with Providence-type lifeguards, known locally as grelhas, and these lasted until about 1950. Replicas have been fitted to museum cars 283, 330, 444 and 508, and tourist cars 1 and 2. The early liveries are unknown, but if they were not yellow from the start, then the trams soon became yellow with white. That gave them their nickname "Amarelos", from the word for yellow. Older residents advised John Price in the nineteen-sixties that they remembered brown trams but could not elaborate. This may refer to the cars which ran on former NCAML cable routes or it may be because the St Louis Car Company, in the absence of an instruction to the contrary, would paint cars in Tuscan Red, hence the red livery applied to the tourist cars 1 and 2. From 1969-70 an increasing number of cars carried all-over advertising, but the rebuilt Remodelados entered service in the traditional Lisboa livery.

Second generation electric trams

After the First World War, faced with considerably increased prices for imports against earnings depressed by industrial and social unrest, Carris decided to build its own tramcar bodies at Santo Amaro works. The first 24 Carris-built cars were American-type clerestory-roof semi-convertible cars similar to 475, possibly built under J. G. Brill licence, but the later cars had domed roofs and simpler lifting windows. 201 two-axle cars were built between 1928 and 1943, including 85 replacements for 1901 cars.

Seating capacity with 2+2 seating was 24, with space for 15 to stand on the rear platform and six at the front, the resulting total of 45 being presumably supplemented by an unspecified number standing inside, certainly the case latterly. All the Standard cars had six-bay domed-roof bodies, 8.38 metres long and 2.38 metres wide, and most weighed 10.73 tonnes.

The Standard cars built as replacements for American-period

A close up of the controls of a hilly route car. The K33 controller is surmounted by an interlock for the air track brakes. The hand track brake is controlled by the larger (nearest the windscreen) of the two wheels on the right; the smaller wheel is for the hand wheel brake. The air brake pedestal is to the right of the instrument panel. *B. King*

A close up of the controls of 233 taken on 12 July 1982, the hand track brake wheel is on the left and the hand wheel brake wheel on the right. *L. Folkard*

cars had inherited Brill 21E trucks and GE59 motors but those of the "new build" programme had swing-link trucks; at first Brill 21 ESL, and later from Maley & Taunton, with 1.98 metre wheelbase. These were fitted with Metrovick MV 115 motors, rated by Carris at 45 hp (continuous), giving improved performance on Lisboa hills. Controllers were either Dick, Kerr DB1 Form K3 or K33, or BTH B510, each one fitted with a Maley & Taunton brake interlock. They had a white rectangle painted at the centre of the truck side-frames; cars thus marked had 45-hp motors, back sanders; and six separate braking systems, viz: air-wheel, hand-wheel, air-track, magnetic-track, hand-track, and run-back. They were the only cars permitted to work the steepest routes.

508-531: Inspired by car 475, these were the first electric cars to be built (rather than assembled) at Santo Amaro works, which explains the choice of 508 for the museum fleet. They were true semi-convertibles, with clerestory roofs containing window pockets. Several cars were later fitted with Metrovick 45-hp motors, and Dick Kerr DB1K33 controllers. All 24 cars were latterly single-ended and operated with trailers.

Disposals: 508 restored in 1991 as a Carris tourist car, now in the museum; 525 to Alvito Children's Park in 1981, replacing 283; sixteen sold to USA. The other six were scrapped by 1980 and their trucks and equipment sold separately.

601-612: These larger domed-roof vehicles were known as "Estrela" cars and were built by J. G. Brill in 1927 for the

802 is seen on a special trip at Praça Comércio on 18 May 2008. *O. Brison*

ex-cable route (the present 28). They were later prohibited from operating on this route. Possibly their greater length than standard cars presented clearance issues on tight curves when the track was doubled. They had large sliding side windows (as on some South African trams). Interestingly the LET was at that time owned by a South African corporation that also owned several South African tramways and similar windows were found on trams in Port Elizabeth. Like 532-551 the platform floors were at level with the saloon floor. These cars were unpopular because only half of the windows could be opened at the same time. 601 and 612 (at least) were reposted as eight bay cars with lifting windows.
Disposals: 609 and 610 received a Caixote body in 1952, the other ten were withdrawn and scrapped by 1972.

532-551: These cars were built in 1927-8 and were the first examples of what was to become the Standard Carris car body, having domed roofs and lifting widows which do not fit into roof-pockets. They differed from later batches in that the platforms are level with the saloon floors: to compensate for this, the cars have smaller wheels. After some years of service on hilly routes, they became single-ended for trailer haulage.
Disposals: Withdrawals began in 1980 with 538 and 541, and the last were 536 and 539 in 1988. 535 and 549 are in the Santo Amaro museum, 533 was sold in 1991 to Kochi in Japan. The others were scrapped.

Bogie cars with Standard bodies, ten trams in the series 305, 327, 328, 329,338, 340, 345, 353, 355, 356:
With the use of the bogies and motors of several existing cars a small series of bogie cars was created with new bodies of the same design as the Standard cars. These bodies had ten windows at each side instead of the six of a 4-wheel car. 353 was fitted with a pantograph by December 1995.
Disposals: 345 and 356 scrapped after accidents before 1995. 305 withdrawn in 1994, the other seven in 1995/6. 329 is in the museum. 355 exhibition car (no motors), later information office for Carristur on Praça do Comércio, now in storage. All others sold, 328 and 353 are owned by a Lisboa enthusiast.

552·571: These were the first true Standard cars (i.e. with lower platforms) and were built in 1931. They were not equipped with full "hilly route" braking systems. After the use of trailers ceased in the 1980s, four were scrapped (552, 567,569, 570) and the other sixteen were upgraded for hilly route service and renumbered 771-785 and 763.

203-282 + 415, 455, 467, 468, 483: When the J. G. Brill open crossbench four-wheel cars were withdrawn in 1932-7, their Brill 21E trucks and GE59 motors were used in new Carris standard-type cars, with air brakes and English Electric K3 or BTH B510 controllers. This did not mean that a new car received the truck and equipment of the old car with the same number.

Instead it received (freshly overhauled) truck and equipment from the float. Five similar cars were built to replace trams lost in accidents.

Entering service:
- 1932: (November-December) 204, 205, 207, 208, 211, 217, 222, 227, 230, 231 (10 cars)
- 1933: 237, 243, 248, 249, 252, 254, 256, 259, 264, 273 (10 cars)
- 1934: 209, 210, 213, 215, 216, 219, 220, 223, 229, 232, 233, 238, 240, 241, 246, 250, 253, 258, 260, 261, 262, 263, 265, 266, 269, 274, 275, 277, 278, 282 (30 cars)
- 1935: 203, 206, 212, 214, 218, 221, 224, 225, 226, 228, 257, 415, 468, 483 (14 cars)
- 1936: 234, 239, 255, 268, 270, 271, 272, 276, 281 (9 cars)
- 1937: 235, 236, 242, 244, 245, 247, 251, 267, 279, 280 (10 cars, 280 was officially finished on 31 December 1937 but entered service only on 20 January 1938)
- 1940: 455, 467 (2 cars replacing two São Luís cars that collided with each other)

Eighteen cars (203-220) were modified to single-ended and worked with the trailers 101-200.

Disposals:
- 203-220 (18 cars) were withdrawn in the early 1970's when trailer operations were reduced.
- 224, 248, 253, 269 (information kiosk at Belém), 271/9, 455, 467 & 468 (9 cars) were withdrawn by 1979.
- 260 (1 car) joined the museum fleet in 1983.
- 231-234, 250, 259 & 268 (7 cars) replaced the original Caixote bodies of 737, 738 and 741-745 around 1985 receiving the numbers of the latter.
- 226 & 276 (2 cars) were modified to single-ended around 1985 and placed on the trucks of 736 and 739 receiving the numbers 761 and 762.
- 229, 239, 249, 251/8, 263, 274, 280 & 282 (9 cars) were withdrawn in 1980-1989.
- 221-223, 225/7/8, 230, 235-238, 240-247, 252, 254-257, 261/2, 264-267, 270/2/3/5/7/8, 281, 415 & 483 (39 cars) were modernised in 1995/6 becoming Remodelados series 541-585.

613·617: Known as the "New Estrela" cars, these were built by Carris as standard-bodied "hilly route" cars, with the equipment described at the start of this section. They had Maley built type 21E swing link trucks, air brakes on the wheels, electro pneumatic brakes on the rails and manual brakes on both the wheels and rails. To reduce the incidence of broken axles, the gears are mounted on sleeve axles. These cars replaced 601-612 on the Estrela route on account of their superior braking system.

701-735: These were standard four-wheel cars built by the workshops for the hilly routes. Car 711 took part in a very thorough series of brake trials in January and February 1936, in the presence of A. W. Maley of Maley & Taunton Ltd, as a result of which cars 711-735 were fitted with various refinements in respect of their braking systems, and were designated universal service cars. Similar equipment was retrofitted to 613-617 and 701-710 and their class became the only class of car allowed to work the Estrela route (28).

Eight cars are still in the fleet for special services or in storage: 617 (numbered 717), 713, 722/3/6 & 732/3/5.
Disposals: Six cars (701/2/7, 714/9 & 724 provided bodies for Remodelados series 541-585. The others withdrawn in the mid 1990's and sold. 703/9 were sold to Sintra. 615 is owned by a local enthusiast, regauged to 1000 mm and stored at Sintra. 704 and 728 went to La Coruña in addition to 712.
801-810: In the late 1930s a need was felt for further bogie cars for the riverside route, and orders were placed for five sets of Maley & Taunton hornless equal-wheel bogies, Metrovick MV109 25-hp motors and English Electric DB1 Form K33 controllers. They were used for five 1939 built bogie cars with domed roofs with roof windows, rounded ends and flush sides. These cars were numbered 801 to 805. In 1943 five cars more were built getting the numbers 806-810. They were completed with Brill 22E trucks, GE59 motors and GE K10 controllers from the float.
Disposals: 802 was withdrawn in 1983 and added to the museum fleet; 801 was withdrawn by 1985, 803-805 were held in reserve from 1989 until withdrawn in 1991. 806 was withdrawn in 1980 to provide trucks and equipment for crossbench museum car 283; 807-810 were withdrawn by 1994. 807 sold to Sóller (Mallorca) but not used and now preserved in Spain.

Third generation electric trams

When tramcar construction was resumed at Santo Amaro in 1947, a new and simpler design was adopted, with flush sides and straight dashes. The first ten 4-wheel cars and the ten bogie cars were double ended, traditionally constructed and had new trucks and equipment. The later batches were single-ended and had aluminium body panels, fixed double transverse seats and latterly fibreglass roofs. They had equipment and in some cases also trucks from earlier series.

Production was not continuous, few trams being built during the initial build-up of bus services (1948-51) or up to the time of opening of the metro (1956-60), but the process was resumed in 1961 and continued for three years. All together Santo Amaro works produced 110 motor cars and assembled 100 trailers between 1947 and 1963.

736-745: Double ended four-wheel cars 736-745 were built in 1947, with Maley & Taunton swing-link trucks, MV 115 45-hp motors, English Electric K33 controllers and "hilly route" braking systems. They were traditional in layout with full saloon bulkheads. From 1983 these ten bodies were replaced by reconditioned bodies from withdrawn Standard cars, fitted with power operated doors and provision for one-man operation, which was introduced in 1984. 736, 739 and 740 became single-ended and were renumbered to 761-763, the others retained their numbers (737, 738, 741-745) and became basically equal to the existing series 613-617 and 701-735.
Disposals: The 1947 body of 741 was set aside and was re-trucked and restored in 1993-4 for the museum; the other 1947 bodies were scrapped.

901-910: These double-ended bogie cars were built using Maley & Taunton equal wheel bogies. The bodies were a "stretched" version of 736-745. They had the same type equipment as 801-5.
Disposals: 903 in 1978; 905/7/8/9 in 1983; 901 in 1984; 902/4/6 & 910 in 1989. 910 sold in 1991 to Kochi (Japan) and regauged to 1067 mm: 904 kept for Carris museum.

Car 744 is one of the cars stored for many years, and now back in service on tourist duties. It is seen near Largo Santo António De Sé in September 2014. *L Vieira*

283-304, 306-322, 403<21 cars>470, 472-482, 484-499, 506, 609, 610: In 1951 a start was made on building unidirectional Caixote type bodies with aluminium body panels and later with fibreglass roofs for the old trams of the São Luís and Brill types series 400-499. The prototype was 492, which initially had bulkheads and windows similar to 736-745 and 901-910. The windows of the other cars of this series were larger and had narrower posts. These cars had bus-type fixed seats and no bulkheads. Driver-controlled air-operated folding doors were provided at the front only for the cars built until 1957. These cars were also equipped to work with trailers. The cars built in 1961-1963 had air-operated doors at both front and back but were not equipped to operate with trailers.

As the cars were considered as replacements, they received the numbers of abandoned cars, partly old open bogie cars and partly cars lost in accidents. This did not mean that a new car received the motors or the truck and equipment of the old car with the same number. Instead it received (freshly overhauled) motors or truck and equipment from the float. The class totalled 90:

- 1951: 492 (prototype, 1 car)
- 1952-1953: 462, 473, 475-480, 482, 506 & 609-610 (12 cars)
- 1954-1957: 284-304, 306-320 (36 cars)
- 1961-1963: 283, 321/2, 403/8/9, 422/8/9, 430/2/3, 442/3/5/9, 451/3, 461/3/5/6, 470/2/4, 481, 484-491, 493-499 (41 cars)
- 283 was built in 1962 (the year after the withdrawal of instruction car 283) and was renumbered 292 when bogie car 283 was reinstated.

Car 498 was used in 1980 for trials with one-man operation, with "deadman" control equipment.

Disposals: Withdrawals began in 1974 and were completed with 422, 429, 442 and 490 in 1994. 506 is in the museum. 442 is owned by an enthusiast.

Standard cars created in the 1980s using parts of other cars.

With the reduction of the network and the ageing of the fleet it became necessary to modify the fleet to be best suited to the remaining network. The oldest 4-wheel trams and smallest series had already been withdrawn by 1980. New articulated trams were considered for the level lines, but were never ordered. The newest trams, the Caixotes, were less successful than their predecessors, the Standard cars; the lower bodies of the Caixotes were too wide for prevailing traffic conditions. The Standards had also become emblematic for the city. In the 1980's a new fleet of Standard cars was created using parts of older trams. These "new" trams were better suited for the hilly lines than the cars from which they received their bodies.

761-763, 771-785: These eighteen cars were created in 1982-4. They were all modified for single-ended one-man operation and had a blind left side. For 761-763 the trucks of the Caixote cars 736, 739 and 740 used with bodies from 276, 226 and 568 respectively. 771-785 were the result of modifying the best cars of series 552-571 for single-ended one-man operation, and upgrading their braking systems to full "hilly route" specification.
Disposals: All are withdrawn; 777 in the museum. 785 is owned by an enthusiast.

737, 738, 741-745: From 1983 the bodies of the Caixote series 736-745 were replaced by reconditioned bodies from withdrawn Standard cars with power operated doors and provision for one-man operation, which was introduced in 1984.
These seven cars became double-ended and are basically equal to the existing series 613-617 and 701-735. 741/4/5 remain in the fleet for special services. The others were withdrawn with the modernisation of the fleet in the mid 1990's. 743 was sold to A Coruña. 738 is owned by a Lisboa enthusiast.

Trailers
Open cross-bench trailers 153-202
The first fifteen trailers were former mule trams delivered in 1898 by Brill. They got the numbers 188-202 and had seats for 32 on eight cross-benches. Carris built 35 more cars of the same design but with nine cross-benches. The last twelve (153-164) had Maley & Taunton pedestal-type running gear.

Disposals: 195 in 1939, 183 in 1947, 194 in 1948, 189 in 1949, 187 and 200 in 1950, 181, 192, 196 and 201 in 1951 and the remaining 40 trailers in 1952-1953.
Except for the Brill cars probably about twenty more other mule cars had a career as trailer behind the electric trams. The exact number as well as the types and numbers are unknown. The annual reports from 1912 until 1917 specified nine closed trailers, three with capacity for 27 and six for 25 (including standing) A photo shows both an open cross-bench type and a closed type. They were all withdrawn by 1930.

Closed trailers
101-200: The withdrawal of the cross-bench trailers by 1953 did not bring the end of trailer operation, for 100 new trailers were built in 1950-5. These modern trailers were assembled by Carris from parts supplied by Metal Sections Ltd of Oldbury (near Birmingham) whose main product was metal bus body framework; their other tramway contracts included cars for Bangkok, trailers for Hong Kong and parts for Blackpool 761/2. Running gear was that originally supplied by Maley & Taunton in 1926 for trailers 153-164, with further sets presumably manufactured locally.
Disposals: Trailer withdrawal started in the early 1970's with the closure of several routes they were used on (Benfica, Lumiar, Areeiro, Carnide). In 1979 there were still 34 available. The last were withdrawn in 1988; 101 is a museum car, and 173 has replaced motor car 269 at Belém.

For some years Lisboa has operated a special service for children at Christmas. In this view taken in December 2017, "Santa" is changing the points in Praca do Comércio to allow the car to gain access to the stop.
L.Vieira

Works cars

389-399: These numbers were used for works cars. Their history however is quite obscure. It appears that until the 1930's only the numbers 395-399 were occupied.

J. G Brill supplied two sprinklers in 1901, nicknamed Pipas, which had a weight of about 20 tons when full. They were withdrawn in 1929. Their trucks were used for other purposes and the tanks were used from 1940 for gasoline storage. They might have had the numbers 398-399.

Records of 1932 show the cars 389-399 being 1901 built freight cars with a weight of 9995 kg, 21E trucks, B18C controllers, two GE 59 motors, manual brakes on tracks and wheels and rheostatic brakes. Records of 1940 show them having a length of 8.25 m. Records of 1979 show them still as zorras, but now with a weight of 11,785 kg and 7000 kg capacity. Again 21E trucks and two GE59 motors, but now with EE DB1K33 controllers, manual track and wheel brakes, air wheel brakes, electro-pneumatic and electro-magnetic track brakes and rheostatic brakes.

In 1906 Brill delivered a car with two cranes. This might have had the number 397. In 1932 this number belonged to a small zorra with weight of 5610 kg and only one GE59 motor. It seems not have existed by 1940. Later the number 397 was used for open car 247 until this car was sold to Detroit.

In 1907 Brill supplied a freight motor that had no cranes and perhaps (records are not clear) a general-purpose "construction-car". These might have the numbers 395-396. Records of 1932 show these numbers belonging to freight cars with a weight of 9025 kg, modified 21E trucks, two GE59 motors, K10D controllers and manual track and wheel brakes. Records of 1940 show 395 as a freight car with a length of 8.25 m and 396 as a tank car with a length of 5.72 m. The latter had disappeared by 1979. In the same year 395 was recorded as a freight car identical to 398-399.

The first records found for 393-394 are of 1940 with them being 1933 built 8.25 m long freight cars. 393 had been withdrawn by 1979 when 394 was recorded as a 15,000 kg freight car with capacity of 5000 kg, 21E truck, two GE59 motors, BTH B510 controllers and manual track and wheel brakes, air wheel brakes and electro-pneumatic and electro-magnetic track brakes.

Records for 1979 show 392 being a 1952 built freight of 10,495 kg and 7000 kg capacity, 21E truck, two GE59 motors, K10 controllers and manual rail brakes and rheostatic brakes. The car was withdrawn after 1979. No records were found about 391, probably it was a freight car.

390 is a 1953 built car for shunting. It weighs 7535 kg, has a 21E truck, two GE59 motors and manual wheel brakes and rheostatic brakes. This small open uncovered car has only one controller as the driver has an easy view in all directions.

389 is a 1956 built rail-grinder car. It weighs 10,410 kg. The truck is 21E and it has two GE59 motors. It has manual and air wheel brakes while the rail-grinder also works as electro-pneumatic rail brake.

There were also a number of rail and sleeper wagons, of which in 1988 ten remained numbered 62-65, 68-71 and 73-74. In 1988 all the works cars and their trailers were renumbered in a single series from Z01 to Z16, and some have been remounted on steel underframes from scrapped passenger trailers. Until the early 1990's the works vehicles were painted grey, but they are now painted yellow to make them more conspicuous. Wagon 07 (ex-68) is a museum car.

The current fleet
The articulated cars (Articulados)

The busiest tram route in Lisboa is route 15 from Praca da Figueira along the industrialised riverbank to Belém and Algés. Most of it is street track, but there is about 2.5 km of paved reservation suitable for high-speed running, in an area formed by land reclamation in the last decades of the 19th century (the Aterro). The route has always had special rolling stock (three generations of bogie cars) and the Swiss consultants recommended in 1978 that it should be retained and modernised. The result was a decision to buy 30 government funded six-axle double-articulated cars.

Carris had no wish to incur development costs for a new design, and waited until one of the car-builders offered a suitable "off-the-peg" product of proven reliability that could be adapted to Lisboa's 900-mm gauge. The answer was found in Spain, where Siemens and Düwag had an order in 1992 for 21 cars for València; 12 were subcontracted to CAF (Beasain) and 9 to Meinfesa of Albuixech. Lisboa placed a run-on order for ten cars, six from CAF and four from Sorefame of Amadora, with an option for 20 more, which could not be taken up due to political changes. The CAF cars (501-6) were delivered between February and June 1995, the first Portuguese-built car (507) arriving in October. After driver training, the first entered public service in April on route 15.

The Lisboa Articulados, designed for a route without gradients, are a clever and economical design that gives the advantages of 62% low floor layout without the cost penalty of external power units for single-wheel drives. Only the two conventional end-bogies are powered (each by two three-phase 103-kW ac motors). Above each of them is a 12-seat saloon with a floor height of 700 mm (two steps up from the low floor portion). The floor in the driving cab is slightly higher (720 mm), presumably to allow for the coupler pocket. At the rear end (the cars are single-ended); the cab is replaced by a seven-place curved seat, of which the centre unit conceals a plug-in back-up controller. The rest of the car has a low floor (350 mm) with six single and 28 double transverse seats, of which eight pairs of double-seats fill the central body section which rests on a bogie with four independent wheels (without linking axles). The centre gangway is of normal width, fitting snugly between the 900-mm-gauge wheels. Access is through four pairs of plug doors, with an entry level only 30 cm above the stop platforms.

Power at 600 volts dc is taken by a one-arm pantograph above the centre section, and is fed through inverters and GTO thyristors with SIBAS-16 microprocessor control, planned to give a 35% economy in power use. Top speed is 70 km/h, which is regularly achieved along the Aterro (Avenida 24 de Julho). Service braking is electric (blended regenerative), with an electro-hydraulic disk holding-brake and four electromagnetic track shoes (two in each end bogie). The driving seat is displaced towards the offside, in an enclosed cab with a sliding rear door. The driver is no longer involved in fare collection. Alongside each entrance is a validator.

All the cars are air-conditioned. Windows are tinted throughout the car, to protect riders from the subtropical sun. Bodywork is stainless steel. Livery is yellow with grey skirting, pale blue roof, and a white panel above the windows. The fleet number is in small grey figures below the offside headlight, and is nearly invisible.

The first car (501) arrived from Beasain on 9 February 1995

Lisboa
Santo Amaro
Depot, Works and Museum

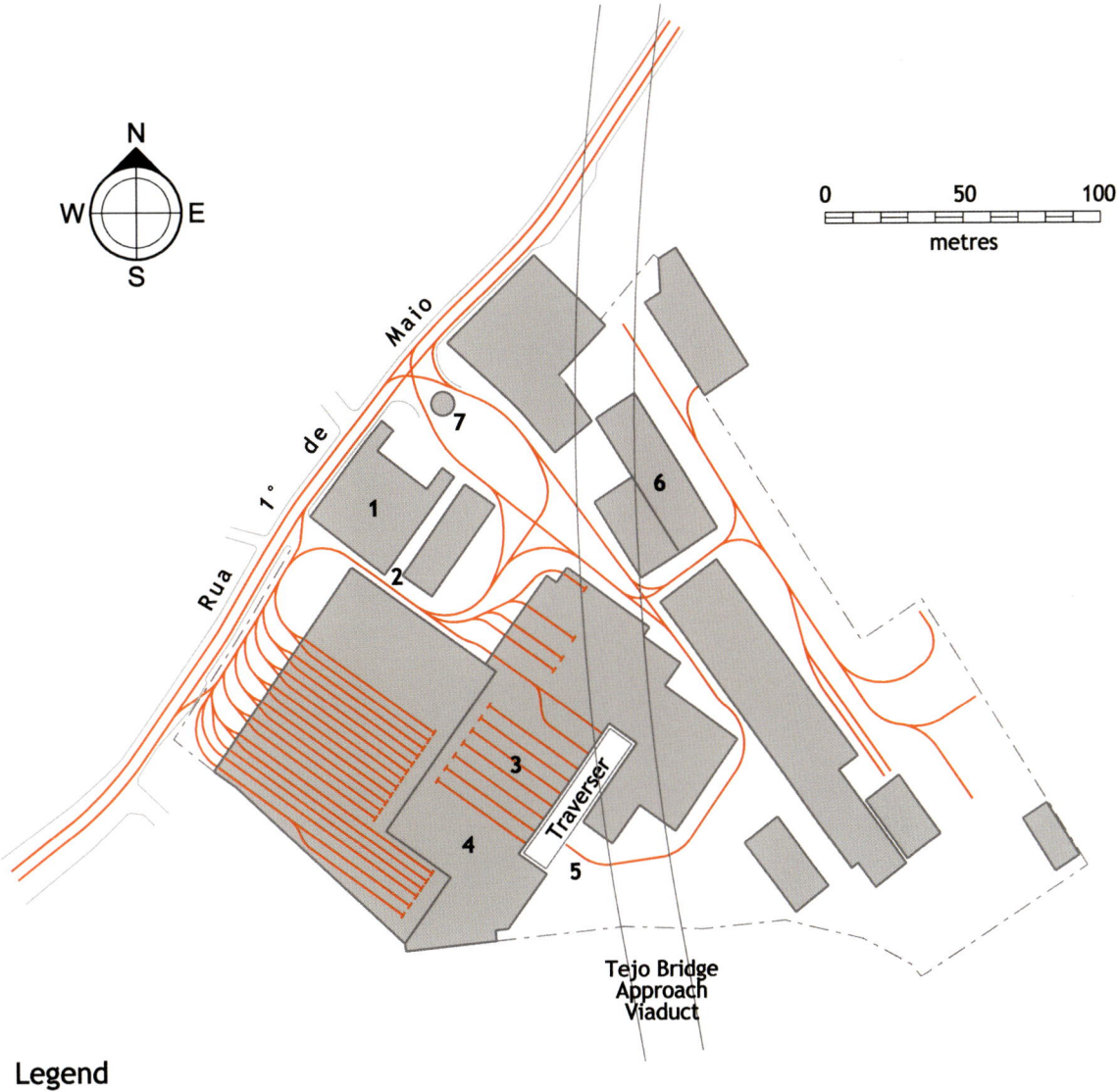

metres

Rua 1° de Maio

Traverser

Tejo Bridge
Approach
Viaduct

Legend

1 Entrance and small exhibit displays.

2 Pick-up point for visitors to travel by historic tramcar to the main exhibits section.

3 Hall containing those fully-restored tramcars that from time to time operate over the city system.

4 Hall containing tramcars that do not (or not regularly) operate on the system. Not rail connected, so tramcars stand on their own short sections of free-standing track.

5 Historic tramcar waits here while visitors visit the exhibits section.

6 Hall 3, containing various motorbuses, tower wagons and any vehicles undergoing restoration.

7 Gatehouse.

© M.J.Russell & R.A.Smith, December 2014. No. 1636, v1.0.

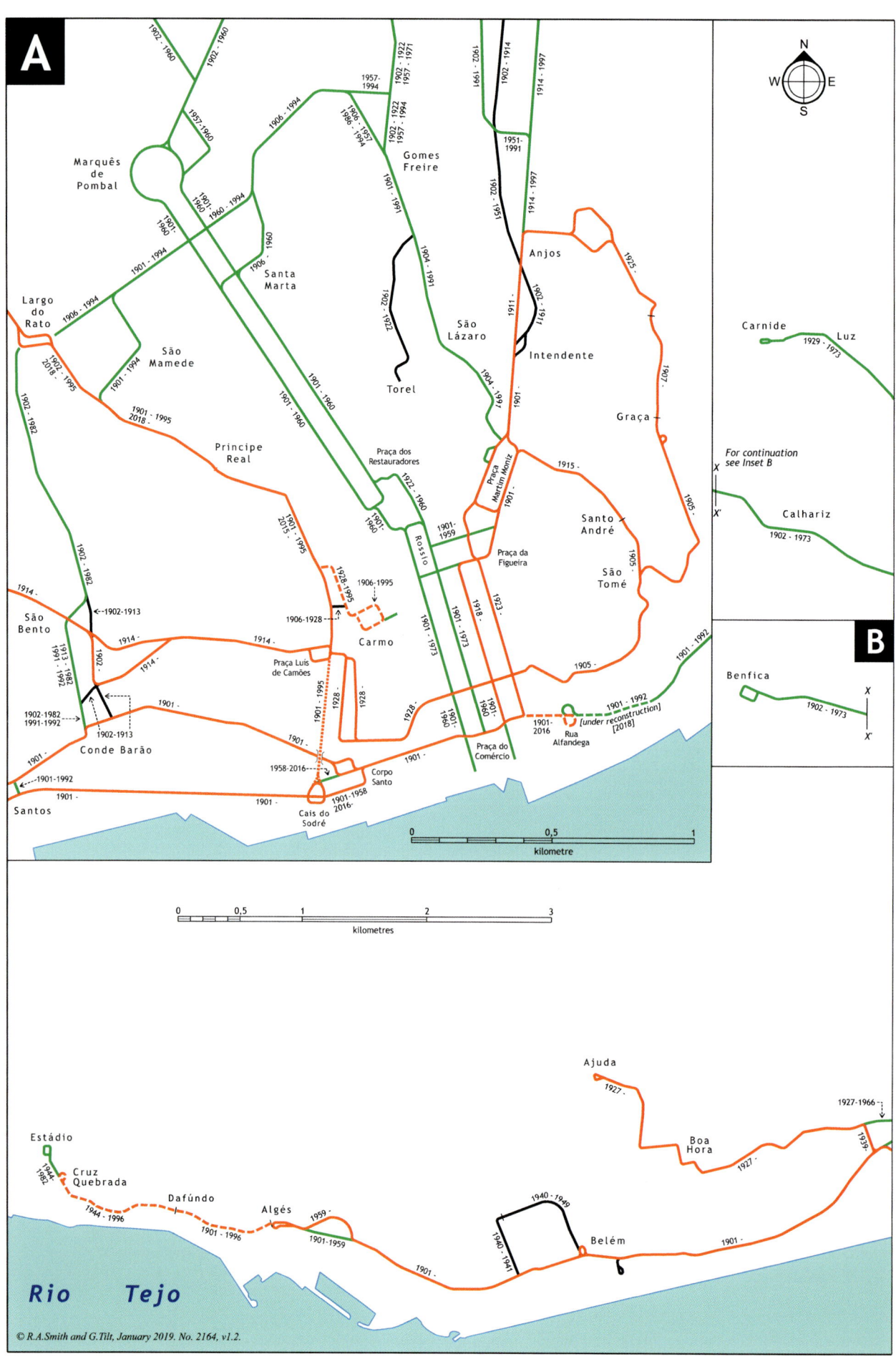

A

N
W—E
S

Marquês de Pombal

1902 - 1960
1957 - 1960
1906 - 1994
1957 - 1994
1902 - 1922
1957 - 1971
1902 - 1991
1902 - 1914
1914 - 1997
1951 - 1991
1902 - 1922
1957 - 1994
1906 - 1957
1986 - 1994

Gomes Freire

Anjos

1925 -

Carnide   **Luz**
1929 - 1973

1901 - 1960
1901 - 1960
1960 - 1994
1906 - 1960

Santa Marta

1901 - 1994
1902 - 1922
1901 - 1991
1904 - 1991
1902 - 1951
1914 - 1997

São Lázaro

Intendente
1901 -
1911 -
1901 -
1907 -

1904 - 1991

Graça

For continuation see Inset B

X
X'

Largo do Rato
1906 - 1994
1902 - 1995
2018 - 1995

São Mamede

1901 - 1994
1901 - 1995
2018 - 1995

1902 - 1982

Principe Real

1901 - 1960
1901 - 1960

Torel

1902 - 1922

Praça dos Restauradores

Praça Martim Moniz
1901 -

Santo André

São Tomé

Calhariz
1902 - 1973

B

Benfica
1902 - 1973

X
X'

1902 - 1982

1922 - 1960
1901 - 1960

Rossio

1901 - 1959

1906-1995
1928-1995
1906-1928
1906-1995

Carmo

1901 - 1973
1901 - 1973
1918 -
1923 -

1915 -

1905 -
1905 -

1905 -

1901 - 1992
1901 - 1992

1914 -
←1902-1913

São Bento
1913 - 1982
1991 - 1992
1902 -
1914 -
1914 -

1914 -

Praça Luís de Camões
1901 - 1995
2015 - 1995
1928

1928

1928

1901 -
1960

1901 -
1960

1905 -

1901 - 1992
1901 -
2016

[under reconstruction] [2018]

Rua Alfandega

1902-1982
1991-1992→
1902-1913

1901 -

Conde Barão

1901 -
←1901-1992

1901 -

Santos

1901 -

1901 -
1958-2016←

Cais do Sodré

Corpo Santo

1901-1958
2016-

1901 -

Praça do Comércio

0   0,5   1
kilometre

0   0,5   1   2   3
kilometres

Ajuda
1927 -

1927-1966

Estádio

Cruz Quebrada
1944 -
1982

Boa Hora

1939 -

1944 - 1996

Dafúndo

Algés
1959 -

1901 - 1996

1901-1959

1940 - 1949

Belém
1940 -
1941

1901 -

1901 -

Rio Tejo

© R.A.Smith and G.Tilt, January 2019. No. 2164, v1.2.

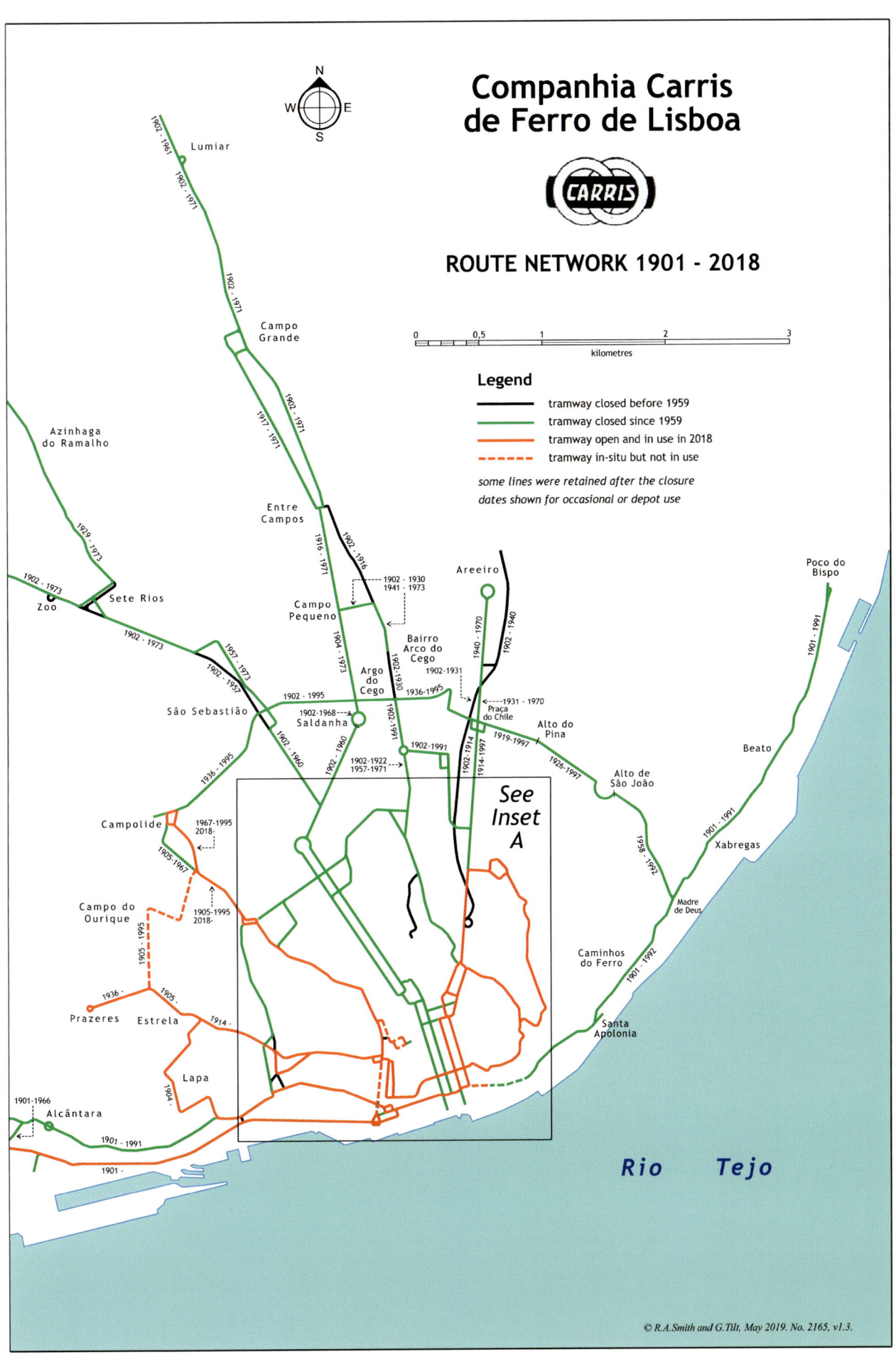

Companhia Carris de Ferro de Lisboa

ROUTE NETWORK 1901 - 2018

0   0,5   1   2   3
kilometres

Legend

— tramway closed before 1959
— tramway closed since 1959
— tramway open and in use in 2018
--- tramway in-situ but not in use

some lines were retained after the closure
dates shown for occasional or depot use

Lumiar

Campo Grande

Azinhaga do Ramalho

Entre Campos

Areeiro

Poco do Bispo

Zoo   Sete Rios

Campo Pequeno

Bairro Arco do Cego

Beato

São Sebastião   Argo do Cego

Saldanha

Praça do Chile

Alto do Pina

Alto de São João

Xabregas

Campolide   See Inset A

Madre de Deus

Campo do Ourique

Prazeres   Estrela

Caminhos do Ferro

Lapa

Santa Apolonia

Alcântara

Rio   Tejo

© R.A.Smith and G.Tilt, May 2019. No. 2165, v1.3.

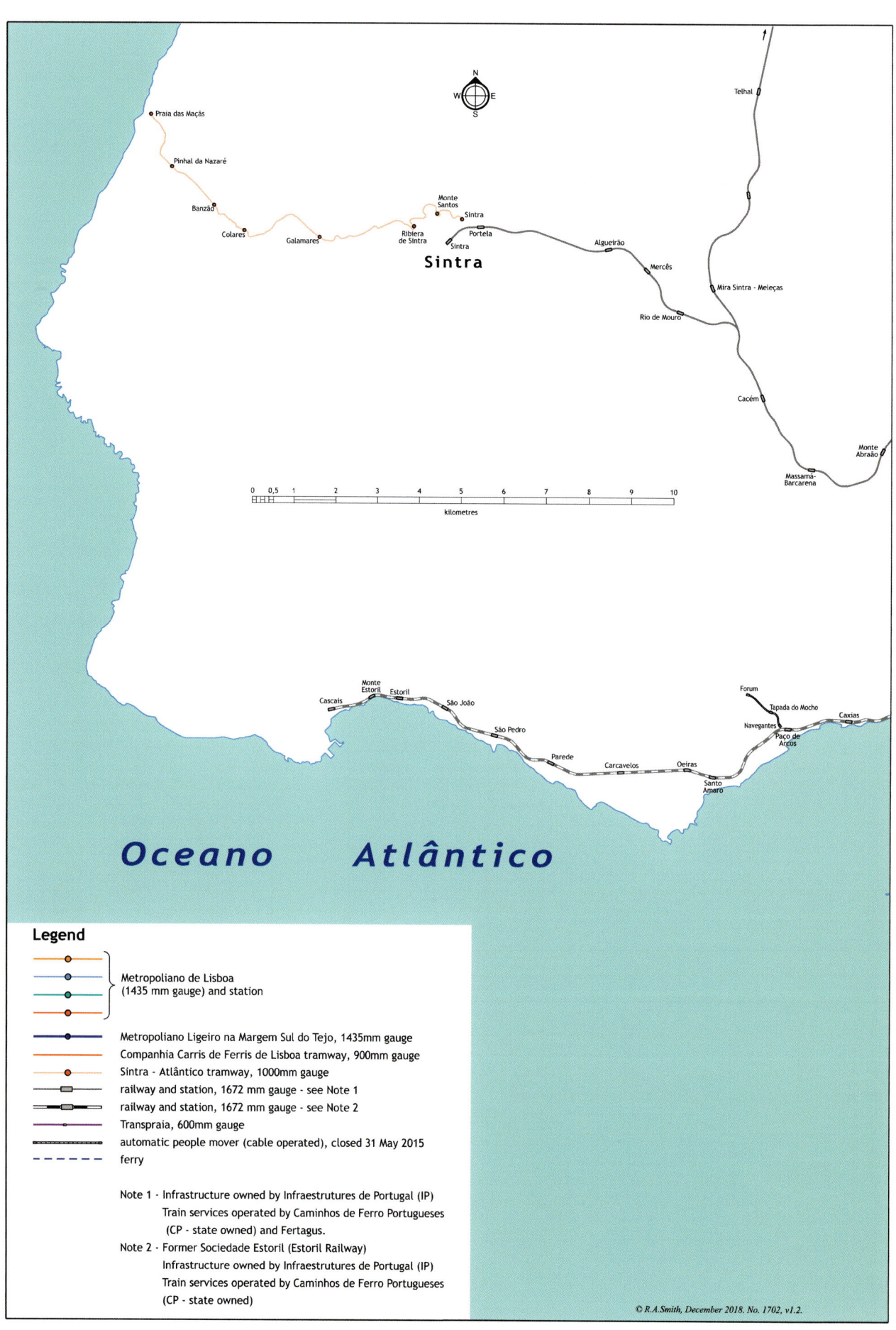

Sintra

Praia das Maçãs
Pinhal da Nazaré
Banzão
Colares
Galamares
Ribiera de Sintra
Monte Santos
Sintra
Sintra
Portela
Algueirão
Mercês
Rio de Mouro
Cacém
Mira Sintra - Meleças
Massamá-Barcarena
Monte Abraão
Telhal

| 0 | 0,5 | 1 | 2 | 3 | 4 | 5 | 6 | 7 | 8 | 9 | 10 |
kilometres

Cascais
Monte Estoril
Estoril
São João
São Pedro
Parede
Carcavelos
Oeiras
Santo Amaro
Navegantes
Paço de Arcos
Forum
Tapada do Mocho
Caxias

Oceano   Atlântico

Legend

Metropoliano de Lisboa
(1435 mm gauge) and station

Metropoliano Ligeiro na Margem Sul do Tejo, 1435mm gauge
Companhia Carris de Ferris de Lisboa tramway, 900mm gauge
Sintra - Atlântico tramway, 1000mm gauge
railway and station, 1672 mm gauge - see Note 1
railway and station, 1672 mm gauge - see Note 2
Transpraia, 600mm gauge
automatic people mover (cable operated), closed 31 May 2015
ferry

Note 1 - Infrastructure owned by Infraestrutures de Portugal (IP)
            Train services operated by Caminhos de Ferro Portugueses
               (CP - state owned) and Fertagus.
Note 2 - Former Sociedade Estoril (Estoril Railway)
            Infrastructure owned by Infraestrutures de Portugal (IP)
            Train services operated by Caminhos de Ferro Portugueses
               (CP - state owned)

© R.A.Smith, December 2018. No. 1702, v1.2.

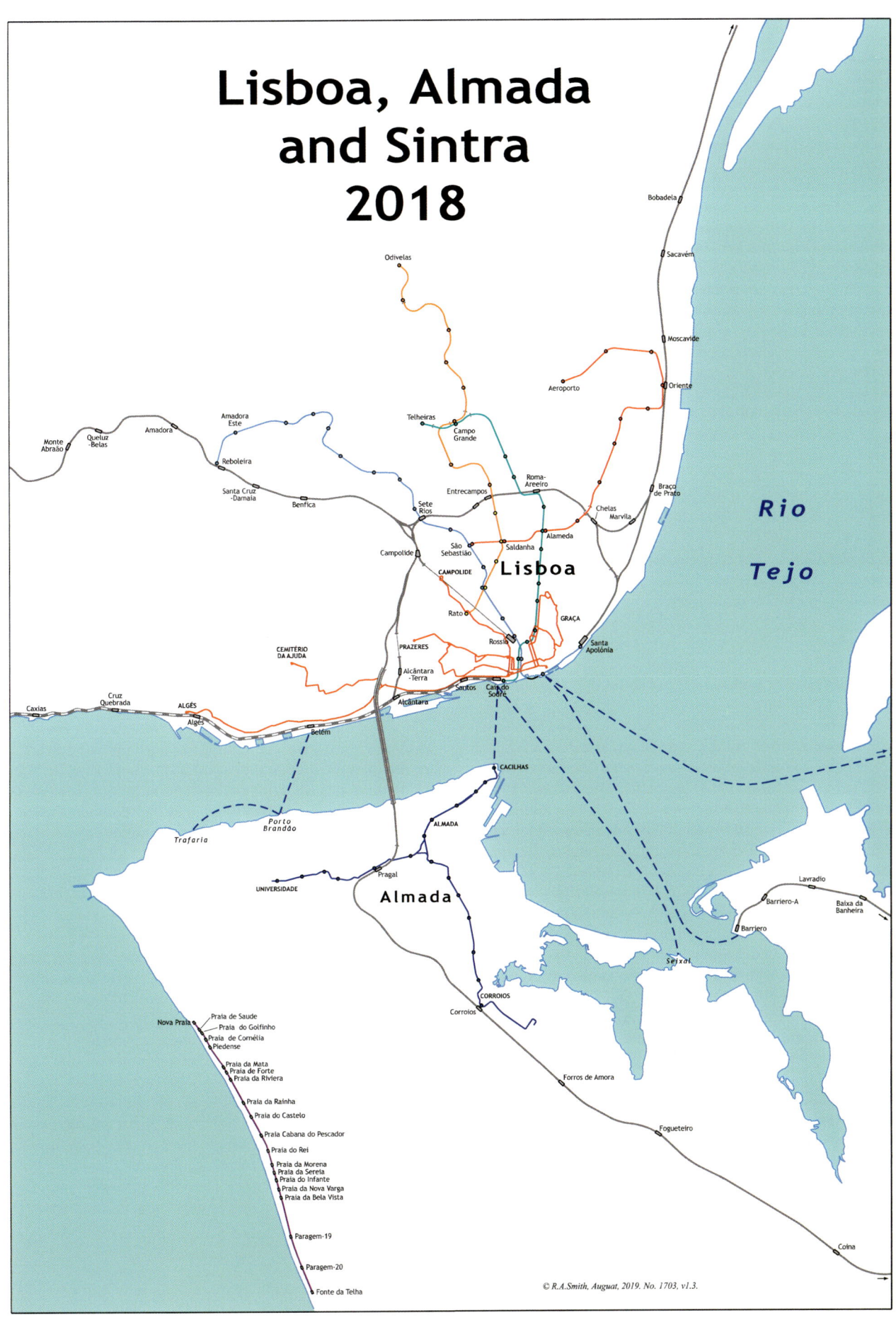

Lisboa, Almada
and Sintra
2018

Bobadela

Sacavém

Odivelas

Moscavide

Aeroporto

Oriente

Amadora
Este

Telheiras

Campo
Grande

Braço
de Prato

Rio

Monte
Abraão

Queluz
-Belas

Amadora

Reboleira

Entrecampos

Roma-
Areeiro

Tejo

Santa Cruz
-Damaia

Benfica

Sete
Rios

Chelas
Marvila

São
Sebastião

Alameda

Saldanha

Campolide

CAMPOLIDE

Lisboa

Rato

GRAÇA

CEMITÉRIO
DA AJUDA

PRAZERES

Rossio

Santa
Apolónia

Alcântara
-Terra

Caxias

Cruz
Quebrada

ALGÉS

Algés

Santos

Cais do
Sodré

Alcântara

Belém

CACILHAS

*Porto
Brandão*

ALMADA

Trafaria

Pragal

Lavradio

UNIVERSIDADE

Almada

Barriero-A

Baixa da
Banheira

Barriero

Seixal

CORROIOS

Corroios

Nova Praia

Praia de Saude
Praia do Golfinho
Praia de Cornélia
Piedense

Forros de Amora

Praia da Mata
Praia de Forte
Praia da Riviera

Fogueteiro

Praia da Rainha

Praia do Castelo

Praia Cabana do Pescador

Praia do Rei

Praia da Morena
Praia da Sereia
Praia do Infante
Praia da Nova Varga
Praia da Bela Vista

Coina

Paragem-19

Paragem-20

Fonte da Telha

© R.A.Smith, Auguat, 2019. No. 1703, v1.3.

and was shown to the press five days later. Public service began in April, limited to the section Praça da Figueira-Belém, while curve clearance was enlarged between Belém and Algés. Operation to Algés began on 1 December 1995.

From 2014 these cars were subjected to a mid-life overhaul.

The remodelled cars (Remodelados)

This title is used by Carris for a fleet of 45 four-wheel trams selected to form the future rolling stock of the four surviving hilly routes (12, 18, 25, 28). The first eight (541-8) entered service on 14 July 1995 on route 25, followed in the autumn by 549 to 553. Further cars were completed at the rate of four per month, with programme completion by 1996. It was intended that eventually these and the articulated cars of route 15 would be the only Lisboa trams in regular public service.

The cars selected were drawn from the 1932-7 series of Standard cars, of which 221 to 282 re-used the trucks and equipments of the 1900 J. G. Brill open cars and 415/55/67/68/83 were replacements for enclosed cars. Although the fleet includes newer Standard cars (613-7 and 701-35), the lower-powered 221-282 series had not been subjected to the body strain imparted by the steepest hills, 45-hp (about 34-kW) motors and air track brakes, and would be released by planned route closures. Starting in 1984, they were given heavy overhauls with structural renewals, and converted for one-man operation. The choice of Standard as against Caixote post-war cars was also influenced by considerations of stronger construction, greater clearance below the waist-rail, and traditional styling. All 45 were in fleet livery, for Lisboa's yellow trams have become a symbol of the city and a major tourist attraction.

German industry tendered successfully for this unusual contract, and set up the "Carris Consortium", comprising AEG-Schienenfahrzeuge, Kiepe-Elektrik, Knorr-Bremse and Ferrostaal. Car 270 was selected as the prototype, and was altered in Lisboa to single ended configuration. It was then taken by lorry to the AEG (ex-MAN) works at Nürnberg in July 1992, followed by a small team from Santo Amaro works. After serving for many months as a full-size-pattern, car 270 was re-assembled in its new form early in 1994, the following work having been carried out:-

Bodywork: Inspected thoroughly and found to be in good condition and suitable for re-use. The door-windows were modified to rectangular to give a more traditional appearance. Car 270 arrived in Lisboa with its original two steps: these were felt unsatisfactory and the production cars were rebuilt to three steps in Lisboa, requiring rearrangement of equipment below the platform. 270 was then retro-fitted.

Truck: New, welded two-axle truck fitted by AEG with Megi-Eiastomer rubber-to-metal primary hornway springs, with the body carried on self-damping leaf-springs, and with roller-bearing axle boxes.

Wheels: Spoked wheels with shrunk-on tyres retained to accord with Lisboa workshop practice.

Current collection: Designed for easy conversion from trolley to pantograph.

Motors: Two truck-mounted, longitudinal 50-kW motors of type ALS 2840 IN by Škoda of Plzeň with cardan-shaft drive to BSI gearboxes on the axles.

Controllers: Separate new front and rear controllers. The leading end has a Kiepe series parallel controller of type NF51 (a 1951 design) with a traditional upright case and handle, using a 24-volt circuit to actuate driving and braking controllers housed

under the rear seat. The 17 power and 15 braking notches give smoother acceleration than the previous BTH B510 type. For safety reasons there is no reversing switch, all backwards moves having to be made by the rear-end shunting controller of Kiepe Type RSH 201.

Resistors: Highly-rated roof-mounted driving and braking resistors in traditional housings; auxiliary saloon-heating resistors beneath forward seats.

Electric brakes: Service braking by highly-rated rheostatic brake, plus 44kN battery-operated 24-volt electromagnetic track brake, operating on a single level of excitation, and applied by a separate push-button. Because of the heavy load on the battery, conventional rotary charging equipment has been replaced by a more powerfull static converter.

Air brake: The air-wheel brake is retained, and takes over automatically from rheostatic brake below 3 km/h to stop and hold the car.

Handbrake: The vertical-shaft handbrake wheel is retained, actuating the wheel-brake through cables. There is no mechanical track brake.

Auxiliaries: Static converter plus battery supplying 24-volt dc supply for control circuit, doors, lighting, electric sanders and magnets.

Objectives: Better performance, greater comfort, quieter smoother ride, lower maintenance costs, traditional appearance.

Car 270 returned to Lisboa in May 1994 for three months of tests. The order was confirmed and deliveries of the fully-assembled trucks began early in 1995. The first few cars (renumbered into series 541-585 in order of completion, 270 becoming 541) were used initially for driver-training prior to 541-8 entering service on route 25 in July 1995.

On 22 April 2021 a contract was signed with CAF for the acquisition of fifteen new articulated low-floor cars for delivery between November 2021 and August 2022. The cars will be 28.5 m long in five sections, single ended with a maximum speed of 70 kph with wire free capability. The contract includes maintenance for five years, the contract value is Euro 43 million.

Sightseeing tours

On 18 April 1965, Carris introduced a sightseeing tour of Lisboa by restored tram. The cars used were St Louis 1901-built saloons 437 and 435 (now 1 and 2) in an ornate red livery. In the 1990's the tourist service was allocated more trams, the first two being from series 700 and later cars being Remodelados. They all received a new number and were painted red. Bogie tram 355 received the same livery and was for many years used as information office on Praça do Comércio, but is now stored in Santo Amaro.

Only the Remodelados are in regular tourist service. 1 & 2 are used as shuttles at the museum. 3, 4 & 10 are in storage.

In 2015 a second, shorter and cheaper tourist service was introduced between Graça and Principe Real. For this service six older trams series 700 were painted green to distinguish them from the Colinhas service. This service operated only during the Summer and was in 2018 replaced by a tourist service Praça do Comércio - Praça da Figueira - Belém using the same green painted cars series 700.

Gallery 2

Lisboa Route 28

Above: Car 701 is at the bottom of Rua do Voz do Operário at Calçada de São Vicente passing the recumbent signalman on 3 May 1964. *L. Folkard*

Left: "Caixote" 497 is in Rua das Escolas Gerais on 3 May 1964. *L. Folkard*

Overleaf: Lisboa 227 threads its way through the narrow Rua das Escolas Gerais Graça on route 11 passing washing and children, in 1969 at a time when more "real" people lived in Graça. *P. Haseldine*

565 on the interlaced track taking the tight corner at Rua das Escolas Gerais on 3 May 1964. The red bat shaped indicator held by the signalman at this point may just be seen. *L. Folkard*

The diminutive two-axle tram 210 looks a giant as it swings around the curves dissecting the alleyways of Graça in 1969. *P' Haseldine*

553 and 574 are seen in Travessa de Sao Tome/Rua des Escolas Gerais on 5 May 2004. *Michael Russell*

A dramatic view of "Caixote" 473 negotiating the junction at São Tomé on 3 May 1964. *L. Folkard*

"Caixote" 429 turns from Rua da Conceição into Rua da Prata on 19 May 1980. Due to the tight clearances, trams crossed to the opposite track to make the turn. This turnout was removed after the closure of Graça circle route 11. When route 12 became a circular the curve had to be relaid to encroach instead on to the pavement in Rua da Prata. *Michael Russell*

741 is seen at the summit of Calçada da San Francisco on 4 May 2016. *Michael Russell*

732 is seen descending the Calçada de San Francisco on 19 May 1980.

Michael Russell

717 at Rua Vitor Cordon with daylight visible under the car on 3 May 2005. These cars returned to service after a period in store when building works necessitated the use of double ended cars on route 28. Note the interlaced curved layout in the background. *Michael Russell*

717 about to descend the Calçada da San Francisco on 3 May 2005. These cars returned to service after a period in store when building works necessitated the use of double ended cars on route 28. Note the reverse curves in the foreground. *Michael Russell*

Estrela at some time between 1904 and 1914, showing 2 cable trailers, the one on the right being manhandled on the turntable outside the depot. They are dwarfed by a crossbench 4-wheeled electric car of the 203-282 class. *Bibliothèque Nationale de France*

Cable Trams, Funiculars and Lifts

By João Firmino, Owen Brison and Ernst Kers

Cable trams

Lisboa is a city built on hills. Mule or horse trams could only go up the steep gradients with the use of extra animals. As the mules and horses were the largest cost of the whole tram operations, it was a too costly affair to have many or long steep inclines within the network. Because of that few mule tramlines went into the hilly areas. In 1873 A.S. Hallidie introduced in San Francisco a system employing a cable running continuously in a conduit below the track, which could be picked up or released by a gripper with metal jaws suspended from the car and passing through a slot in the road. Many such systems were built in America and other countries. Three cable tramlines, three funiculars and three vertical lifts were built in Lisboa. For all these projects, except for the Chiado vertical lift, the promotor and designer was Raúl Messnier du Ponsard, an engineer born in Porto of French parents. All three cable tramlines had a gauge of 900 mm. The funiculars and vertical lifts are described separately.

Estrela

The Nova Companhia de Ascencores Mechanicos de Lisboa (NCAML) which already operated the Lavra and Gloria funiculars, opened its first cable tramline on 15 August 1890. The route was from Largo de Camões via Rua do Loreto, Largo de Calhariz, down the steep (12% over about 200 m) Calçada do Combro, Rua dos Poias de São Bento and up the steep (10% over about 500 m) Calçada da Estrela to Largo da Estrela. The length was 1.7 km. The depot and winding house were located at Estrela. The installations came from the Maschinenfabrik Esslingen. Photos, including a works photo of Esslingen, show small open grip cars with a small closed trailer with four windows at each side. There were six of these combinations with capacity for 10 on the grip car and 24 in the trailer. These combinations were replaced in 1896 by larger closed cars which were operating as single units with the gripman on the front platform. They looked quite similar to the trailers but had five windows. The line was double track over most of its length, but single tracks at the narrowest locations. The system used a side slot in the left rail. The trams were unidirectional which meant a solution had to be found for turning at the termini. Originally tight turning loops were installed. The one at Camões was in the north-east corner of the square. The one at Estrela had a trailing switch for the track to the depot located shortly beyond.

Estrela gripper car 6 with unidentified trailer. (Benoliel collection) *J. Firmino*

Car 3 on turntable at Largo da Estrela. *J. Firmino*

Estrela in 1913 when the line closed. *J. Firmino*

However a tramcar pulled by a cable with normal speed (Estrela line 2.5 m/s = 9 km/h) for straight sections doesn't perform well in tight curves. Raul Messnier wrote in an article in the Gazeta dos Caminhos de Ferro of 16 October 1890 that dedicated equipment had been ordered to get the trams slowly through the end loops. He didn't give details about how this worked but the ship with the equipment sank on its way to Lisboa. The NCAML decided to use small turntables instead which were installed before the opening of the line. The turntable at Camões was located in a crossover in front of the Rua do Norte, about 50 m before the loop. As the grip car + trailer combinations were far too long for the turntables, this must have resulted in extensive shunting, probably using manpower. Apparently neither the CML nor the NCAML were happy with it. Messnier designed first for the Camões a system with an auxiliary cable that could move the cars slowly through the turning loop. But in the end a more simple solution was chosen by taking advantage of the local existing light incline. The tramcars went through the loop by gravity. The proposal for this solution was sent to the CML a month after opening of the line and apparently implemented soon after. The crossover with the turntable was removed. At Estrela both tracks were converging into the turntable located at the same place as the turning loop. In November 1891 the NCAML proposed to make a new turntable (perhaps the one from Camões?) in the track to the depot in front of the entrance which would also accommodate the track foreseen for extending the line. It's unknown if this change was made. A 1905 drawing shows there was no turning loop anymore but both tracks converging into a stub track with in the left (arrival) spur a very small turntable to turn the cars and to give access to the depot. According to the contract of 1882 for the concession, an extension had to be made to Rua de São João de Bemcasos (now Rua Silva Carvalho) before February 1905. It was never realised. Instead the NCAML bought two Dion-Bouton buses in 1904 for a service between Estrela and Amoreiras. Carris protested on 28 November 1904 and the bus service was ended after only a few hours. The buses were sold to the fire brigade. The cable tramline was closed on 3 July 1913. On 1 March 1914 the replacing electric tramline was opened.

Graça

The Graça cable tramline was opened by the NCAML on 26 March 1893. It had its downtown terminus on the Largo Fernandes da Fonseca where now is the northern end of Martim Moniz at the corner of Rua da Palma. The route, about 900 metres long with a gradient of at least 1 in 10 for almost 700 metres, went up through Rua dos Cavaleiros, Calçada de Santo André, turned to the left passing under the Arco de Santo André (later demolished), again to the left into Calçada da Graça and then to the right to terminate at the north end of Largo da Graça. Tracks and cable however went on for almost 200 m more into the Rua da Graça to the location of the winding house and depot. These were located where in 1928 the Royal Cine was built. The building of this cinema is now in use by a supermarket. The line had double track, but interlaced at the narrowest points. The system used a central slot and the cars were bi-directional. This avoided complex solutions for the

Graça 1909 picture in Rua Fernandes de Fonseca/Carreirinha do Socorro. *J. Firmino*

termini as were made for the Estrela line. However the two tight curves between the Calçada the Santo André and the Calçada da Graça couldn't be passed with use of the relative fast moving main cable. The solution was found by installing an auxiliary slow moving cable between the top of the Calçada de Santo André and the bottom of the Calçada da Graça. Photos show two types of cars. It seems the originals were closed cars with six top curved windows at each side, longitudinal benches and roofs with a clerestory. Apparently these were between 1897 and 1899

Graça – Calçada de Santo André – passing through arch of same name. *J. Firmino*

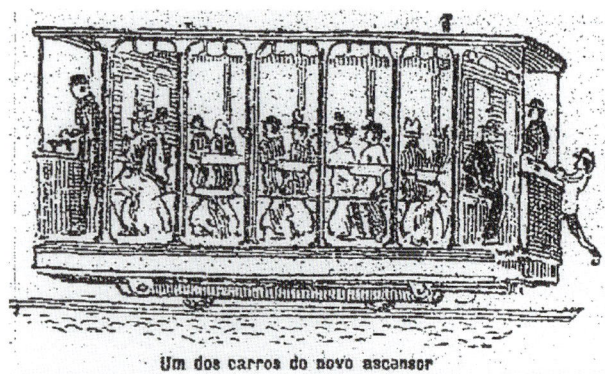

Ascensor de São Sebastião – engraving from the periodical d'O Seculo. *J. Firmino*

replaced by a half open type with cross-benches and flat roofs. A photograph exists which shows a mystery car numbered 5, with a 5-window body similar to the Estrela cars, but with a different truck and gripper controls. It may have been transferred from the Estrela line. The line was closed on 25 May 1913. Works for removing it to make construction of the electric successor possible started the next day. The replacing electric tramline was opened on 1 January 1915. It didn't go up all the way to Graça but connected at the top of the Calçada de Santo André with the already existing short branch at São Tomé of the electric line via Rua do Limoeiro and Escolas Gerais to Graça.

After the NCAML cable tramlines were closed (Graça 25 May 1913 and Estrela 3 July 1913), there was an advertising of the NCAML in the Diario de Notícias of 8 July 1913 offering for sale 13 closed and several open cars. The ad said they were suited for use as e.g. beach house or guards shelter.

São Sebastião

The third cable tramline, the ascensor de São Sebastião, was operated by another company, the Companhia de Viação Funicular (CVF). The concession was given by the CML already in 1888 together with concessions for a number of other but never realised lines, to the Syndicato Sanches de Baena. During the nine years that followed the concessions were transferred twice to another company, the winding house was built and equipped and material to construct the line was bought and stored. But the money to finish the works failed until 1897. The line was finally opened on 15 January 1899. The downtown terminus was on Largo de São Domingos aside to the Teatro Dona Maria II. From there it followed the route Rua das Portas de Santo Antão - Rua de São José - Rua de Santa Marta - Rua de São Sebastião da Pedreira - Largo de São Sebastião da Pedreira - Estrada de Palhava (now Rua Dr. Nicolau Bettencourt) with the terminus just across the Estrada de Circunvalação (now this part of Circunvalação is the Rua Marques de Fronteira). Here was also the winding house and the depot. The length was about 2.6 km and the line had double track except for a short section in Rua de São José with interlaced tracks. Like the Graça line a central slot was used for the cable and the cars were bidirectional. Although no photos are known, it appears there were on the moment of opening in 1899 four open cross-bench cars with six benches. Two closed cars were added to the fleet in the first half of 1900. All six cars were locally built. While both the Estrela and Graça lines had long gradients of at least 10% over large

parts of their routes, the São Sebastião line did only have a few short sections of about 7% but was for about half its length close to only 2 to 3 %. During its existence the line suffered major technical problems causing complete shutdowns during February and March 1899 and again in February 1900. On 3 January 1901 the cable broke. The CVF was financially broken too. The tramcars remained paralysed on the street for many months and the line was never reopened.

Funiculars

Although not real trams, the Lisboa funiculars are part of the Lisboa public transport and operated first by the same company that operated the cable tramlines and are now operated by the CCFL. As the funiculars are in public streets, the cable is in a conduit slot. The original installations were, like with the cable trams of the NCAML, delivered by Esslingen.

Lavra 1884-1914

The funicular do Lavra opened on 19 April 1884 by the Companhia de Ascencores Mechanicos de Lisboa (CAML). It's about 175 m long and has and average gradient of about 23%. For (emergency) braking the cars had at the cog-wheels running on a Riggenbach ladder rack instead of normal rails at the outer sides of the line. A photo taken in or short before 1909 proves that at that time the ladder racks still existed. Probably this system was only changed with the electrification. Working was

Lavra car1 seen at the lower terminus on 3 May 2005. *Michael Russell*

Lavra car 2 at passing point, 28 April 2014. *C. F. Isgar*

originally by water counterbalance whereby tanks on the cars
were filled with water at the upper terminus and emptied at the
lower one. Since 1888 steam engines were used to propel the
cable. In October 1884 the CAML transferred all its concessions
and belongings, included the Lavra funicular, to a new company,
the Nova Companhia de Ascencores Mechanicos de Lisboa
(NCAML) which was later also going to operate the Gloria and
Bica funiculars and the Estrela and Graça cable trams.

Gloria 1885-1914

The Gloria funicular opened on 24 October 1885. It's about
250 m long and has an average gradient of about 17%. Also here
a Riggenbach ladder rack was used for emergency braking. The
cars had four normal wheels running on rails. There was only one
cogwheel on each car mounted outside on the normal wheel
at the lower end of the line. Here ladder rack(s) and cogwheels
were at the "inside" of the line. The cars had knife-board benches
on the roofs, which could be accessed by a winding staircase on
the platform at the hill side. The Gloria changed to steam power
in 1887.

Bica 1892-1914

The Bica funicular opened on 28 June 1892. It's about 260 m long
and has an average gradient of almost 18%. It was steam powered
from the beginning and had no ladder racks for emergency
braking. The original cars were open of stepped type.

Modernisation

In 1914 all three funiculars were closed for electrification. The

Gloria, probably on the day of inauguration. *J. Firmino*

propulsion was going to be electric and seven (apparently one
spare) new cars were ordered in 1913 from United Electric Car
Co. The Gloria funicular reopened on 3 September 1915 and
the Lavra on 5 December. The Bica however suffered a serious
runaway accident when the new system was tested in October
1916. Due to this incident, and probably the war circumstances
didn't help either, the Bica was not reopened for over a decade.

Gloria car 2 is almost at the top of the incline on 4 May 1964 with two pedestrians taking advantage to assist their ascent. *L Folkard*

Gloria car 2 has just commenced the descent on 12 July 1982. *L Folkard*

Lavra and Gloria 1915 until now

The Gloria and Lavra lifts were rebuilt to the powered counterbalance system, and as such are unique in the world (the only other examples, in Seattle, Providence and Sydney, closed many years ago). The Lavra and Gloria cars are connected by a cable, but this merely runs over an idler pulley at the top and is not powered. Instead, the cars themselves are fitted with 2 x 45 hp electric motors, drawing current through an overhead wire and a GE simple five notch controller, and are driven up and down the hill in a similar manner to a tram, the descending car thus helping to pull up the ascending car. In effect, the ascending

Gloria car 1 setting off from the lower station on 13 September 1995. *Michael Russell*

The two Gloria cars, 1 and 2, parked during the night-time lull in service and seen at the intermediate passing loop at 02:05hrs on 16 September 1995. *Michael Russell*

Glória car 2, close up of cranked truck, 26 April 2014. *C. F. Isgar*

Bica cars 1 and 2 passing, 4 May 1964. *L Folkard*

car's controller regulates the power supply to both cars, an arrangement which necessitates four overhead wires. The four cars involved have horizontal bodies and inclined under-frames like a cliff lift, but ride on cranked Brill type trucks. Current is taken (or supplied) through wheel collectors on pantograph mountings, and a light in each cab glows to indicate to the driver of the ascending car that the "pilotman" of the descending car has moved his controller (or equivalent control gear) to "full power", this serving as a starting signal to the driver of the lower car. In response, he notches up in the classical manner, powering the motors of both cars simultaneously. The technical term

for this is "double traction" (not to be confused with multiple-unit), and it means that power for both cars is regulated by the controller and resistors of one. Conventionally there would be a "train line" between two mechanically coupled cars, but in the case of these lifts the extra overhead wires serve as the train line, and the cable replaces the mechanical coupling.

Bica 1927 until now

On 6 December 1924 the CML ordered to electrify and reopen the Bica funicular within a year. Instead as originally foreseen and as the Lavra and Gloria with motorised cars, the Bica was

Bica car 2 at upper station, 28 April 2014. *C. F. Isgar*

converted to a true funicular with external propulsion. It got two rather ugly shed type cars built by Theodore Bell which arrived in 1926. A newspaper of that time described them as looking like a "Casa do Banho". Perhaps they had in mind the small outside shed types of bathrooms that were often used in the old days. On 2 June 1927 the Bica was finally reopened. The current cars probably date from 1933. Or perhaps only the bodies were replaced.

Midsummer visitors will usually find one of the lifts closed for overhaul and repainting which is carried out in the street. The cables are renewed about every four years.

Vertical lifts

Three vertical lifts have provided public transport in Lisboa, of which one still exists. First was the one operated by the Empresa do Elevador do Chiado, which opened on 15 February 1892. The lift was inside the building of the Hotel Universal. The upper entrance was on the T-crossing of the Rua do Carmo, Rua Garrett and Rua Nova do Almada, a very busy commercial area. The lower entrance however was in the Rua do Crucifixo, a narrow back-street. The height difference between the two levels was about 14 metre. The Elevador do Chiado closed in 1912. The building was absorbed by the Armazéns do Chiado. Technical details about the working of this lift are unknown, but it might have been using water counter balance.

The second vertical lift was operated by the Empresa do Ascensor Município-Bibliotheca and opened on 12 January 1897. The lift was known as the Elevador da Biblioteca, the Elevador do Município or the Elevador de São Julião. It was built as an open iron tower behind the building Largo de São Julião 13, which was also the lower entrance. This is very close to the city-hall (Município). From the top a footbridge connected to the terrace of the Palácio do Conde Visconde de Coruche on the Largo da Biblioteca (now Largo da Academia Nacional de Belas Artes). The footbridge crossed high above the Calçada de São Francisco. The level difference between "Município" and "Biblioteca" was about 30 metre. Water counter balancing was used to work it. The lift was closed in 1915.

Last of the lifts was the still existing Elevador de Santa Justa. It was opened on 10 July 1902 by the Empresa do Elevador do Carmo and covers a level difference of about 30 metre. Originally it was steam powered. The 45 metre high tower is on the flight of steps at the western end of the Rua da Santa Justa. It's a most improbable pseudo-Moorish iron structure affording fine views over the Baixa and it bears a fine brass plate which proclaims that it was supplied by R. Waygood & Co Ltd. of London, Lift Makers to His Britannic Majesty. From the top level a 25 m long footbridge crossing high above the Rua do Carmo connects to the alley (Travessa Dom Pedro de Menezes) next to the ruins of the Igreja do Carmo.

The Carmo church was destroyed by the earthquake of 1 November 1755. With this disaster many thousands were killed. The death-toll isn't known, but the economic historian Álvaro

Car 530 passes the Carmo lift on 3 May 1964. Note the magnificent display of greenery being carried on the head of the lady standing near the front exit. *L. Folkard*

Pereira estimated that of Lisbon's population of approximately 200,000 people, some 30,000–40,000 were killed by the earthquake and tsunami that was caused by it. On the moment of the earthquake at about 9:40 in the morning, the church was full of people because of All Saints Day. Many were killed by the collapsing roof and walls and the ruins are retained as a monument in the memory of that most terrifying day in the Lisboa history.

There was a tram track in the alley until a few years ago, perhaps dating from the opening of the branch line to Carmo on 7 November 1906. The turning loop of Carmo came in service only on 28 August 1948. The lift was a direct rival to the by Carris wanted but never realised line from Rossio via Ruas do Carmo and Garrett to Chiado. On 20 November 1905 the Empresa do Elevador do Carmo leased with the option to buy its concession and all tangible property to the Lisbon Electric Tramways Ltd. The option to buy was exercised in 1913. In 1906 Carris and the EEC introduced transition tickets and the LET started to buy EEC shares. On 4 July 1907 works started to electrify the lift, which was reopened on 2 November 1907.

LISBOA | Metropolitano de Lisboa

By Owen Brison

In 1953 it was decided to build an underground railway, funded largely by the state and operated by a concessionary company, Metropolitano de Lisboa SARL. The first lines were officially inaugurated on 29 December 1959, with public service starting the next day. The (standard gauge) lines formed a Y-shaped system from Restauradores to Sete Rios and Entre Campos. The operating company was nationalised in 1975 and became Metropolitano de Lisboa EP in 1978. It shares a Chief Executive with CCFL, but Lisboa still has two quite separate transport systems (Metro and CCFL) and through ticketing is confined to the passe social, zapping and certain tourist tickets.

The Metro, unusually for Portugal, is of standard 1435-mm gauge, with left-hand running. Direct current at 750 volts is taken from a top-contact lateral third rail, and most of the original lines are in shallow double-track cut-and-cover tunnel following the ground contours of the streets, sometimes with gradients as steep as 4% (1 in 25); most later lines are in bored tunnels. The working sites required many tramway diversions during the initial construction. Some stations are deep enough to require long escalators, and most have been, or are being, provided with lifts for accessibility. The original system was wholly in tunnel except for a covered bridge at Rua São Sebastião. There are now a number of locations where the lines are on viaducts including Campo Grande station and its approaches. The Yellow Line between Senhor Roubado and Odivelas is on viaduct while between Olaias and Bela Vista the Red Line crosses over a valley which contains the CP "Linha da Cintura" (Belt Line).

The Metro is open between 06.30 and 01.00 hrs each day, essential maintenance work being carried out at night. Signalling is automatic, actuated by ac track-circuits, with two aspect colour light signals actuating electro-magnetic train stops fitted to the trains. Top speed was 60km/h except where there are speed restrictions; however, since 2011 and on account of the financial crisis, top speed has been reduced to 45km/h to save energy. On weekdays trains have 6 cars, reduced to 3 late at night and on weekends. At times of expected heavy traffic, for example for a football match, 6-car trains run "out of hours". On weekdays between 10hrs and 17hrs, all Yellow Line trains cover the Rato to Campo Grande section, while alternate trains go on to Odivelas: at other times of the day and on weekends, all Yellow Line trains serve Odivelas.

Extensions:

Extensions were built from Restauradores to Rossio (1963), Rossio to Anjos (1966) and Anjos to Alvalade (1972), after which the Government decreed that no more should be built until all the existing stations had been lengthened. This took all the available funds for the next 15 years, but 1988 brought extensions from Sete Rios to Colegio Militar and from Entre Campos to Cidade Universitária. 30 March 1993 saw lines opened from Cidade Universitária to Campo Grande and from Alvalade to Campo Grande, with a depôt at Calvanas. Campo Grande is operated as two terminal stations side-by-side. The original Y shaped system of one line that branched into two (with alternate trains on the two branches) has been replaced by a system of four independent lines. These are the Blue Line from Santa Apolónia to Reboleira, the Green Line from Cais do Sodré to Telheiras, the Yellow line from Rato to Odivelas and the Red Line from São Sebastião to Aeroporto. Opening dates are given below. On 9 May 2017 it was announced that the current Yellow Line will be extended past Rato, with stations at Estrela and Santos, to Cais do Sodré, where it will join the Green Line end-on. The layout at Campo Grande will be modified to give a circular Green Line while the Yellow Line will run between Telheiras and Odivelas.

Station Lengths

When the Metro was built, the planned capacity was a mere 10 000 passengers per hour and direction, requiring two-car trains at 2½-minute intervals. As traffic built up, this could be augmented to over 16 000 per hour by reducing the headway to 1½ minutes. Ten of the twelve original stations were built with platforms only 40m long (2½ car lengths), a curious lack of foresight that took no account of experience elsewhere. In 1972, work began on lengthening the short stations to 70m, to take four-car trains, while the new stations on the Anjos - Alvalade extension were built for 4 cars. Due partly to the return of many Portuguese citizens from Angola and Mozambique, traffic growth was such that 4-car trains had to be introduced on weekdays while seven stations still had short platforms; these were served by the front two cars only, and some rush-hour trains with a Directo board by-passed them. At some rebuilt stations the opportunity was taken to provide platforms of six-car length (105m), a portion being fenced off. In time it became clear that six-car trains would be needed, and work began to lengthen the four-car stations. The final four car station, Arroios, was closed to passengers on 19 July 2017; it reopened on 14 September 2021.

Depots

The original depôt and overhaul works (Parque de Material e Oficinas I, or PMO I) was at Sete Rios, reached by a single-line bored tunnel from Palhavã station. Two four-wheel diesel locomotives, one Ruhrtaler and one Moyse, were used to move cars around non-electrified parts of the site, while a turntable able to turn one car at a time was provided. Running maintenance and cleaning was done underground at Sete Rios station and the other termini. Although a suburban railway runs past the depôt and had a siding into the yard, physical connection was ruled out by the difference in gauge. In 1993 the second depot and works complex, PMO II, was inaugurated at Calvanas, with access from Campo Grande station by viaduct. In 1999, PMO III was inaugurated at Pontinha, with access by a tunnel from Pontinha station. Both heavy and running maintenance, as well as cleaning, is performed at PMO II and III; there is a turntable at PMO II. Following the opening of PMO III at Pontinha, the original depot PMO I at Sete Rios was closed and is now a terminus for long-distance coaches.

Interchanges

Tram interchanges were originally built at Sete Rios and Entrecampos, at a time when it was intended to maintain the outer portions of the Benfica, Carnide and Lumiar tram routes as feeders. There are now interchanges with CP and Fertagus trains at Roma-Areeiro, Entrecampos and Sete Rios, with the CP at Oriente, Santa Apolónia and Restauradores, with the CP (Estoril Line) and ferries at Cais do Sodré and with ferries at Terreiro do Paço. There is a major bus interchange, for urban and suburban routes, at Campo Grande, while many other stations are close to bus or tram stops. In addition, the Metro Red Line now serves Lisbon Humberto Delgado Airport. Passenger interchange between Metro lines is possible at Campo Grande, Saldanha, Alameda, Baixa Chiado, Marquês de Pombal and São Sebastião. There are physical connections between the respective lines near the latter four stations, while both lines at Campo Grande share a common access to PMO II.

Opening and Other Dates

29 December 1959: Official Inauguration, Restauradores to Sete Rios and Entre Campos
30 December 1959: Start of public service
27 January 1963: Restauradores to Rossio
28 September 1966: Rossio to Anjos
18 June 1972: Anjos to Alvalade
14 October 1988: Sete Rios to Colegio Militar and Entre Campos to Cidade Universitária

03 April 1993: Cidade Universitária to Campo Grande and Alvalade to Campo Grande
15 July 1995: Rotunda II station opened. This marked the start of operation on two independent lines: the Yellow Line from Campo Grande to Rotunda II and the Blue Line from Colégio Militar to Campo Grande via Rotunda I and Rossio.
18 October 1997: Colégio Militar to Pontinha on the Blue Line.
19 October 1997. In the early hours of the morning, fire broke out at Alameda station, at the time closed for rebuilding and the building of Alameda II on the future Red Line. There were two fatalities: the Station Master at Areeiro and a security guard. The line between Areeiro and Arroios through Alameda remained closed for several months.
29 December 1997: Rotunda to Rato on the Yellow Line. On the same day, Rossio station was closed for rebuilding and for the line to Restauradores to be lifted. The Blue line then operated between Colégio Militar and Restauradores.
01 March 1998: The names of four stations were changed: Palhavã, Rotunda, Sete Rios and Socorro became Praça da Espanha, Marquês de Pombal, Jardim Zoológico and Martim Moniz, respectively.
03 March 1998: Service reinstated after the Alameda fire between Campo Grande and Martim Moniz on what then became the Green Line; this marked the start of operation on three independent lines.
18 April 1998: Rossio to Cais do Sodré on the Green Line; the relevant part of Baixa Chiado station opened a week later.
19 May 1998: Red Line from Alameda to Oriente; stations Cabo Ruivo and Olivais opened several months later.
08 August 1998: Restauradores to Baixa Chiado on the Blue Line.
09 June 2000: During construction work on the future Terreiro do Paço station, water and mud entered the tunnel; there were no injuries. Completion of the Blue Line to Santa Apolónia was significantly delayed.
03 November 2002: Campo Grande to Telheiras on the Green Line.
27 March 2004: Campo Grande to Odivelas on the Yellow Line.
15 May 2004: Pontinha to Amadora Este on the Blue Line.
19 December 2007: Baixa Chiado to Santa Apolónia on the Blue Line.
28 August 2009: Alameda to São Sebastião on the Red Line.
17 July 2012: Oriente to Aeroporto on the Red Line.
13 April 2016: Amadora Este to Reboleira on the Blue Line.

Rolling Stock

The original cars, of type ML7, Nos A1-A24, were built in 1959 in Germany by Linke-Hofmann-Busch with Siemens-Schuckert equipment and were tested on the Hamburger Hochbahn. They were 16.4m long and 2.70m wide, each axle having its own 100-kW motor on account of the steep gradients. Tare weight was 35 tonnes per car. Service braking was rheostatic down to 10 km/h and air below, and for emergency each car had four low-tension electro-magnetic track brake shoes. Each car had a full width cab (with side and end doors) at one end only and three air-operated double sliding doors each side. Service trains were made up of one or (later) two back-to-back pairs, although the cars could operate singly. Cars 6, 9, 12 and 16 were destroyed by fire near Arroios on 21 May 1976. ML7 cars 25-38 were built to the same design by Sorefame in 1964 under LHB licence, followed by 39-70 in 1971-2 and 71-84 in 1973-4, all with 36 seats and standing space for 137. Nos 71-84 were the last to enter service in the

1959 red-and-grey livery. These cars were finally withdrawn from service on 30th January, 2000. The first two cars, A1 and A2, have been retained in working order and were returned to original condition in 2002 (the seating arrangement having been changed over the years). Car A1 carries a plate inside proclaiming a total of 2,969,418 km during its 41 years of service.

In 1982-5 Sorefame supplied 56 cars of a new design, type ML79, Nos M101- M156. The first cars entered service on 01/01/1984. They were built in stainless steel with polyester glass-fibre reinforced ends. Again, each car had a full-width cab (with side and end doors) at one end only and three air-operated double sliding doors each side. They ran in back-to-back pairs, with equipment distributed between the cars of a pair. In practice, the pairs were 101 with 102, 103 with 104 and so on. The average tare of the cars in a pair was 30.8 tonnes. They seated 40 and had standing room for 124. Again, all axles were powered. They had Siemens control equipment, Paris Metro type MF77 monomotor bogies by MTE, and 241kW traction motors by Alstom (20 cars) and EFACEC (36 cars). Normal service braking was rheostatic but with 20% provided by air-operated brake blocks for tread-cleaning, while stopping was by disc brakes. Again, each car had magnetic track brakes. The final cars were withdrawn from service on 11/07/2002.

Two prototype three-car trains, type ML90, were produced by Siemens and Sorefame and entered service on 29/03/1993, and so far all further cars, of types ML90 (production), ML95, ML97 and ML99, have evolved from these prototypes. Service trains are normally made up of one or two three-car sets, each of which is made up of two motored cars with one trailer between them, although they can (and have) run as two- or four-car sets without the trailers. Each motor car has a full-width cab (with side doors) at one end only while both motor and trailer cars have three double plug doors each side. The prototype cars have an end door in each cab, but later cars (including the production ML90 series) do not. ML90 and ML95 sets have locked inter-car doors for emergency use. ML97 and ML99 sets have open, concertina, connections between the cars of a set, although each car retains two bogies: they are not articulated. The motor cars have all axles powered; in the case of ML90 cars, each bogie has two 195 kW motors, while the later series all have 175 kW motors. Service braking is rheostatic or regenerative with air-operated disc brakes to stop, while all cars have electromagnetic track brakes. In ML90 and ML95 sets, motor cars seat 40 and trailer cars seat 48, with 122 and 130 standees respectively. In ML97 and ML99 sets, the respective figures are 38, 40, 127 and 141. Recently, some changes have been made to the seating. A small number of sets, including the first ML90 prototype, have been given experimental longitudinal seating, while a number of "non-concertina" motor cars have had a block of four seats removed at the inner end to make room for bicycles, push chairs and bulky luggage. ML99 cars differ from previous builds in that their transverse seats are cantilevered out from the car walls, providing an easier-to-clean floor. Tare weights for this family of cars vary between 29 and 30.4 tonnes for the motor cars and between 18.8 and 19.6 tonnes for the trailers. Cars in each three-car set are numbered consecutively, with the number preceded by an M for the motor cars and an R ("Reboque") for the trailer cars. Thus the first prototype set has cars M201, R202, M203. while M712, R713, M714 make up the final ML99 set.

Building dates and number sequences are as follows, built by Sorefame/ Adtranz/ Bombardier at Amadora:

ML90, prototypes 1993, production 1995/96, cars 201 - 206 (prototypes) and 207 - 257.

ML95, 1997/98, cars 301 - 414.

ML97, 1999, cars 501 - 554.

ML99, 2000/02, cars 601 - 714.

It was announced on 9 May 2017 that a further 33 sets are to be acquired.

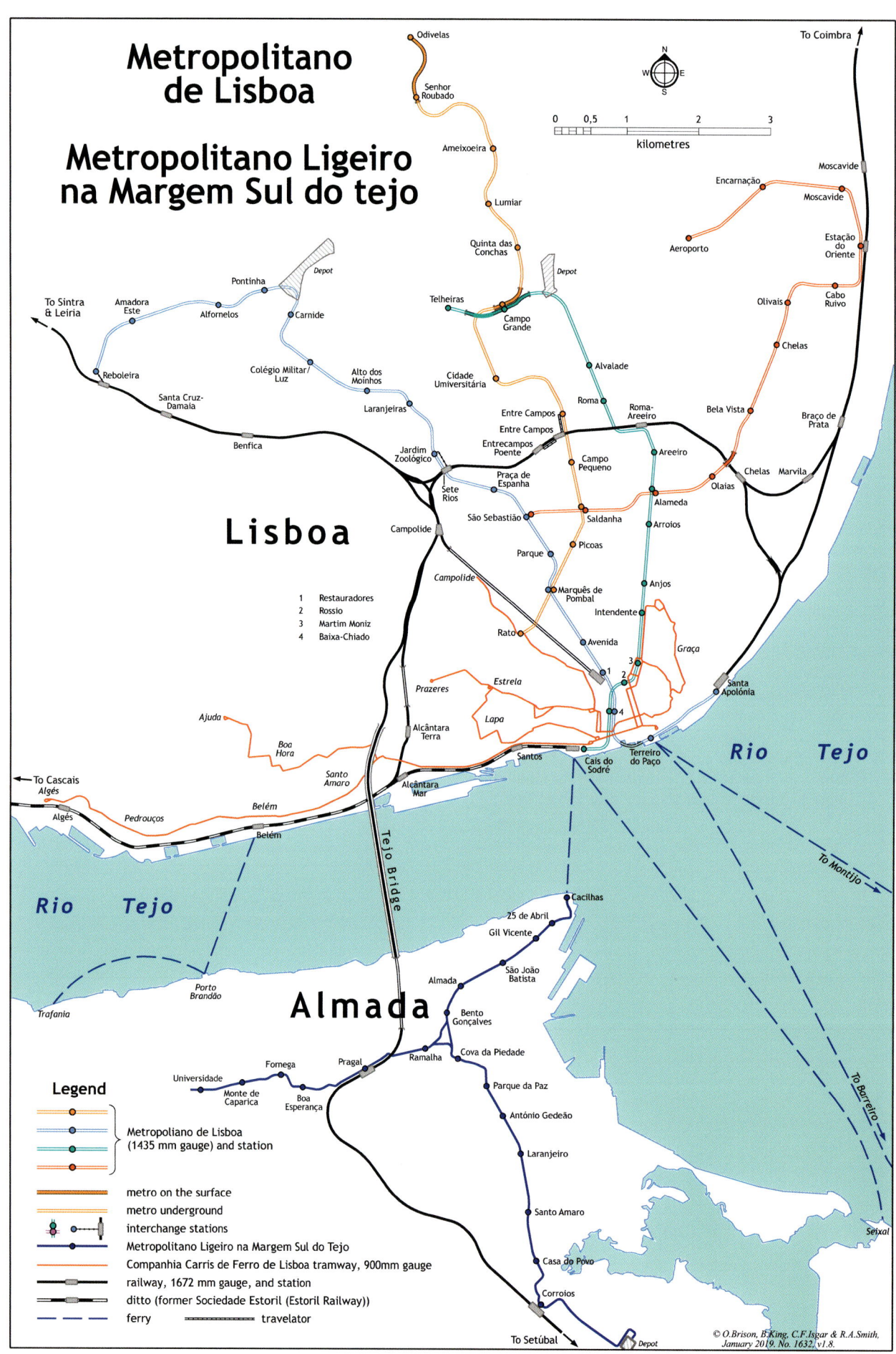

Metropolitano de Lisboa

Metropolitano Ligeiro na Margem Sul do tejo

Lisboa

Almada

Rio Tejo

1 Restauradores
2 Rossio
3 Martim Moniz
4 Baixa-Chiado

To Sintra & Leiria

To Cascais

To Coimbra

To Setúbal

To Montijo

To Barreiro

N
W E
S

0   0,5   1   2   3
kilometres

Legend

Metropoliano de Lisboa (1435 mm gauge) and station

metro on the surface
metro underground
interchange stations
Metropolitano Ligeiro na Margem Sul do Tejo
Companhia Carris de Ferro de Lisboa tramway, 900mm gauge
railway, 1672 mm gauge, and station
ditto (former Sociedade Estoril (Estoril Railway))
ferry   travelator

© O.Brison, B.King, C.F.Isgar & R.A.Smith,
January 2019. No. 1632. v1.8.

ALMADA

MTS – Metro Transportes do Sul

C020 is seen at Ramalha on 26
April 2014. The steep section at this
point is clearly evident. *C.F.Isgar*

By Owen Brison

The Metro do Sul do Tejo, operated by the Consortium MTS - Metro Transportes do Sul, S.A., is a standard gauge light-rail system in the towns on the south bank of the Tejo River opposite Lisboa.

The first serious proposals for building the system were made in 1985, but it was only in 1994 that a study into the economical and technical feasibility was undertaken. The following year the three municipalities involved, Almada, Seixal and Barreiro, came to an agreement with the government to develop the new system. The local municipalities were to contribute 20% and the government 80% of the investment costs. However some years marked by impasses and setbacks followed. It was only in 1999 that the municipalities and the government came to an agreement to launch a competition for the construction and operation of the system.

Progress remained slow and it was 2002 when the MTS consortium was founded and granted the contract for the design, building and maintenance of the system and for the supply of rolling stock. The consortium is made up of:

Barraqueiro Transportes, S.A; Ascendi Group SGPS, S.A.; Ensulmeci, S.A.; Teixeira Duarte – Engenharia e Construções, S.A.; Sopol, S.A

Construction work started in 2003 and the first tram arrived in 2005. The first section between Corroios and Cova da Piedade was inaugurated on 30th April 2007, and entered public service the following day. On 15 December 2007 the line between Cova da Piedade and Universidade opened and finally on 27 November 2008 the first phase of the project was completed when the branch to Cacilhas was opened.

Network

The network is T-shaped with a triangular junction, bounded by the stops of Cova da Piedade, Ramalha and Bento Gonçalves, at the centre. There are three routes:

1. Cacilhas - Cova da Piedade - Corroios
2. Pragal -Cova da Piedade - Corroios
3. Cacilhas - Pragal - Universidade

The length of the network in public service is 12 km. The depot is another 1.5 km beyond the Corroios terminus, on the route of a planned future extension to Fogueteiro. There are 19 stops, six each on the Cacilhas and Universidade branches and seven on the Corroios branch. Of these stops, sixteen are within the municipality of Almada; two are in the municipality of Seixal while the Santo Amaro stop is on the border between Almada and Seixal.

Corroios and Pragal are interchange stations with Fertagus trains (which run between Roma - Areeiro in Lisboa and Setúbal). Pragal also allows interchange with the CP "Intercidades" services between Lisboa Oriente, Alentejo and the Algarve. Cacilhas is the interface with ferries across the Tejo to Cais do Sodré in Lisboa. Most of the system is on reserved track at the side, or in the centre, of roads. In the central area of Almada and in the Ramalha area there is some street running on roads with limited access by private cars. There are island platforms at the stops of Bento Gonçalves and Almada, as well as Cacilhas and Corroios.

Tram services on all three lines are every 15 minutes on Mondays to Saturdays with every 5 or 10 minutes during peak hours. On Sundays the frequency is every 20 minutes.

In order to facilitate construction of the new tramway to Universidade at Ramalha it was necessary to lower the ground level adjacent to an apartment building as evidenced by this view taken in September 2007. *O. Brison*

C017 at Largo da Costa Pinto on 13 May 2009. *Michael Russell*

C022 at Cova da Piedade on 13 May 2006. *Michael Russell*

Rolling stock

Siemens supplied 24 vehicles of the Combino-Plus model, numbered C001 to C024. The original order was for the Combino type, but faced with the structural problems occurring with the Combinos delivered previously to other systems; Siemens built the Combino-Plus type instead. These double-ended, 4-section articulated 100% low-floor trams have eight axles and are powered by six motors of 100 kW each. They are 36.36 m long and 2.65 m wide. The trams carry notices proclaiming a total capacity of 311 passengers, made up of 74 seated and 237 standing places.

In early 2007, tram C008 spent several months in public service in Melbourne, Australia, as a demonstrator. It was thus the first of the MTS cars to carry fare-paying passengers. It arrived in Melbourne on 2 January 2007, first ran in public service on 15 March and made its final run in Melbourne public service on 16 June 2007. After returning to Portugal, C008 was modified as a test bed for the Siemens "ultracapacitor" system.

Depot

The depot (or "PMO": Parque do Matérial e Obras), mentioned above, has a total of 11 roads, several of which have pits and are reserved for maintenance. Equipment includes a wheel lathe and a mechanical washing plant for the vehicles. The depot also includes the Traffic Control Centre and an electrical substation.

Ticketing

Ticketing is via pre-purchased contactless smartcards: either the "Lisboa Viva" photocard for 30 day passes, or the cardboard "Viva Viagem" (or alternatively, "Sete Colinas") cards for sporadic travel. The Viva Viagem cards may be bought (currently for €0.50) from machines at each stop, or from one of a small number of manned MTS kiosks; they are valid for one year. These cards may be loaded with multiple journeys, with a discount if 10 are loaded at once. On first entering a tram, a passenger must "touch-in" the card on one of the machines available for the purpose; the passenger may then travel on multiple vehicles, touching-in on each, provided that the final touch-in is within an hour of the first (even if this final journey goes past the hour). The machine states how much of the hour is left, and also the number of unused journeys still on the card.

Since 26th October 2015 MTS and Fertagus have been added to the Lisboa "zapping" system, whereby the cardboard "Viva Viagem" or "Sete Colinas" tickets, or even the Lisboa Viva plastic card, can be loaded with up to €40, which is then deducted a bit at a time when the holder travels on Carris/Metro/CP/Transtejo/Soflusa.

Revenue Protection is carried out by roving teams of inspectors. At first, these teams comprised MTS employees, but more recently private security contractors have been used.

Future

The original plans foresee a second phase with the extension of line 1 from Corroios to Fogueteiro and a third phase with a new line Fogueteiro - Seixal - Barreiro. According to these plans the total length of the system will become 28 km with 37 stops. However, except for agreement about the routing between Corroios and Fogueteiro, no progress has been made. In addition to these plans, the municipality of Almada supports extensions to Trafaria and Costa da Caparica. If and when these projects will be realized is unclear.

C019, C004, C023 at Ferry terminus, Cacilhas, 26 April 2014. *C. F. Isgar*

C021 at Cacilhas on 13 May 2009. *Michael Russell*

C018 at Cova da Piedade on 26 April 2014. *C. F. Isgar*

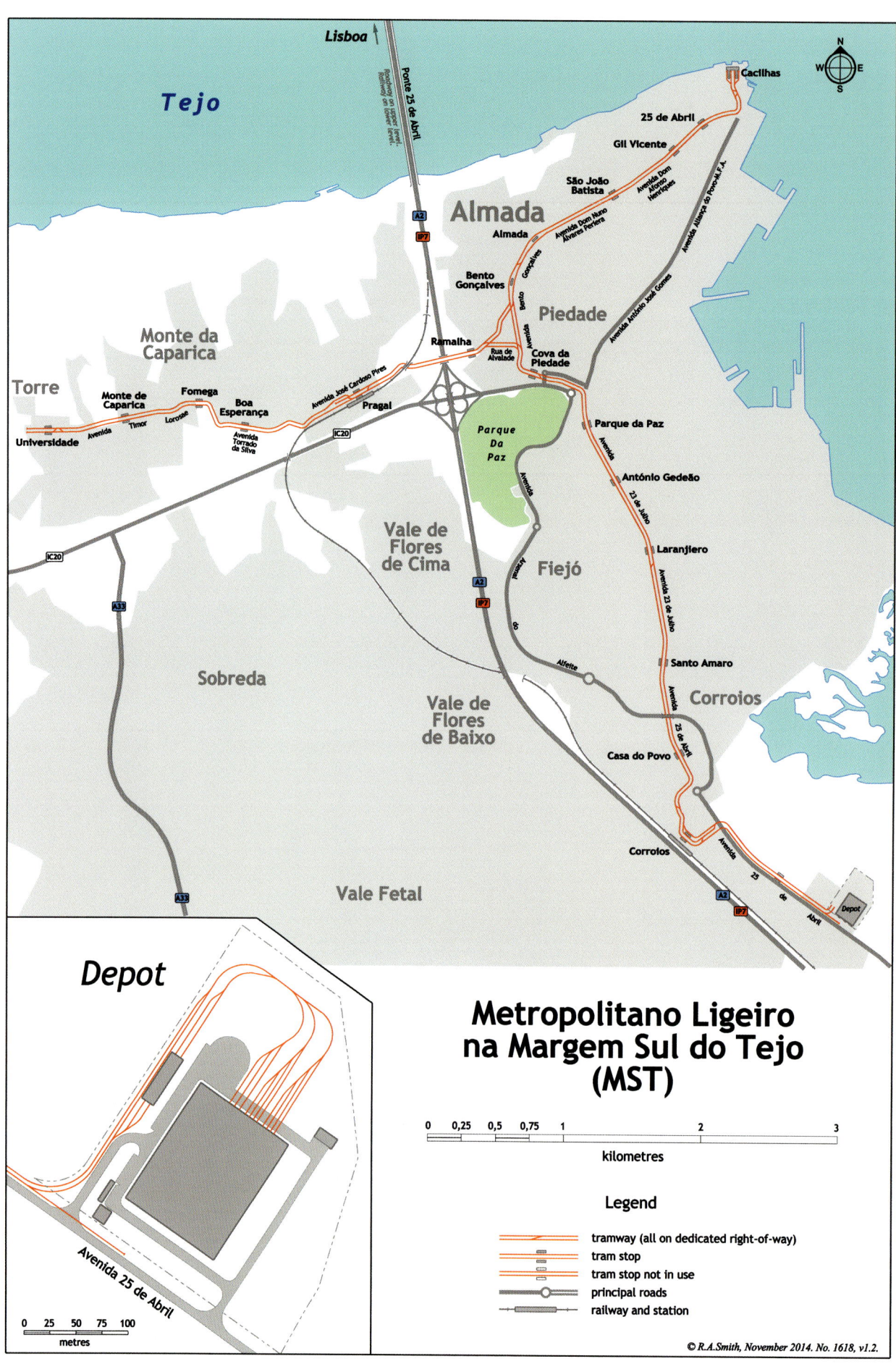

Tejo

Lisboa

Ponte 25 de Abril
*Roadway on upper level.
Railway on lower level.*

Cacilhas

25 de Abril

Gil Vicente

São João
Batista

Avenida Dom Afonso Henriques

Avenida Aliança do Povoal.F.A.

Almada

Avenida Dom Nuno Alvares Pereira

Almada

Bento
Gonçalves

Bento Gonçalves

Piedade

A2

IP7

Ramalha

Rua de Alvalade

Avenida António José Gomes

Cova da
Piedade

Monte da
Caparica

Avenida José Cardoso Pires

Pragal

Parque da Paz

Parque
Da
Paz

Torre

Fomega

Monte de
Caparica

Boa
Esperança

Timor   Lorosse

Avenida

Avenida
Torrado
da Silva

IC20

Universidade

António Gedeão

Avenida

23 de Julho

Laranjiero

Avenida 23 de Julho

IC20

Vale de
Flores
de Cima

Avenida do Rossy

Fiejó

A2

IP7

A33

Santo Amaro

Avenida

Corroios

Sobreda

Vale de
Flores
de Baixo

Alfeite

Casa do Povo

25 de Abril

Corroios

Vale Fetal

Avenida 25 de Abril

A2

IP7

Depot

Depot

Avenida 25 de Abril

0   25   50   75   100

metres

Metropolitano Ligeiro na Margem Sul do Tejo (MST)

0   0,25   0,5   0,75   1   2   3

kilometres

Legend

⟶	tramway (all on dedicated right-of-way)
	tram stop
	tram stop not in use
○	principal roads
	railway and station

© R.A.Smith, November 2014. No. 1618, v1.2.

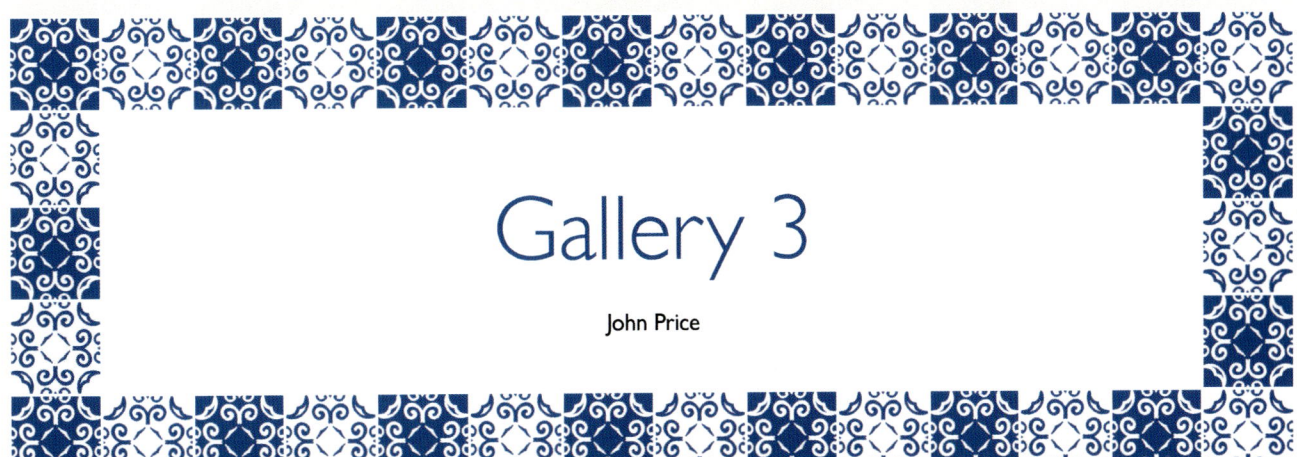

Gallery 3

John Price

Portugal saw very few tourists before the 1960s, even fewer tram enthusiasts. Norman Forbes visited Porto in 1949, and was amazed to find early Starbuck cars still in use as trailers. John Price, author of the first two editions of this book and leading author of the third and fourth editions, visited in 1959, and was astonished to encounter one of them still in service, almost certainly the oldest trams in use anywhere. He instantly took this picture, in case there was no other opportunity, then hastened to the street corner and just managed to secure a second photograph of trailer 10 before it disappeared from view. He resolved there and then that one of these cars, of the type used by G. F. Train for his pioneering line in London must be acquired for the nascent museum at Crich. Thus it came about that number 9, which was in better condition than 10, was acquired in 1964, probably triggering interest within STCP and providing the initial impetus to establish a museum in Porto.

The side view of trailer 10 taken in October 1959 at Bolhaõ, one of the last of these trailers to remain in service.

Courtesy of the National Tramway Museum, photographer J.H. Price

The second view of trailer 10 taken at Bolhaõ in October 1959. *Courtesy of the National Tramway Museum, photographer J.H. Price*

Starbuck trailer 9 cosmetically restored in a livery typical of the period is seen making a rare appearance outdoors at Crich on 10 September 2016.
Kath Lomas King

PORTO

Tramways

By Pedro Milheiro and Ernst Kers

Sunset along the Atlantic in October 1978
starring Belga 284. *Tim Boric*

Porto is the second city of Portugal, and is the natural capital of the fertile north of the country. The population of the greater Porto conurbation is in excess of one million. The city is dramatically located on a granite outcrop high above the river Douro, on the north bank near the river mouth. The oldest part of the city is a UNESCO World Heritage Site described as "an outstanding urban landscape with a two thousand year history of continuous growth, linked to the sea". The former Roman name of Portus Cale led to the naming of Portugal, as the nascent country emerged from foreign domination. Houses near Infante are amongst the oldest continuously inhabited buildings in Europe. Porto has also given its name to the justly celebrated fortified wine produced in the upper Douro valley, and traditionally blended and matured on the south side of the river in Vila Nova da Gaia. Matosinhos on the nearby coast is a major centre in the world of fish, and the adjacent port of Leixões has replaced the Douro as the focus of the shipping industry.

Mule tram, perhaps in Matosinhos. *MCE*

Dawn and rise of an era

In the 2nd half of the 19th century, Porto had no public transportation, apart from a few coaches and wagons that were slow and cumbersome. The best way for people to circulate, was simply to walk! So by 1858 Albino Francisco de Paiva Araújo saw an opportunity and applied for a licence with that aim, but it was denied; only 12 years later, on the 25 August 1870, granted by royal decree, another entrepreneur, Barão (Baron) da Trovisqueira was to succeed in gaining a licence for a line to Matosinhos. The King granted this enterprise exemption from taxation.

In spite of this demonstration of goodwill, the terms of the licence were however very severe, because the system had to be working and providing regular services by the end of June 1872, or all the goods, cars, rails, equipment, etc. that had been bought without taxes would become property of the Crown, without any compensation. These hard conditions would not be applied later to the other companies.

The works had begun by July 1871, and were a true challenge for Companhia Carril Americano do Porto á Foz e Matosinhos (CCAPFM) often called Carril Americano. As the name indicates this line was from Porto to Foz and Matosinhos. In Portugal tramcars were called carros americanos as they were invented in America. There was no experience in laying rails, curves and

col. John R. Stevens
Mule tram built for the CCFP around 1874. *Science Museum*

operating light tram cars. Derailments were frequent and the deadline was just a few months away, 30 June 1872! But "where there´s a will there's a way", and by 9 March 1872, services began, and were officially inaugurated by 15 May.

Services began in the Rua dos Inglezes (nowadays Rua Infante Dom Henrique). Although the terminus was in an important commercial area with the Bolsa, the Feitora Inglesa, the market and Alfândega, the line would need to be extended for about 800 m to reach Praça Dom Pedro, the true centre.

On the 20 March 1873, CCAPFM received a licence from the Camara Municipal do Porto (CMP) to install a branch line from Massarelos via Rua da Restauração to the Campo dos Martires da Patria, about the location of the current bandstand, still some distance from the true centre.

CCAPFM had also applied for a licence to extend their lines to Praça and to the future railway station Pinheiro (nowadays Campanhã), which was at that moment projected and would be opened in 1875. However on 27 March 1873 the members of CMP voted with six against two in favour of the plans of a new company, the Companhia Carris de Ferro do Porto (CCFP) for the construction and exploitation of an urban tram network and a second line to Foz, where the terminus would be at Cadouços.

When, in 1873 CCAPFM wasn't granted permission to extend its rails in the city, the company was doomed to an existence in poverty. Although CMP allowed the spur from Alameda de Massarelos to Cordoaria via Rua da Restauração, it was a short one, preventing the growth of the system within the city.

In 1874 CCFP started its operations. Both companies used mules for traction as in Latin and South America instead of horses as was usual in North America and most other European countries.

On the day CCFP opened its line to Foz, 14 August 1874, CCAPFM lowered their fare from Porto to Foz from 100 Reis to 80 Reis, equal to that of CCFP. With the competition of CCFP to Foz and after 1882 also to Matosinhos, and unable to reach the centre or any other destination except Matosinhos, the revenue was severely limited.

On 27 June 1878 both companies received authorisation to use steam locomotives. For CCFP this permission was for their line from Boavista to Cadouços. That same year CCFP bought four tram locomotives from Henschel. CCAPFM received permission for the part of their line between the

A tram in front of the Igreja de Santo Ildefonso was always a popular subject for photographers. Today it can be done again as line 22 passes here. This former mule tram is seen around 1900 on the way to Campanhã. *Commercial postcard*

A steam tram pulled by the Merryweather locomotive has just arrived at Boavista. Just above the locomotive a team of mules is waiting, probably to bring one of the carriages to the city centre and Campanhã. A small mule tram is visible in the background. The photo was made by Domingos Alvão around 1897 (1893-1901). This is one of the few photos with the Merryweather locomotive in Porto. *CPF.*

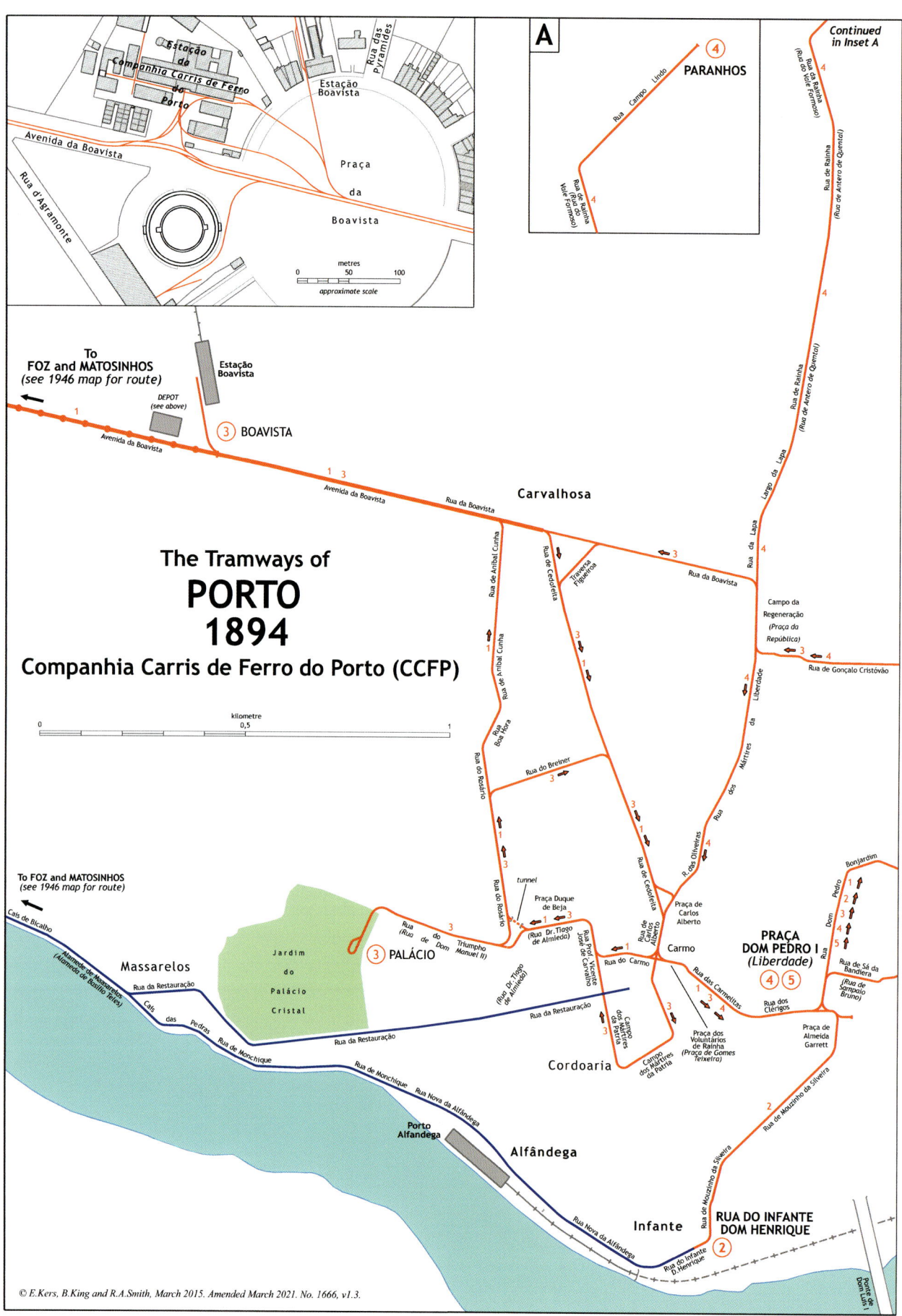

© E.Kers, B.King and R.A.Smith, March 2015. Amended March 2021. No. 1666, v1.3.

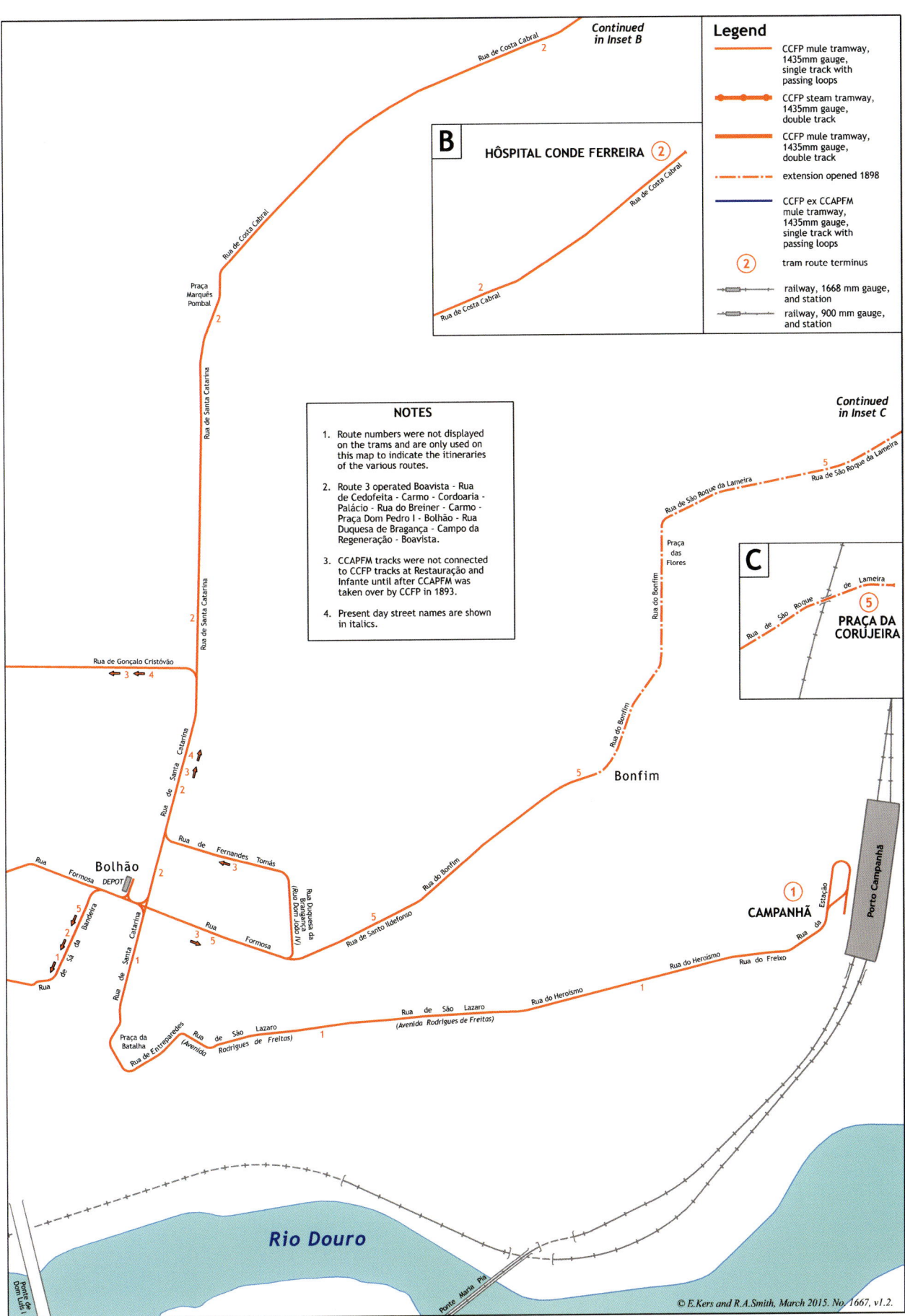

Legend

CCFP mule tramway, 1435mm gauge, single track with passing loops

CCFP steam tramway, 1435mm gauge, double track

CCFP mule tramway, 1435mm gauge, double track

extension opened 1898

CCFP ex CCAPFM mule tramway, 1435mm gauge, single track with passing loops

② tram route terminus

railway, 1668 mm gauge, and station

railway, 900 mm gauge, and station

B

HÔSPITAL CONDE FERREIRA ②

Rua de Costa Cabral

Rua de Costa Cabral

C

PRAÇA DA CORÚJEIRA ⑤

Rua de São Roque

de Lameira

Continued in Inset B

Rua de Costa Cabral 2

Continued in Inset C

NOTES

1. Route numbers were not displayed on the trams and are only used on this map to indicate the itineraries of the various routes.

2. Route 3 operated Boavista - Rua de Cedofeita - Carmo - Cordoaria - Palácio - Rua do Breiner - Carmo - Praça Dom Pedro I - Bolhão - Rua Duquesa de Bragança - Campo da Regeneração - Boavista.

3. CCAPFM tracks were not connected to CCFP tracks at Restauração and Infante until after CCAPFM was taken over by CCFP in 1893.

4. Present day street names are shown in italics.

Praça Marquês Pombal

Rua de Costa Cabral

Rua de Santa Catarina

Rua de Santa Catarina

Rua de Gonçalo Cristóvão

Rua de Santa Catarina

Rua de São Roque da Lameira

Rua de São Roque da Lameira 5

Praça das Flores

Rua do Bonfim

Rua do Bonfim

Bonfim 5

Rua de Fernandes Tomás 3

Rua do Bonfim

Rua Formosa

Bolhão DEPOT

Rua de Santa Catarina

Rua Duquesa da Bragança (Rua Dom João IV)

Rua de Santo Ildefonso 5

Rua Formosa 3 5

CAMPANHÃ ①

Estação

Porto Campanhã

Rua da

Rua de Sá da Bandeira 5

Rua de Sá 2 1

Rua

Praça da Batalha

Rua de Entreparedes (Avenida)

Rua de São Lazaro Rodrigues de Freitas)

Rua de São Lazaro (Avenida Rodrigues de Freitas)

Rua do Heroismo

Rua do Heroismo 1

Rua do Freixo

Rio Douro

Ponte de Dom Luís I

Ponte Maria Pia

© E.Kers and R.A.Smith, March 2015. No. 1667, v1.2.

Ouro depot and Matosinhos, but apparently decided that the necessary investments in locomotives and track reinforcement or replacement were not justified. At that time the services to Matosinhos were still limited to once an hour, which means that for CCAPFM on the busy part of their line to Foz, the use of locomotives was only allowed on a short section. For about eight to ten years CCFP used the steam locomotives

only during daytime in the summer period. They were however making the position of CCAPFM weaker. The 1879 travel times for CCAPFM were 35 minutes from Rua dos Ingleses to Foz (Rua Senhora da Luz) and 60 minutes to Matosinhos. For CCFP the times between Praça D.Pedro and Foz (Cadouços) were 35 minutes when steam locomotives were used between Boavista and Foz, and 40 minutes to Foz or 50 minutes to Praça in case of

Left: Six mules were needed to pull a single tram-car uphill from Praça Dom Pedro to Carmo. In the foreground a closed car is just leaving Praça, in the background an open car with longitudinal benches is preparing for the ascent. In the background, in Rua de Santo António a track is visible. This long steep street was never used by mule trams, which means that the tracks for the electric tram were already in place, but the overhead lines were not yet installed. The photo was taken by Aurelio Paz dos Reis, probably in the spring or summer of 1901. *CPF.*

Below: The entry into service of the first electric tramcar in Portugal, and indeed the Iberian peninsula, does not appear to have attracted much in the way of news coverage, and the archive does not seem to include any official photographs. We therefore offer this image of transformado 1 at Infante, the exact location from which heritage trams still depart 125 years later. The image was copied from "Os Velhos Electricos do Porto" by Manuel Castro Pereira (1995). He credits it to "Colleccao de Sr. Daniel Gonsalves"

The only known photograph of Arrabida depot. In the foreground is an example of the original type of transformado truck. Beyond the simple traverser the car on the extreme right is one of the former Larmanjat trailers. Second from the left, with twin headlamps is the Siemens & Halske car 14, which was renumbered 26 in 1901. The number on the dash is indistinct, but is not 26, so the picture dates from between 1898 and 1901. Boavista had no electric car works facilities until 1904. *Memorias do Reino de Portugal*

animal traction. The difference without doubt caused by the level difference that existed on CCFP line.

CCFP extended the urban network and in 1882 also their Foz line to Matosinhos. In 1886 they constructed a timber bridge across the river Leça to make it easier for inhabitants of Leça da Palmeira to reach the terminus of their tramline in Matosinhos. CCAPFM decided on their only extension since 1873 by extending their line in 1887 over an iron bridge to Leça da Palmeira with a new terminus at the northern end of the wooden bridge of CCFP. Perhaps CCAPFM won the competition for the travellers going to or from Leça da Palmeira, but it's questionable if the numbers of those justified the costs of building the bridge. It might have pushed the company from poor to technically bankrupt.

Ripert cars

In 1883 Empresa Portuense de Carros Ripert started to compete using Ripert cars. This type of car was able to run on the normal road surface as well as using tram tracks. On July 12 a line was opened following the route Cedofeita - Carmo - Clerigos - Praça Dom Pedro V (now Praça de Liberdade) - Sa

da Bandeira - Formosa - Sta. Catarina - Batalha - São Lazaro - Heroismo - Campanhã. The frequency was every 30 minutes. On July 23 a second line was opened from Praça Dom Pedro V to Álfândega. In the direction Álfândega the route was via Rua das Flores, in the opposite direction via Rua Mousinho da Silveira. In contrast to Lisboa this appeared not to be long lasting. On 15 November 1884 the Ripert company stopped working in Porto. All cars, other equipment and most of the mules were sold to the Lisboa Ripert company.

A further Ripert service appears to have been operated by CCFP to S. Mamede since 1883 until opening of the electric tramline in 1910. No more details are known.

The End of one and a new start of another

Since even before the start of CCFP operations in 1874 the subject of a merging of both companies was discussed a number of times, though without results. By 1893 the financial situation of CCAPFM had become unsustainable. CCFP took the opportunity in 1893 to purchase the shares of their competitor and merged both. CCFP was to last until the 14 July 1949, when

Porto—Estação do Ouro

The old Ouro depot of the CCAPFM around 1903 (1902-1904). Tramcar no.2 with two trailers is just arriving from Porto. At the left is another tramcar with three open freight trailers on a side track. This location is now called Largo António de Calém. *Commercial postcard.*

While A Constructora was building a series of 24 tramcars with six windows each side and the numbers 43-54, 57-66 and 68-69, two smaller trams appeared with the numbers 55-56 of the type which was built earlier. They are believed to be the original 33-34 dating from 1902 with new trucks and motors. In 1907 the numbers of all electric trams were increased by 100. The photo was taken around 1909 (1907-1911). *Grainy reprint of a postcard.*

the shareholders would vote for its dissolution. The records of CCAPFM have not survived.

Porto is a city of hills. While a light tramcar could easily be hauled on a horizontal plane by two mules, four or six were needed to climb the slopes. This was a very expensive

operation and surviving correspondence shows that the use of the Mekarsky compressed air system using twelve cars was investigated in 1876, as was cable operation in 1886. Meanwhile in the USA electric traction had demonstrated its advantages and reliability and in 1895 the first electric trams were introduced in Porto. By 1904 the last mule tramline was converted. The steam tram was retained until 1914.

When CCFP established the first electric traction lines, this was done as a necessary technical solution and up-grading, but simultaneously as a trial; Porto city centre steep gradients were a constant challenge for animal traction; not being able to use other forms of mechanical traction, mainly in the Carmo-Praça-Batalha section where the severe gradients were located, electric traction seemed the best solution. After the 3 initial overhauled and motorised mule trams / steam trailers, followed by another 3 cars in 1896 and a 3rd set of 3 in 1897, CCFP became apprehensive concerning its legal rights. What legal title did they have as a safeguard for the investment? Just a mere licence, or permission that could be revoked by CMP whenever it wished? As an answer to this issue was of the utmost importance, CCFP instructed a lawyer to study it, anticipating what could happen for instance, if they were ordered by CMP to remove the tracks and shutdown the entire operation. Did they have the right of an indemnity? And could they fight back and go on with the services? Now, instead of seeking a technical solution, CCFP was seeking a legal one, a new contract.

The turn of the century and the following years brought some peace between the parties, but many periods of dispute. CMP had long disagreed with CCFP behaviour, considering some of the acts offensive to the public, stating the only aim was profit,

A Brill tram of the 1909 order in front of the São Bento railway station. At the far right the tracks leading to Rua de Loureiro are just visible. The photo can be dated around 1913. *Source postcard*

at the expenses of the greater good, i.e. the public interest and needs. After a period when it seemed CCFP was being successful in persuading CMP for a new contract with previously accorded conditions and an understanding was being achieved, an open tender was unexpectedly launched in 1906. For this new contract CMP imposed different conditions, which CCFP was not willing to accept, claiming a 99 year period with the minimum period of 50 years, contradicting CMP wishes for a 75 year period, with the possibility of takeover after the first 35 years of operation. CMP refused to yield its position in these two most important issues, and so CCFP decided not submit a bid.

In 1906 the concession was won by Mathieu Lugan and Paiva Irmãos, and passed within two months to the newly formed "Companhia Viação Eléctrica do Porto" (VE). This new company would construct a complete new tram network of metre gauge to replace the old CCFP standard gauge network (1.435 mm / 4'8½"). VE made contacts with several suppliers of tramcars in the USA and Europe and acquired the first new rails and a site in Massarelos to construct a new power plant, depot and workshop. VE were to inform CMP a month in advance for every 5 km of track laying, as it would be necessary to notify CCFP to remove (at their own expense) the now superfluous existing tracks. The decision of VE to build a metre gauge rail network, instead of keeping the existing standard gauge meant that the purchase of the old system was out of the question, although it could be an economical and time saving solution. Meanwhile CCFP just ignored all these developments, and went on exploiting the system.

From 1902 to 1906 both the rail network and the services were improved, as CCFP didn't anticipate that someone else would get the contract. But when this actually happened in 1906, no more improvements or track building were made in the network, until the ownership of the contract was solved.

A legal dispute was in the air. CMP knew this could happen, and took precaution, by forcing the successful concessionaire to deposit 1,000 contos (one billion réis, since 1911 this was equal to one million escudos). Should the Town hall be prosecuted by CCFP and lose the legal action, this bail would be used to compensate CCFP, and if it wasn't enough, the concessionaire would pay the rest, with no other compensation. Portuguese law was clearly insufficient, without the courts' intervention, to establish what legal rights CCFP had, if any. The amount of the compensation was another major issue, and this could lead the

newly born company to bankruptcy, should the courts grant CCFP a larger sum then the expected one. An expropriation law would be passed only 6 years later, in 1912. So, even CCFP claiming public interest to acquire the existing rail network would be a major problem. Most important of all, public transportation could come to a halt in the country's second city.

Meanwhile, parallel negotiations were being carried on by CCFP and VE: a merger was agreed, the name chosen "Companhia Carris Eléctricos do Porto", but suddenly and unexpectedly CCFP just bought VE (24 August 1908), received the charter, and the land VE bought to construct the depot and workshops at Massarelos, and shut VE down. What CMP fought against had just happened, and now, for the first time since its inception, CCFP had a solid legal instrument and became the legal operator with exclusivity inside the Porto area, until the 22 December 1981.

The hard life of a tram company

Now CCFP could continue with its own agenda. The change of gauge was cancelled as well as the acquisition of a complete new fleet of rolling stock. They were obliged to construct, but not obliged to operate, all the lines stipulated in the concession within five years, except the Freixo line via Campanhã railway station. It couldn't be constructed, because the low railway bridges over the Rua do Freixo didn't allow clearance for tram use. Although completing the lines, and making them ready for operation, CCFP never intended to exploit all of them. Low estimated revenues in thinly populated areas were the main cause. With CMP agreement they would be lifted without ever being used. On 20 December 1911 CCFP reported to CMP about the status of the lines they had to construct according the concession. Ready were:

The line of Rua da Boavista to Estrada de Circunvalação (Monte dos Burgos), the future line 6;

The line from Praça Carlos Alberto via Trindade to Rua Sta. Catarina, part of the future circular line 19;

The line from Bonfim to Campanhã, part of the future line 11;

The line from Alameda de Massarelos to Rua de Cedofeita via Rua D. Pedro V, Piedade and Torrinha, part of the future circular line 16;

Extensions to the Estrada de Circunvalação (this road is also the municipal border) of the old Linhas de Paranhos (Circunvalação -Amial, future line 7), S. Roque de Lameira (Circunvalação-S. Roque, future line10) and Costa Cabral (Circunvalação-Areosa, future line 9);

The route from the old Linha da Rua Duqueza de Bragança via the Rua d'Alegria and Constituição until the Rua Antero Quental, part of the future line 20.

Not yet ready were:

The line from Av. Boavista to Ouro via Lordelo, the never opened line 4;

The extension of the Campanhã line to the Estrada de Circunvalação (Freixo);

The line from Infante via Ribeira to the lower deck of the Ponte Luís I, the never opened line 15. This line was made in 1912 but CCFP said they would open it as soon as they had extended it over the bridge and in Vila Nova de Gaia. However they had asked for the concession, but never received it. It is likely that the concession had to be given by the State, but the responsible authorities had at that time, shortly after the revolution turning Portugal into a republic, more urgent matters to deal with;

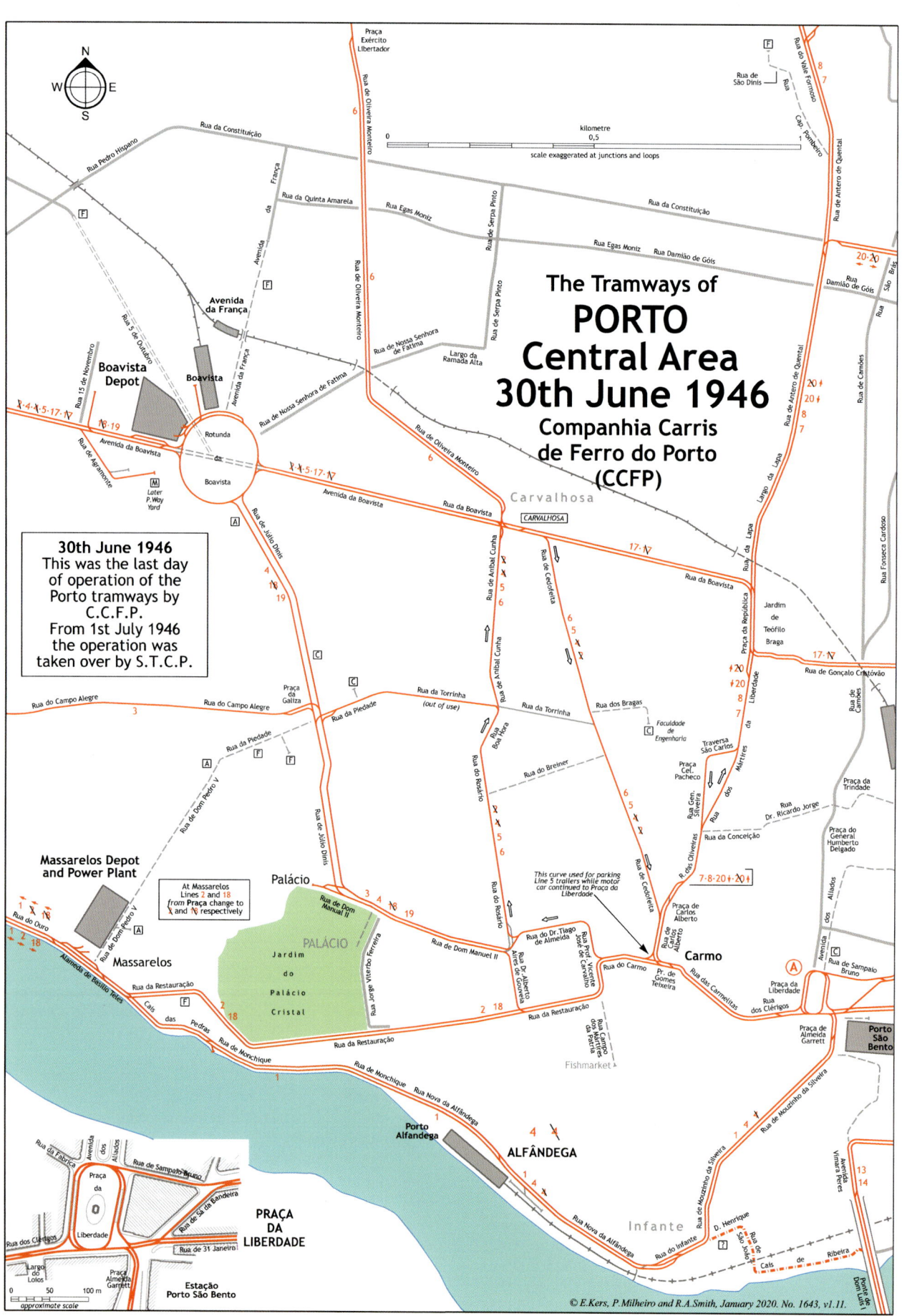

The Tramways of
PORTO
Central Area
30th June 1946
Companhia Carris
de Ferro do Porto
(CCFP)

30th June 1946
This was the last day
of operation of the
Porto tramways by
C.C.F.P.
From 1st July 1946
the operation was
taken over by S.T.C.P.

At Massarelos
Lines 2 and 18
from **Praça** change to
X and N respectively

This curve used for parking
Line 5 trailers while motor
car continued to Praça da
Liberdade

Massarelos Depot
and Power Plant

Boavista
Depot

Later
P.Way
Yard

PALÁCIO

Jardim
do
Palácio
Cristal

PRAÇA
DA
LIBERDADE

Estação
Porto São Bento

Porto
São
Bento

Carmo

Palácio

Massarelos

Alfândega

Porto
Alfandega

Fishmarket

Infante

Carvalhosa

CARVALHOSA

© E.Kers, P.Milheiro and R.A.Smith, January 2020. No. 1643, v1.11.

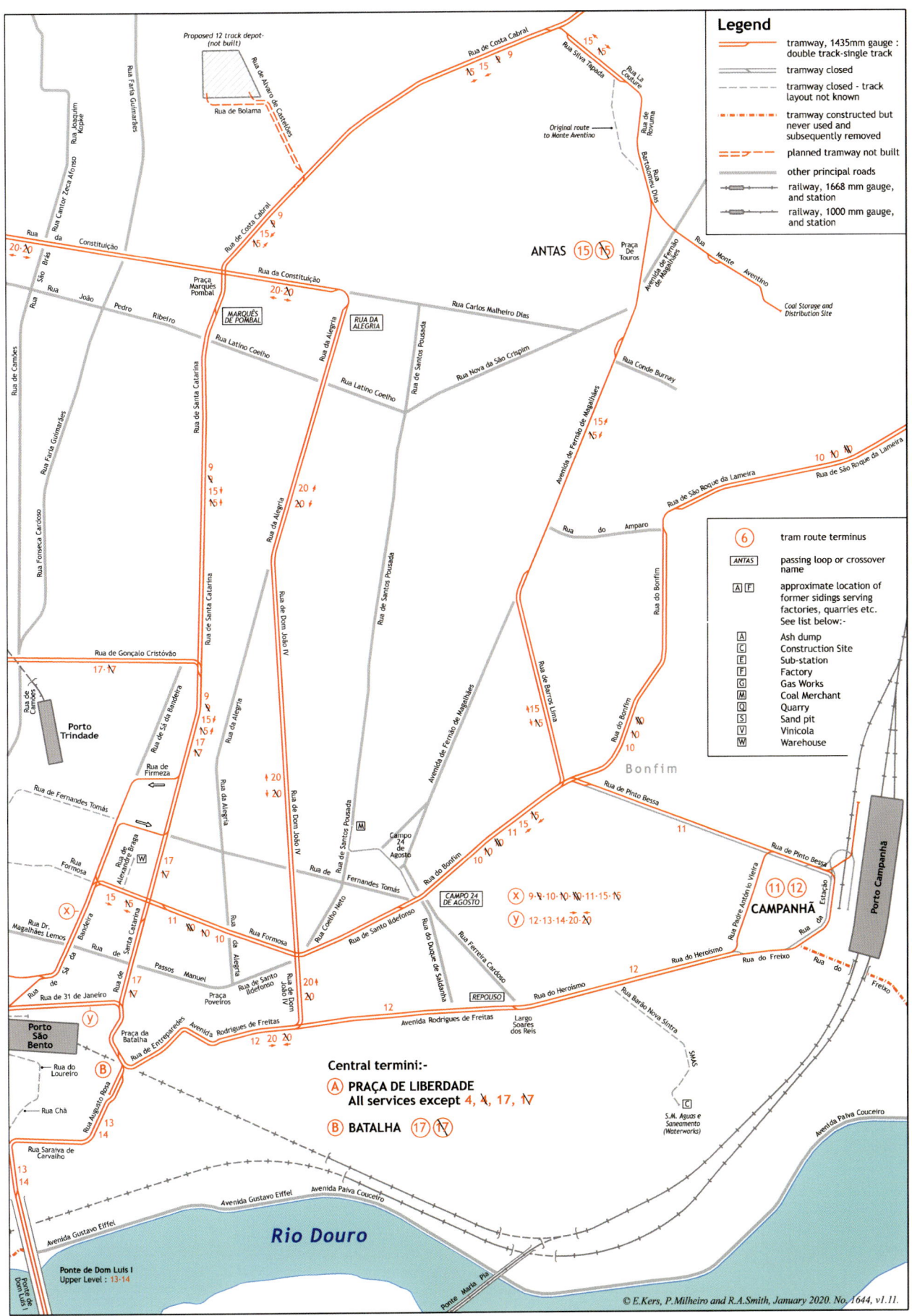

The Tramways of
PORTO
1946
Companhia Carris de Ferro do Porto
(CCFP)

30th June 1946
This was the last day of operation of the Porto tramways by C.C.F.P.
From 1st July 1946 the operation was taken over by S.T.C.P.

0   0,5   1   2   3
kilometres
scale exaggerated at junctions and loops

LEÇA DA PALMEIRA
① ⑤ ⑲

MATOSINHOS ⑰

Praia da Leça
Castelo
Av. do Dr. A. Guimarães
Leixões

Matosinhos

See Note A

Note A
Original route prior to harbour extension in 1938

Matosinhos

Porto de Leixões

Doca Mercado

GODINHO

Prado

Senhora da Hora

Real

Estrada da Circunvalação

Circunvalação

Avenida Meneres

Praça Cidade do Salvador

CIRCUNVALAÇÃO

Biquinha

Site for planned 12 track depot (not built)

Castelo de Queijo

Avenida da Boavista

FONTE DA MOURA ④ Ⓧ

ANTÓNIO AROSO

2 Ⓧ 5·17· Ⓝ 18· Ⓝ 19

Avenida Dr. Antunes Guimarães

Campinas

Ramalde

PRELADA

MONTE DOS BURGOS ⑥

Rua Monte dos Burgos Carvalhido

Sub-station No.2
Rua da Telhe ira

AMIAL

Rua do Amial

Rua do Amial

Rua Vale Formoso

Igreja de Paranhos

Rua Delfim Maia

C (?)

PARANHOS ⑧

Rua do Campo Lindo

PONTE DE PEDRA ⑦

São Mamede

Rua de Silva Brinco

Conde

São Mamede de Infesta

Rua Godinho de Faria

At Gomez da Costa Lines Ⓧ and Ⓝ from **Praça** change to **18** and **2** respectively

P. Way Yard

FOZ-MOLHE

Avenida Marechal Gomes da Costa

Cadouços

Foz do Douro

Esplanada do Castelo

Rua da Senhora da Luz

FOZ ⑰

Avenida Dom Carlos I

PASSEIO ALEGRE

Avenida da Boavista

Ⓧ·4·5·17· Ⓝ 19

BESSA

Rua Azevedo Coutinho

F

LORDELO ③

Rua do Campo

Rua Azevedo Coutinho

Rua Guerra Junqueiro

ARRÁBIDA

Largo António Calém

BICALHO

Rua do Ouro

Rua do Ouro

LARGO DO CALÉM

1 2 Ⓧ 18 Ⓝ

See Central Area Map

Francos

Rio Douro

Oceano Atlântico

General Torres

Vila Nova da Gaia

Rua do Barão do Corvo

Rua Conselheiro Veloso da Cruz

Avenida Republica

13
14

FC.

AC.

Coimbrões

DEVESAS ⑭

Vila Nova de Gaia

Jardim Soares Reis

Rua de Dr. Soares dos Reis

SANTO OVIDIO ⑬

© E.Kers, P.Milheiro and R.A.Smith, July 2015. No. 1641, v1.9.

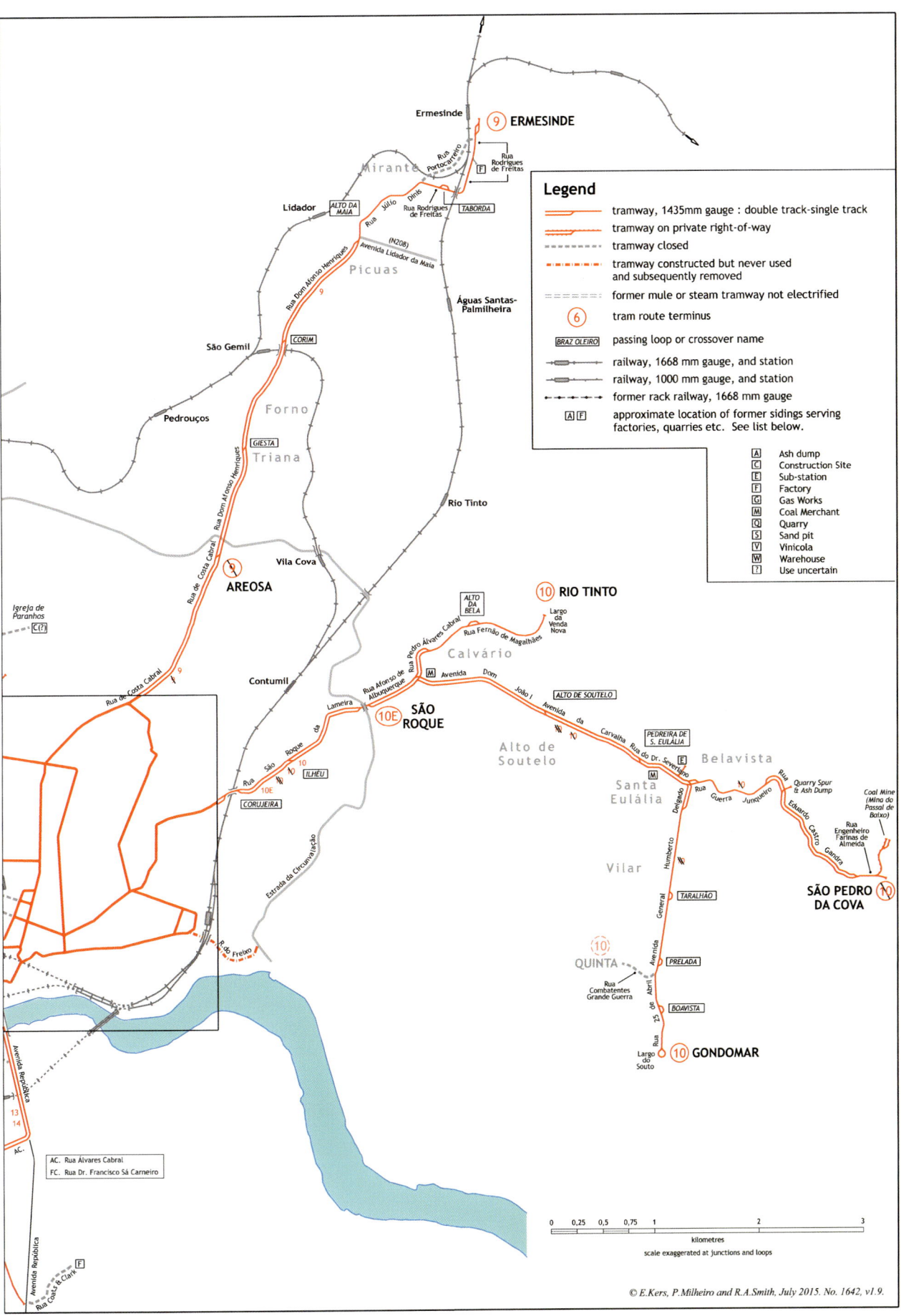

Legend

———	tramway, 1435mm gauge : double track-single track
	tramway on private right-of-way
- - -	tramway closed
—·—·—	tramway constructed but never used and subsequently removed
=======	former mule or steam tramway not electrified
⑥	tram route terminus
BRAZ OLEIRO	passing loop or crossover name
	railway, 1668 mm gauge, and station
	railway, 1000 mm gauge, and station
	former rack railway, 1668 mm gauge
Ⓐ Ⓕ	approximate location of former sidings serving factories, quarries etc. See list below.

Ⓐ	Ash dump
Ⓒ	Construction Site
Ⓔ	Sub-station
Ⓕ	Factory
Ⓖ	Gas Works
Ⓜ	Coal Merchant
Ⓠ	Quarry
Ⓢ	Sand pit
Ⓥ	Vinicola
Ⓦ	Warehouse
?	Use uncertain

Ermesinde ⑨ **ERMESINDE**

Mirante

Lidador — *ALTO DA MAIA*

Rua Portocarreiro

Rua Rodrigues de Freitas Ⓕ

Rua Rodrigues de Freitas *TABORDA*

Rua Júlio Dinis

Rua (N208)

Avenida Lidador da Maia

Picuas

Águas Santas-Palmilheira

Rua Dom Afonso Henriques

9

São Gemil *CORIM*

Forno

Pedrouços

GIESTA

Triana

Rio Tinto

Vila Cova

⑨ **AREOSA**

Rua de Costa Cabral

Rua Dom Afonso Henriques

Igreja de Paranhos Ⓒ(?)

Rua de Costa Cabral 9

ⓄⒶⓁⓉⓄ DA BELA

⑩ **RIO TINTO**

Largo da Venda Nova

Rua Fernão de Magalhães

Calvário

Rua Pedro Álvares Cabral

Contumil

Rua Afonso de Albuquerque

Ⓜ Avenida Dom

Lameira

⑩Ⓔ **SÃO ROQUE**

João I Avenida da

ALTO DE SOUTELO

Rua São Roque da

10

Alto de Soutelo

Carvalha Rua do Dr. Severiano

PEDREIRA DE S. EULÁLIA

Ⓔ

Belavista

Rua da 10Ⓔ *ILHÉU*

Santa Eulália

Ⓜ

Rua Guerra

Rua Quarry Spur & Ash Dump

Coal Mine (Mina do Passal de Baixo)

CORUJEIRA

Delgado

Junqueiro

Rua Eduardo Castro Gandra

Rua Engenheiro Farinas de Almeida

Estrada da Circunvalação

Vilar

Humberto

SÃO PEDRO DA COVA ⑩

R. do Freixo

General

TARALHÃO

⑩ **QUINTA**

Avenida 25 de Abril

PRELADA

Rua Combatentes Grande Guerra

BOAVISTA

Largo do Souto

⑩ **GONDOMAR**

Avenida República

13
14

AC.

AC. Rua Álvares Cabral
FC. Rua Dr. Francisco Sá Carneiro

Avenida República

Rua Coals & Clark Ⓕ

0 0,25 0,5 0,75 1   2   3
kilometres
scale exaggerated at junctions and loops

© E.Kers, P.Milheiro and R.A.Smith, July 2015. No. 1642, v1.9.

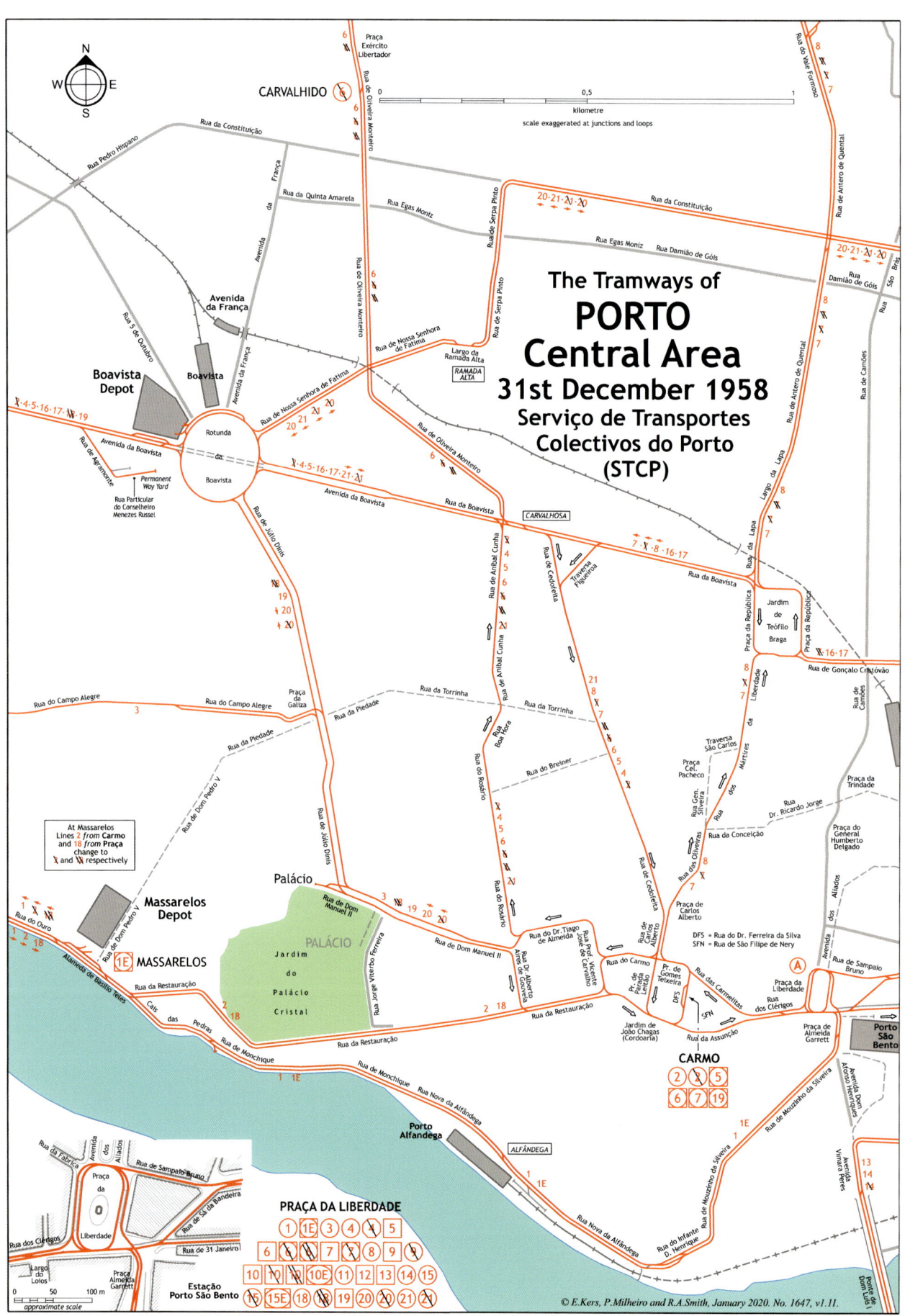

The Tramways of
PORTO
Central Area
31st December 1958
Serviço de Transportes
Colectivos do Porto
(STCP)

© E.Kers, P.Milheiro and R.A.Smith, January 2020. No. 1647, v1.11.

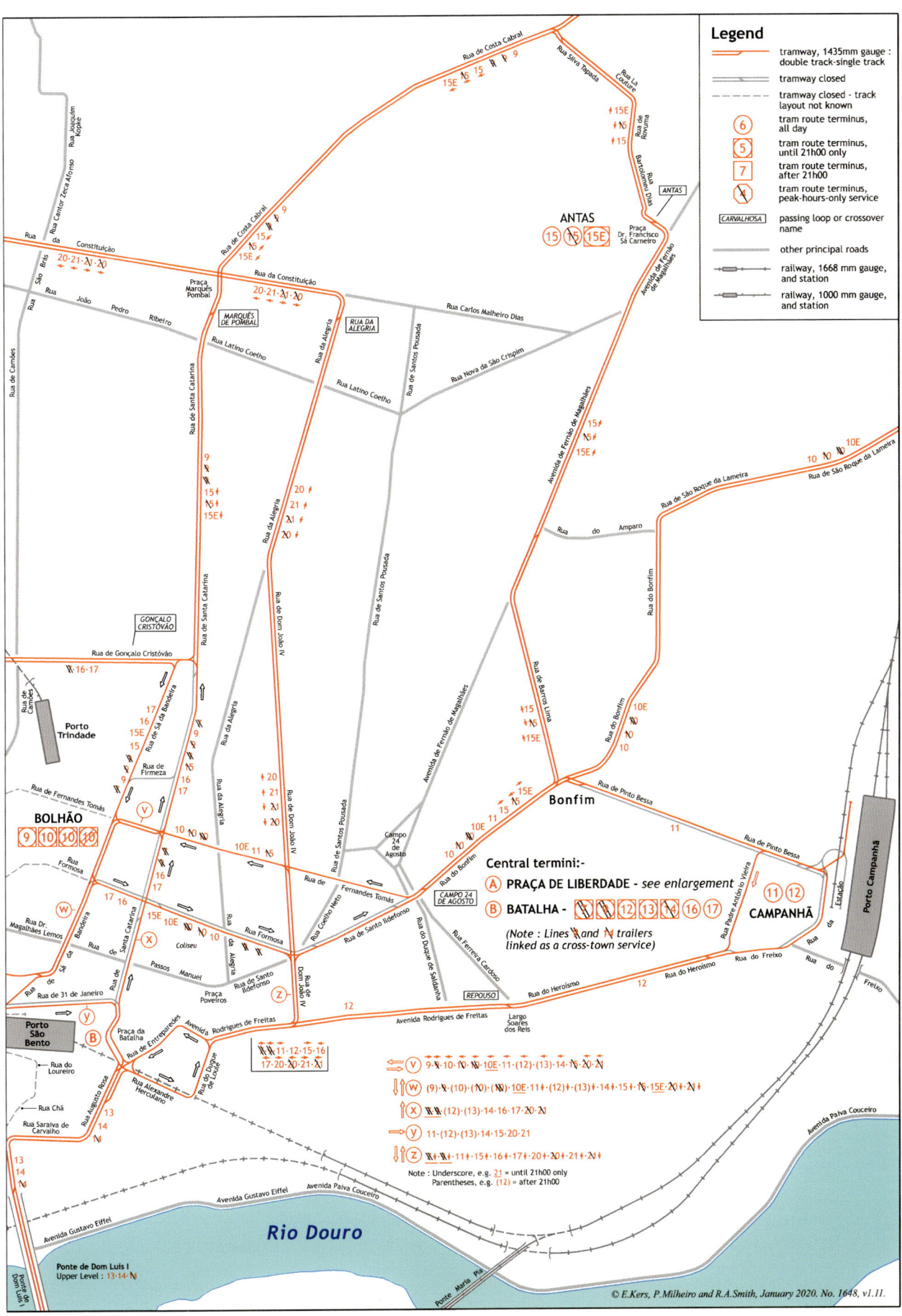

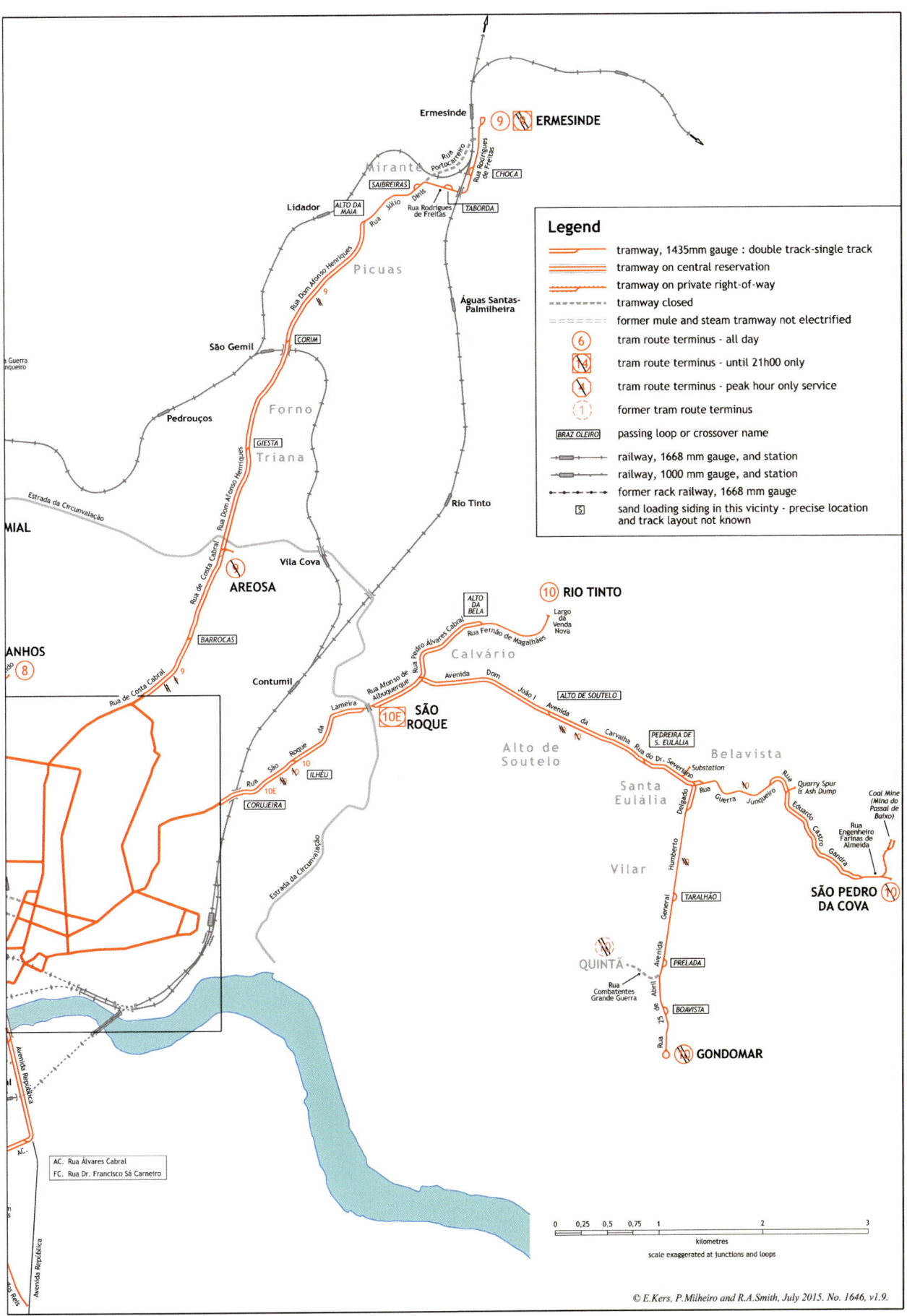

Legend

| tramway, 1435mm gauge : double track-single track |
| tramway on central reservation |
| tramway on private right-of-way |
| tramway closed |
| former mule and steam tramway not electrified |
| ⑥ tram route terminus - all day |
| ⑭ tram route terminus - until 21h00 only |
| ⑭ tram route terminus - peak hour only service |
| ⑴ former tram route terminus |
| *BRAZ OLEIRO* passing loop or crossover name |
| railway, 1668 mm gauge, and station |
| railway, 1000 mm gauge, and station |
| former rack railway, 1668 mm gauge |
| Ⓢ sand loading siding in this vicinity - precise location and track layout not known |

Ermesinde

⑨ ❌ **ERMESINDE**

Mirante

SAIBREIRAS

Rua Portocarreiro

Rua Rodrigues de Freitas

CHOCA

Lidador

ALTO DA MAIA

Rua Júlio

Dinis

Rua Rodrigues de Freitas

TABORDA

Rua Dom Afonso Henriques

Picuas

Águas Santas-Palmilheira

São Gemil

CORIM

Forno

Pedrouços

GIESTA

Triana

Estrada da Circunvalação

Rio Tinto

...MIAL

Vila Cova

⊘ **AREOSA**

Rua de Costa Cabral

Rua Dom Afonso Henriques

ALTO DA BELA

⑩ **RIO TINTO**

Largo da Venda Nova

Rua Fernão de Magalhães

Calvário

BARROCAS

...ANHOS

⑧

Rua de Costa Cabral

Contumil

Rua Pedro Álvares Cabral

Rua Afonso de Albuquerque

Avenida

Dom

ALTO DE SOUTELO

João I

Avenida

da

Carvalha Rua do Dr. Severiano

PEDREIRA DE S. EULÁLIA

Belavista

Lameira

10E **SÃO ROQUE**

Alto de Soutelo

Substation

Rua

Quarry Spur & Ash Dump

Coal Mine (Mina do Passal de Baixo)

Santa Eulália

Guerra

Junqueiro

Eduardo Castro

Rua Engenheiro Farinas de Almeida

Rua São Roque da

Rua São Roque da

⑩ *ILHÉU*

10E

CORUJEIRA

Humberto

Vilar

Delgado

SÃO PEDRO DA COVA ⑩

Estrada da Circunvalação

General

TARALHÃO

⊘ **QUINTÃ**

Avenida

PRELADA

Rua Combatentes Grande Guerra

BOAVISTA

Rua 25 de Abril

⊘ **GONDOMAR**

Avenida República

AC. Rua Álvares Cabral

FC. Rua Dr. Francisco Sá Carneiro

| 0 | 0,25 | 0,5 | 0,75 | 1 | 2 | 3 |

kilometres

scale exaggerated at junctions and loops

© E.Kers, P.Milheiro and R.A.Smith, July 2015. No. 1646, v1.9.

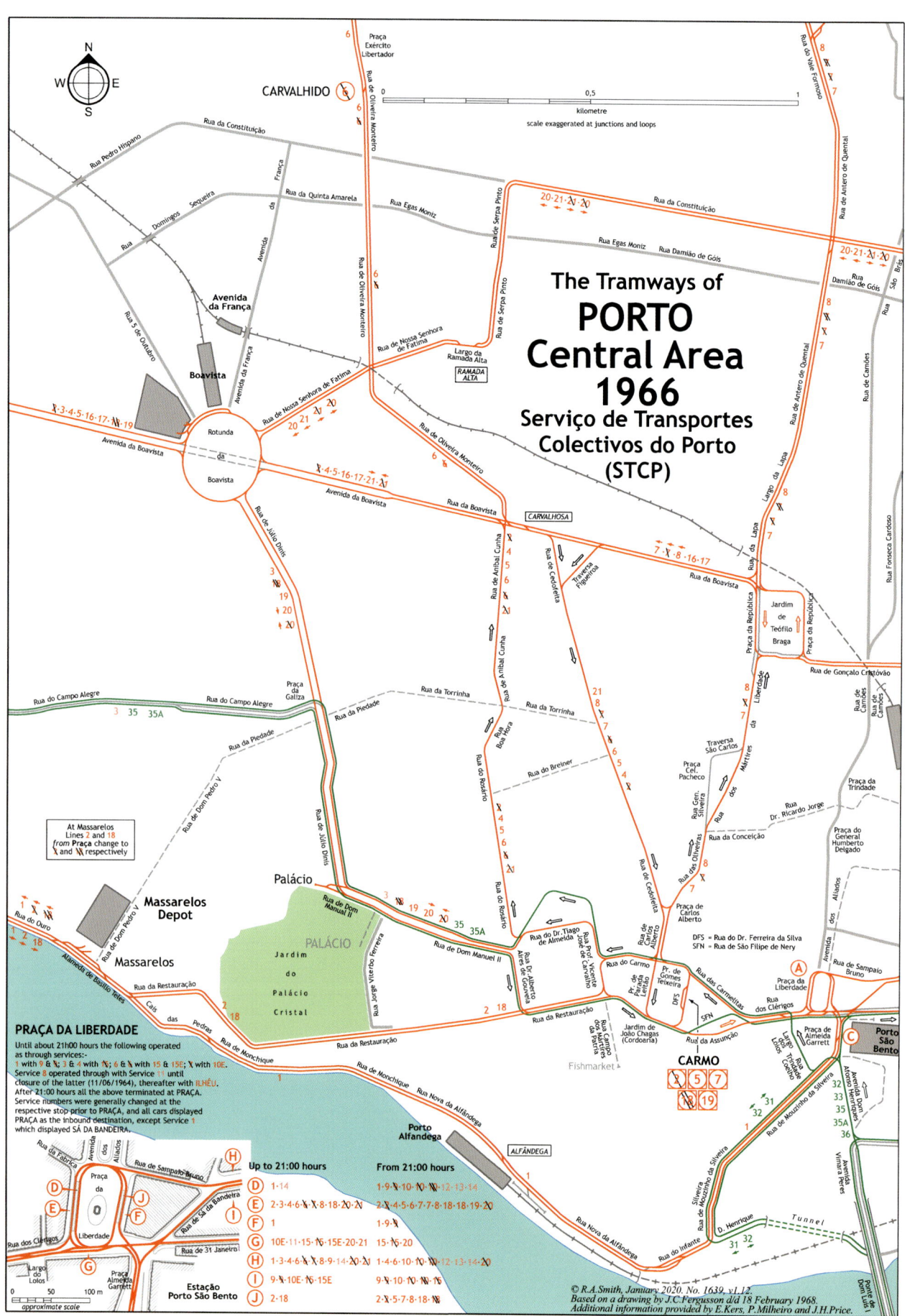

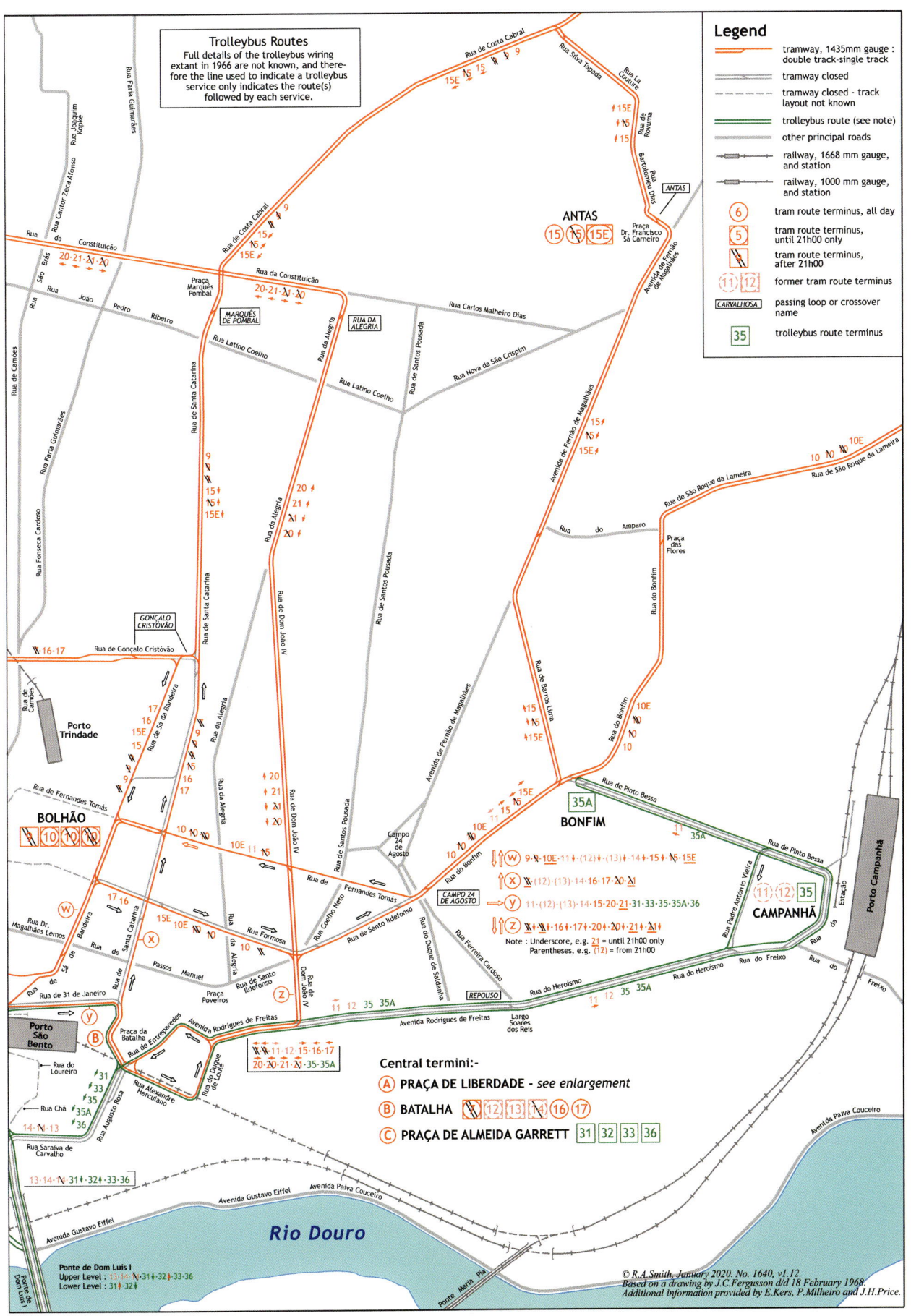

Trolleybus Routes
Full details of the trolleybus wiring extant in 1966 are not known, and therefore the line used to indicate a trolleybus service only indicates the route(s) followed by each service.

ANTAS

BONFIM

CAMPANHÃ

BOLHÃO

Note : Underscore, e.g. 21 = until 21h00 only
Parentheses, e.g. (12) = from 21h00

Central termini:-
Ⓐ **PRAÇA DE LIBERDADE** - *see enlargement*
Ⓑ **BATALHA** 12 13 14 16 17
Ⓒ **PRAÇA DE ALMEIDA GARRETT** 31 32 33 36

Rio Douro

Ponte de Dom Luís I
Upper Level : 13·14· 31 32 33·36
Lower Level : 31 32

© R.A.Smith, January 2020. No. 1640, v1.12.
Based on a drawing by J.C.Fergusson d/d 18 February 1968.
Additional information provided by E.Kers, P.Milheiro and J.H.Price.

Legend

	tramway, 1435mm gauge : double track·single track
	tramway closed
	tramway closed - track layout not known
	trolleybus route (see note)
	other principal roads
	railway, 1668 mm gauge, and station
	railway, 1000 mm gauge, and station
6	tram route terminus, all day
5	tram route terminus, until 21h00 only
	tram route terminus, after 21h00
(11) (12)	former tram route terminus
CARVALHOSA	passing loop or crossover name
35	trolleybus route terminus

The Tramways of
PORTO
1966
Serviço de Transportes
Colectivos do Porto
(STCP)

kilometres

scale exaggerated at junctions and loops

Note A
Original route prior
to harbour extension
in 1938

Railway between Senhore da Hora
and Matosinhos closed to passengers
1st July 1965.
Closed completely by 1st July 1966

PONTE
DE
PEDRA (7)

LEÇA DA
PALMEIRA   (1) (5) (19)

MATOSINHOS   (16)

Leixões

SÃO MAMEDE

São Mamede
de Infesta

MATOSINHOS   (1) (5) (16) (19)

GODINHO

Porto de
Leixões

Doca
Mercado

MONTE
DOS
BURGOS (6)

AMIAL

Prado

Praça Cidade
do Salvador

CIRCUNVALAÇÃO

Biquinha

Real

Estrada da Circunvalação

Circunvalação

Senhora
da Hora

PEREIRÓ
(3) (4)

Ramalde

Campinas

PRELADA

ARCA D'AGUA

PARANHOS
(8)

Sub-station
No. 1

Castelo
de Queijo

ANTUNES
GUIMARÃES

Rua Dr.
Joaquim
Costa

Avenida da Boavista

CASTELO
DO QUEIJO

ANTÓNIO
AROSO

At Gomez da Costa
Lines X and XX
from Carmo change to
18 and 2 respectively

Carcereira
Depot

Francos

See
Central Area
Map

AV. BRASIL

Avenida Marechal Gomes da Costa

BESSA

Avenida
da Boavista

GOMES
DA COSTA

LORDELO
(35) (35A) (3)

Foz do
Douro

Cadoucos

Esplanada
do Castelo

Rua da
Senhora
da Luz

FOZ
(17)

PASSEIO
ALEGRE

Largo
António
Calém

LARGO
DO CALÉM

R. do Ouro

SECIL

Rua do Ouro

Ponte da
Arrábida

Rio Douro

Oceano
Atlântico

Av. Diogo Leite

Vila Nova
da Gaia

Coimbrões

(14) (14) (33)

COIMBRÕES

(a)

SANTO OVIDIO
(31) (32) (36) (13)

footer / copyright

© R.A.Smith, December 2018. No. 1637, v1.10.
Based on a drawing by J.C.Fergusson d/d 18 February 1968.
Additional information provided by E.Kers, P.Milheiro and J.H.Price.

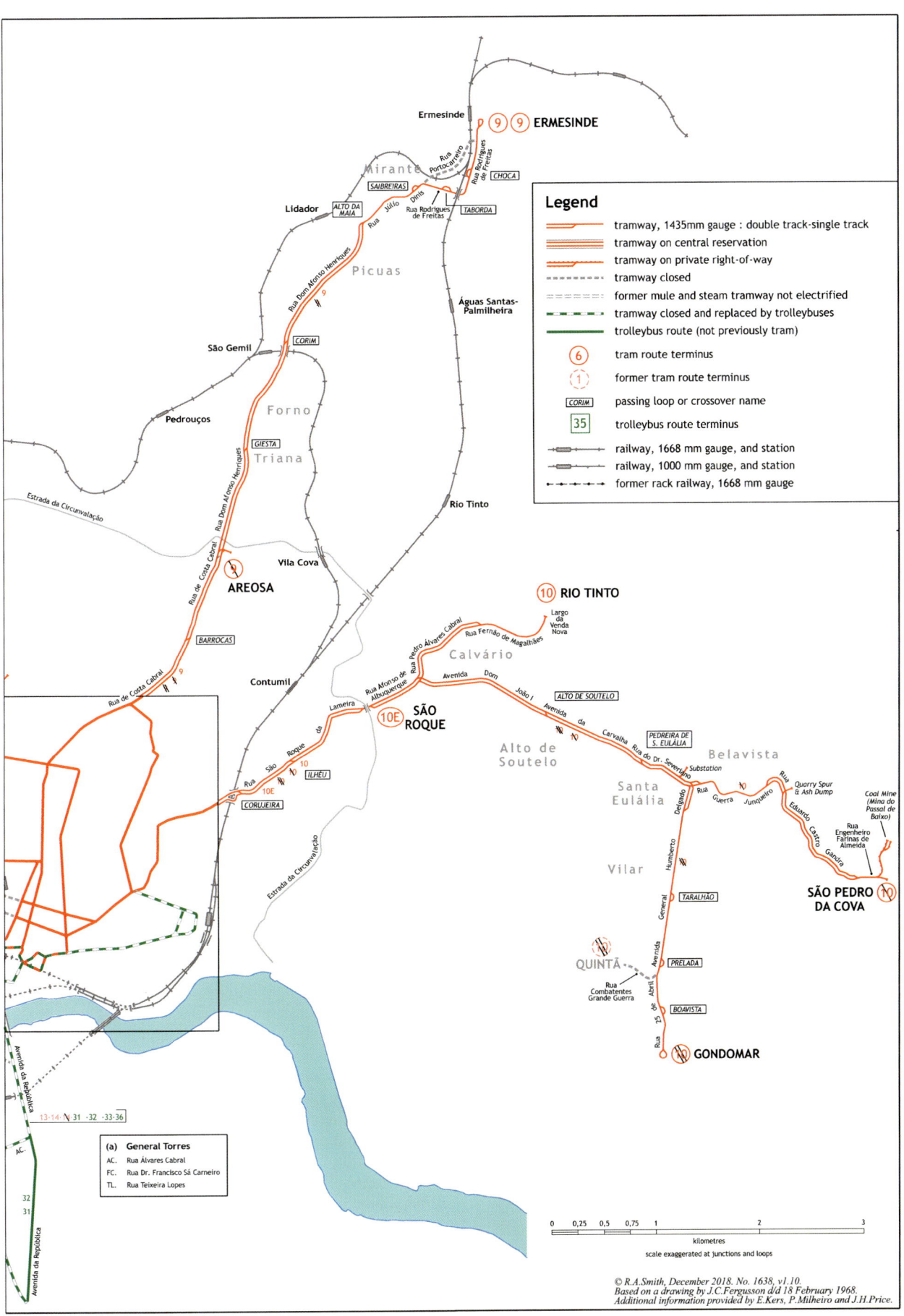

Legend

	tramway, 1435mm gauge : double track-single track
	tramway on central reservation
	tramway on private right-of-way
	tramway closed
	former mule and steam tramway not electrified
	tramway closed and replaced by trolleybuses
	trolleybus route (not previously tram)
6	tram route terminus
1	former tram route terminus
CORIM	passing loop or crossover name
35	trolleybus route terminus
	railway, 1668 mm gauge, and station
	railway, 1000 mm gauge, and station
	former rack railway, 1668 mm gauge

Ermesinde

9 9 **ERMESINDE**

Mirante

SAIBREIRAS

Rua Portocarreiro

Rua Rodrigues de Freitas

CHOCA

Lidador

ALTO DA MAIA

Dinis

Rua Júlio

Rua Rodrigues de Freitas

TABORDA

Picuas

Rua Dom Afonso Henriques

9

Águas Santas-Palmilheira

São Gemil

CORIM

Forno

Pedrouços

GIESTA

Triana

Rua Dom Afonso Henriques

Rio Tinto

Estrada da Circunvalação

Vila Cova

9 **AREOSA**

Rua de Costa Cabral

BARROCAS

10 **RIO TINTO**

Largo da Venda Nova

Rua Fernão de Magalhães

Rua Pedro Álvares Cabral

Calvário

Contumil

Rua Afonso de Albuquerque

Avenida Dom

João I

ALTO DE SOUTELO

Rua de Costa Cabral

9

Lameira

10E **SÃO ROQUE**

Avenida da Carvalha

Rua do Dr. Severino

PEDREIRA DE S. EULÁLIA

Substation

Belavista

Rua

Quarry Spur & Ash Dump

Coal Mine (Mina do Passal de Baixo)

Rua do São Roque da

10

ILHÉU

Alto de Soutelo

Santa Eulália

Delgado

Rua Guerra

Junqueiro

Eduardo Castro Gandra

Rua Engenheiro Farinas de Almeida

6 10E

CORUJEIRA

Vilar

Humberto

SÃO PEDRO DA COVA 10

General

TARALHÃO

Estrada da Circunvalação

Avenida

QUINTÃ

Rua Combatentes Grande Guerra

Avenida 25 de Abril

PRELADA

BOAVISTA

Rua

10 **GONDOMAR**

Avenida da República

13·14 31 ·32 ·33·36

AC.

32

31

Avenida da República

(a) General Torres

AC. Rua Álvares Cabral
FC. Rua Dr. Francisco Sá Carneiro
TL. Rua Teixeira Lopes

0   0,25   0,5   0,75   1   2   3

kilometres

scale exaggerated at junctions and loops

© R.A.Smith, December 2018. No. 1638, v1.10.
Based on a drawing by J.C.Fergusson d/d 18 February 1968.
Additional information provided by E.Kers, P.Milheiro and J.H.Price.

The direct line from Praça to the upper deck of the bridge. CMP first had to make the new Avenida, but only did that 40 years later when CCFP didn't exist anymore.

The line along Av. Boavista until Castelo do Queijo, part of the future lines 2 and 5.

There was no obligation to construct any further lines within CMP area.

But in 1908 and in possession of the contract, CCFP at the same time as building the lines, was also considering a larger extra urban network. In 1906 at the signing of the deed only three lines were in existence outside Porto; the two Matosinhos lines, and the Vila Nova de Gaia line. The Matosinhos lines were the former CCAPFM now electrified (later to become line 1), and CCFP steam line (later to become line 5) in an adjacent street. Not agreeing with the percentage of the revenues due to CMP, but not being able to prevent it, CCFP came up with an idea: explain and make CMP agree with the principle that they should calculate the ratio of the inner (within Porto limits) lines mileage versus total network mileage, and use this figure to reduce the tax; as some revenues were obtained outside Porto, CMP shouldn't collect money where the concession conditions didn't apply. By deliberation of the 28 December 1908, this rule became effective. What seemed to be a minor adjustment would

result in a reduction of taxes in the years to come. From this date on, CCFP would extend the rail network outside Porto, where few tickets were previously sold, and manipulate zones to reduce income on city sections. Passengers were even carried on coal cars to boost fares outside CMP jurisdiction.

In the years 1909-1914 the fleet was expanded, the steam tram replaced by electric traction and a new power plant constructed on the premises acquired by the VE takeover. CCFP intended to use the new line on Avenida da Boavista between Fonte da Moura and Castelo do Queijo, which they were obliged to build in the concession, to provide a faster service to Matosinhos, but CMP wanted the former steam line via Cadoucos to be retained and electrified, although it wasn't included in the concession. CCFP was apparently now in dispute with both councils because Matosinhos would not allow track doubling in Rua Brito Capelo, thus obliging CCFP to electrify the former steam line in Rua Roberto Ivens; Matosinhos relented in 1930. When the line from Boavista to Palacio opened in 1934, CMP wanted CCFP to run from Boavista to Foz via Avenida Marechal Gomes do Costa, but this did not happen until buses appeared in 1948.

During World War I CCFP was not able to maintain tramcars and power supply adequately. A major problem was that Siemens had supplied most electric equipments for the power plant and the major part for the rolling stock. During WWI it was almost impossible to get spare parts for these. Trams were cannibalised to keep others running, effectively reducing the size of the fleet. By 1920 the company was in a crisis, the number of available tramcars was far less than needed and power failures occurred frequently. An order for new tramcars from Brill had to be cancelled because of lack of finance. Attempts were made to get new tramcars from Germany on the WWI reparations, but this also failed. Post war labour unrest and strikes deepened the crisis. In 1921 advantage was taken of a new law allowing contract modification due to "unnatural and unforeseeable" causes to gain financial concessions from CMP. In the second half of the 1920's and 1930's, despite the 1928 fire at Boavista slowly the situation improved. The reliability of the power supply improved. Tramcars received new electric installations and were repaired, overhauled or reconstructed. The workshops also started to construct new tramcars of the semi-convertible Brill design of 1900. CCFP became a competent well-operated company, with all necessary skills to build new, reconstruct and maintain all equipment without the need to import spare parts from abroad. They had workshops, carpentry, painting, electrical and mechanical facilities, tram building and repair works, including two large foundry facilities with an average capacity of 1 ton of molten metal each; even brass or aluminium could be smelted. These furnaces were the largest in the north of the country until 1975. But modernisation hardly took place, trams and other equipment became old-fashioned compared to contemporary systems in other countries.

Legal actions were filed during years to come, all in what was believed to be the company's best interest. Massarelos power station needed a large amount of water. This vast flow was subterranean and was diverted after an intervention made by CMP; predictably a complaint to the courts was filed (and won by CCFP). Another legal action took place when Norte de Portugal (Porto-Póvoa de Varzim Railway Company) decided to extend the metre gauge line from the existing Porto-Boavista railway station to a newly built central one at Trindade. CCFP, according to the 1906 contract claimed the exclusivity of construction

Praça da Batalha with the Igreja de Sto.Ildefonso and Brill-23 no.249, Brill-28 no. 239 and another Brill type tram. Car 249 is working a line 14 Praça - Devezas service and the 239 a line 17 Foz - Batalha service. Car 249 still has a second trolley-pole for working across the Ponte Dom Luís I bridge. The photo is probably taken in 1932 or 1933. *Commercial postcard*

Cars 104 and 76 at Boavista depot in 1955. *Courtesy of the National Tramway Museum, photographer D Trevor Rowe*

of railway lines within Porto. Due to an agreement made while discussing the 1929 takeover plan referred to below, CCFP dropped this action.

However courts didn't always rule in favour of CCFP. In 1917 CCFP constructed a new line to the Passal de Baixo coal mine in São Pedro da Cova, considered essential to supply coal to the new Massarelos power plant, but due to WWI could not acquire the rails and materials. Instead CCFP asked permission from CMP to lift little or unused rail lines, including those of line 19, which was in operation, so the rails could be used for the São Pedro da Cova line. CMP agreed, with a specific condition that as soon as the situation was back to normal, CCFP had to reconstruct it. But time went on, and as CCFP failed to do so, CMP filed a complaint to enforce it. The Supreme Administrative Court ruled on the 14 July 1939 in favour of CMP, compelling CCFP to re-lay the lines. CCFP still thought these lines were not profitable and came up with another solution, constructing two new lines, instead of reinstalling the "old" ones. These became lines 3 and 15.

Another recurrent dispute was the impossibility of CMP to force the concessionaire to construct new lines, in order to help development of areas away from the city centre, which CMP wanted to urbanise. CCFP had the exclusivity, but not the obligation, of constructing new lines. With the lines referred to in the concession contract completed, they would build new lines only if they wished, but could prevent any one offering competitive services, even by bus or any other mechanical means (but apparently not including taxis) within Porto boundaries.

So in 1929, CMP made a proposition to CCFP to buy the entire system, but both parties could not agree about the price CMP had to pay.

In September 1930 CCFP proposed to CMP to start bus lines to parts of the city, like Freixo, which were not served by tramlines. In October 1930 six Ford chassis were acquired. In 1931 the workshops built five twenty-six seater buses plus a spare body and a freight truck. However CCFP never opened a bus line and in 1932 the buses and the spare body were sold.

The end of CCFP

One of the conditions of the contract was the legal right of CMP to claim the concession: "after the first 35 years of exploitation, CMP has the legal right of claiming the concession, and for doing so, shall notify the concessionaire in advance of 5 years of the date…" And they did, on the 3 December 1936, intending the takeover would take place on the 22 December 1941, the very day when the concession would celebrate its 35th anniversary!

Meanwhile WWII began, adding a widespread fear to this complex process, as nobody could predict the conditions of the country when it ended. This worried the Portuguese Government, as economic conditions were in 1941 completely different from the ones in 1936 when the takeover was decided, even though the country was neutral. The solution seemed easy, postponing it by two years, in the expectation the war would end by then. WWII went on, and in November 1943 the Government again postponed the takeover date, until the 22 December 1945, without any other changes.

On the 24 November 1945, with WWII ended, the Government decided to select the 30 June 1946 to implement the takeover, as a new CMP mayor and aldermen would be appointed by the 1 January 1946, and that date also matched an economic cycle as in Porto 6 months passes were issued.

Porto STCP: Ponte de Pedra Tramcar no. 187+119. *Courtesy of the National Tramway Museum, photographer N.N Forbes*

Zorra 67 in the Massarelos power-plant. *MCE.*

And besides the municipality would have enough time to get acquainted with the transition. So with both CCFP and CMP agreement, the suspension period was once more postponed until the 30 June 1946.

According to the 1906 concession the VE/CCFP had to hand over for free after 75 years everything necessary and in good condition for the exploitation of the system. If the concession was ended after 35 years, then CMP had to pay the amount of money that the company would have earned in the remaining 40 years. Some questions arose. How much money would the company have earned? The calculations for that had to be based on the results of the years before. But had the date of 22 December 1941, the intended date of the buy off, or 1 July 1946, the date STCP took over the operations, to be taken for the calculations? WWII caused an enormous increase in passenger numbers, which significantly affected the outcome of the calculations. How much money would CCFP have spent to maintain the system in good condition? Was the Massarelos power plant necessary for the operations of the system and should it be included? etc. The shareholders of CCFP of course wanted as much money as possible while CMP wanted to give as little as possible. All conditions were fulfilled for the Grande Finale of about half a century of continued disputes between CCFP and CMP. This last dispute ended with CMP paying CCFP 143,891,319$86 (one hundred forty-three million, eight hundred ninety-one thousand, three hundred nineteen Escudos and eighty-six Centavos). CCFP ended its existence in July 1949. CMP/STCP published in the 1950's in two volumes, counting together over 1000 pages, a report with the details about the buy off.

STCP and the failed modernisation of the tram system

From 1 July of 1946 the trams were operated by the Servico de Transportes Colectivos do Porto (STCP). Therefore on that date, just prior to the first tram leaving Boavista depot to fulfil its duties at 04.30, a huge crowd saw the Mayor getting a bunch of flowers before tram 278 (nowadays 288 and the Museum "Belga"), left the premises for line 1 operation. The STCP inspection committee in what was an almost impossible task, evaluated in 1946 the condition of the industrial unit five years before, in 1941, worn by heavy use during the war years when the number of passenger almost doubled, concluding that the "good and working condition" requirement was met.

Now rail network layout improvements, new and modern cars, a new depot and new lines were envisaged. But one of the early decisions was to increase the number of trams in daily use, as STCP had 177 trams from CCFP, and at least 60 or 70 cars were kept every day as reserve; but from the director of the Massarelos power station came the "red flag": 125 trams is the highest number we can run without damaging the power station, or we face the risk of a major power failure and breakdown!. This number of cars wasn't absolute, of course, but should be considered prior to operation. Due to constant delays in some lines, innovative steps had to be implemented, in order to offer more seats, mainly for those who live outside Porto. This would be accomplished by ground-breaking decisions and solutions. Plans were made to modernise the system and the first buses were acquired to work routes to parts of the city where no tramlines existed.

When CCFP bought the 1906 concession from VE, the contract had a mandatory list of the lines to be constructed, services to be operated, including the quantity of inbound and outbound trams in any and all of the lines. With few exceptions, they had their origin at Praça, so apart from this hub, no reversing loops were ever built or used for tram operations, only crossovers that proved impractical and impossible to use in daily operation in the town centre, particularly 40 years later in 1946. A decision was taken to use Carmo, Batalha and Bolhão as alternative route destinations whenever it proved difficult to reach Praça. Even the longer routes in daily operation should start and terminate there, releasing the centre from tram congestion. Another major concern was the narrow central streets heavily used by both trams and motor traffic, with double tracks installed, which also made reversing practically impossible should an unexpected event occur. Breakdowns, power failures, derailments, were a nightmare. So works began in 1946 in Batalha, followed by Carmo in 1947 and Bolhão in 1949/50, in order to create alternative reliable layouts with reversing loops, and extending facilities, allowing the creation of new routes and enabling trailer operation without the need for shunting and coupling.

Although the existing fleet was in a fair condition, a significant proportion of the trams (around 90) dated from 1905 to 1912, and around 60 more from 1926 to 1930, giving a high average age in excess of two decades. Even the newest trams were

built to a design dating back to 1900. This didn't fit the concept STCP had of operating a modern and up-to-date standard fleet. The solution was to build or buy a fleet of new 2 axle and bogie cars. Bogie cars were never bought due to, but not only because of, financial problems, but single truck 2 axle cars were constructed in STCP workshops. However with the exception of one prototype, these cars were with their 1920's type of electric equipments and timber seats better characterised as primitive instead of modern. Even the modern prototype had a truck (79E) introduced by Brill in 1919 for use under Birney Safety cars.

In the 1950s it was decided to replace the trams by trolleybuses. The first of these came into operation in 1959. Although several tramlines were replaced by trolleybuses, the tramcars which were displaced from these lines were used to improve the service on the remaining tramlines. Only after 1967 was the number of tramcars substantially reduced, with the post-war cars being among the ones withdrawn. The same year it was decided to replace first all tramcars and subsequently the trolleybuses by diesel buses. But after the Revolution of 1974 and the return from the colonies, the Portuguese economy started to grow and the demand for public transport with it. New buses intended to replace the trams were acquired but needed to extend the network and to replace older buses. It appeared not to be possible to replace all the old trams as was planned and although many tramlines were closed in the late 1970s, other lines were retained until 1993. All tram services on Sundays were suspended in 1978 however. Only in 1995, 100 years after opening of the first electric tram line was the closure of the last line [18] announced. But the old trams had become emblematic of the city. Instead of suppressing the last line, it was transformed into a heritage system. Portuenses and tourists still can enjoy travelling on "eléctricos antigos".

That trams are not old fashioned, but can be vital for travelling to and in a modern but historic city is demonstrated by the new "Metro do Porto". While buses are trapped in the traffic jams, the modern trams run faster and move larger numbers of people than buses are able to do.

Network

Mule and Steam Tram Routes

The mainline of CCAPFM started in Rua dos Inglezes (since 1894 called Infante) and followed the bank of the river to Foz (Castelo). From Foz the Atlantic coast line was followed to Matosinhos, where the tramline was situated in Rua Juncal da Cima, now known as Rua Brito Capelo. In 1873 a branch line from Cordoaria to Massarelos via Rua da Restauração was opened. Although the time-table changed quite often, for many years the frequency on the mainline was every 20 minutes between Inglezes and Foz and once an hour to Matosinhos. In 1887 the mainline was extended to Leça da Palmeira (Rua do Arnado) using an iron bridge built by the Willebroeck company, to cross the river Leça. This bridge was used until 1938, when it was demolished due to the extension of the docks.

On 12 August 1874 CCFP opened its own line from Praça Carlos Alberto via Boavista and Fonte da Moura then on private right of way to Cadouços. In the same year the line was extended in the city to Porta do Olival (Cordoaria near Torre dos Clerigos). In 1875 the first urban lines were opened.

The line from Boavista to Cadouços was converted for the use of steam traction in 1878. In 1882 this line was extended from Cadouços via Rua da Gondarem to Matosinhos. The line reached

During the STCP period until the closure of the tramlines across the bridge small old trailers were used behind narrow pre-WWI tramcars on the combined route of the lines 9// and 14/. Here is such a combination, 118 +13, is at the railway station Vila Nova de Gaia - Devesas. *Source unknown*

the Atlantic coast just south of Castelo do Queijo where it crossed the line of CCAPFM. In Matosinhos the line was situated in Rua Juncal da Baixa, now called Rua Roberto Ivens, a street parallel to Rua Brito Capelo. For passengers to or from Leça da Palmeira CCFP constructed a pedestrian timber bridge in 1886 across the river Leça, which was not used by the (steam) trams. This bridge was demolished in 1922.

The 1883 CCFP network:
1. Matosinhos – Cadouços – Boavista - Praça D. Pedro – Campanhã
2. Infante – Praça D. Pedro - Cruz das Regateiras (now Hospital Conde Ferreira)
3. Boavista – Cordoaria - Palácio de Crystal – Rua de Breyner - Praça D. Pedro – Duquesa da Bragança (now D. João IV) – Campo de Regeneração (now Praça da República) – Boavista (one way)

The total network of CCFP had an extent of 24,183 m with 9,932 m double track and passing loops. During several years between Boavista and Matosinhos in summer the traction was mixed steam and mule, in winter it was mule only until the entry into service of further locomotives.

After the merger with CCAPFM CCFP opened one more mule tramline on 21 June 1894 to Bonfim (Igreja). On 8 May 1898 this line was, still with mule traction, extended to São Roque (Mercado de Corujeira).

Mule Trams Routes CCAPFM
- 9 March 1872 Alfândega - Foz

- 18 May 1872 Foz - Matosinhos Rua Juncal de Cima (Brito Capelo) at the bank of the Rio Leça
- 24 May 1872 Rua dos Inglezes (Infante) - Alfândega
- 1873 (probably summer) Cordoaria - Massarelos
- 1887 (probably May/June) Matosinhos Rua Juncal de Cima (Brito Capelo) - Iron bridge (made by Willebroeck) crossing the river Leça - Leça Rua do Arnado. Rua do Arnado doesn't exist anymore, but it was where the northern quay of the docks in the area is now, a little west of the movable bridge.

Mule (steam) tram routes CCFP
- 11/14 August 1874 (12 August official inauguration) Praça Carlos Alberto - Rua de Cedofeita - Boavista – Cadouços (steam Boavista to Cadouços 1878, extended to Matosinhos 1882).
- 19 October 1874 Porta do Olival (close to Torre dos Clerigos) - Praça Carlos Alberto
- 17 February 1875 Porta do Olival - Palácio (with special event journeys before that date).
- 13 June 1875 (start test running 12 May) Praça Carlos Alberto - Carmo - Praça - Rua D.Pedro - Bolhão - Rua Sta. Catarina - Praça da Batalha - Rua Entreparedes - S.Lazaro - Heroismo – Campanhã.
- 4 August 1875 trams run via Rua do Rosário in the direction of Boavista. Via Rua de Cedofeita now only trams in the direction of Carmo.
- 22 September 1875 Bolhão - Rua de Sta. Catarina - Rua Gonçalo Cristovão - Campo de Regeneração (Praça da Republica) - Rua da Boavista (outward direction, trams inward went via Cedofeita - Praça) (circular in central area).

Car 187 seen at São Pedro da Cova in the late 1950s/early 1960s. Note the conductor turning the trolley pole with retriever attached. The line to the coal mine is to the left. *P Eaton, courtesy P Milheiro*

Car 255 is at Largo da Venda Nova in the late 1950s/early 1960s. *P Eaton, courtesy P Milheiro*

- 1876 Rua Sta.Catarina - Largo de Aguardente (Praça Marques de Pombal).
- 1876 Rua Formosa - Rua Duqueza de Bragança (now Dom João IV) - Rua F.Thomaz.
- 12 May 1877 Largo de Aguardente - Rua Costa Cabral - Cruz das Regateiras (Hospital Conde Ferreira).
- Unknown between 1876 and 1883 the route via Rua Sá da Bandeira - Rua Sampaio Bruno.
- 25 May 1882 (22 May official inauguration) Cadouços -

Matosinhos Rua Juncal de Baixa (now Roberto Ivens) at the boundary of the Rio Leça. (steam).
- 9 July 1882 Rua dos Inglezes (Infante) - Rua Mousinho da Silveira – Praça.
- 1886 (probably) Campo de Regeneração - Lapa - Rua da Rainha - Rua Valle Formosa - Campo Lindo.
- 21 December 1892 Rua Martyres da Liberdade - Rua Coronel Pacheco - Rua Oliveiras used in inward direction by the linha de Paranhos: outward via Praça - Rua D.Pedro - Rua Formosa – Rua Sta.Catarina - Rua Gonçalo Cristovão - Campo de Regeneração - Lapa - Rua da Rainha - Rua Valle Formosa - Campo Lindo: return via Martyres da Liberdade - Rua Coronel Pacheco - Rua Oliveiras - Praça Carlos Alberto - Carmo - Rua das Carmelitas – Praça (circular in central area).
- 21 June 1894 Linha do Bonfim.
- 7 May 1898 Bonfim - Mercado de Corujeira.

Electrification of routes

In 1895 the Restauração line was electrified, including a section of the Marginal as far as Arrábida. Here were the first power station, depot and workshops of the electric trams. Subsequently the Marginal was electrified in stages in 1896-1897 and then the existing urban lines in the period 1898-1904. That last year mule traction was abandoned but the steam tram was retained for ten more years.

In 1905 the electric tram lines across the Ponte Dom Luis I to

The three Starbuck trailers, 8, 9 and 10, probably new to CCAPFM, at Massarelos depot in 1959. *J. H. Price*

Vila Nova de Gaia were opened. On the bridge two trolley poles had to be used to prevent oxidation of the ironwork by stray currents. A number of tramcars were equipped with a second trolley pole for use on the lines to Vila Nova de Gaia. After 1933 the trams could cross the bridge with one trolley-pole. In 1930 the single track (situated at the eastern side) was changed to double track. The distance between the centre of the tracks was only 2.430 m. This close arrangement of the tracks on the bridge meant that on the lines to Vila Nova de Gaia only narrow cars could be used: Carros Antigos, Constructora, Inglês and Brill-23.

Introduction of Electric traction
- 12 September 1895 Campo dos Martyres da Patria (Cordoaria) - Massarelos – Arrábida.
- 3 August 1896 (official inauguration 1 August 1896) Infante - Estação do Ouro.
- 4 April 1897 (official inauguration 26 March 1897) Estação do Ouro - Castelo da Foz.
- 20 July 1897 (official inauguration 29 May 1897) Castelo da Foz - Castelo do Queijo.
- 31 October 1897 Castelo do Queijo - Rua Godinho.
- 28 December 1897 Rua Godinho - Leça Rua do Arnado. This was the location where the mule trams formerly of CCAPFM had their terminus.
- 23 April 1898 Infante - Praça D.Pedro. Initially this service was separate from the Marginal as it had been with the mule trams. From 1902 the trams of the Marginal continued to Praça.
- 22 August 1898 Leça Rua do Arnado - Castelo da Leça.

- 18 May 1899 (official inauguration 12 May 1899) Praça D.Pedro - Marques de Pombal.
- 13 November 1899 Marques de Pombal - Hospital Conde de Ferreira.
- 6 October 1900 Carmo - Batalha. This was when trams started to use 31 de Janeiro / Sto.António.
- 26 November 1900 Batalha - Campanhã
- 23 May 1901 (official inauguration 11 May 1901) Carmo - Boavista. Boavista depot welcomed the 1st electric tram (CE 23) on 23 May.
- 24 October 1902 Rua Sta. Catarina - Praça da Corujeira (São Roque).
- 17 September 1903 Palácio route: Rua do Triumfo (now D.Manuel II) and Rua do Breiner. With this the approximately 500 m long section Carmo - Porta do Olival - around the Jardim da Cordoaria - Hospital de Santo António was abandoned.
- 14 March 1904 Rua da Boavista - Rua C.Cristovão and Rua F.Thomaz - Rua Duqueza de Bragança.
- 18 August 1904 Carmo - Campo Lindo and Boavista – Bessa.
- 28 October 1905 Praça de Almeida Garrett - Ponte Luís I – Vila Nova de Gaia (Sto.Ovídio and Devesas).
- 1909 Carvalhosa – Carvalhido.
- 1909 Massarelos - Rua D.Pedro V - Rua Piedade - Rua Torrinha.
- 1909 Rua das Oliveiras - Trindade - Rua F.Thomás. it is believed that the Constituição route was opened at the same time, but no data has been found to corroborate this.

- 19 February 1910 Rua Vale Formoso - S.Mamede da Infesta.
- 29 May 1910 Hospital Conde Ferreira - Circunvalação/Areosa.
- 1911 Bonfim – Campanhã.
- 1911 Carvalhido - Monte dos Burgos.
- January 1911 Areosa - Aguas Santas.
- November 1912 São Roque - Venda Nova.
- 1913 Devesas - Arco do Prado.
- 7 November 1914 Bessa - Fonte da Moura - Castelo do Queijo (or maybe Fonte da Moura - Castelo do Queijo only in February 1915). Probably at the same time also Rua Brito Capelo - Rua Sousa Aroso - Rua Roberto Ivens until Rua do Conde S.Salvador.
- 1916 Aguas Santas – Ermesinde, in two stages February and August. In 1925 this route was changed to use a new road which crossed under a viaduct of the Minho / Douro railway.
- 25 April 1917 Matosinhos, Rua Roberto Ivens from Rua do Conde S. Salvador to Rua de Sto.Amaro (140 m). It is believed that this part had already been used by the steam tram, but it seems that the electric tram here was delayed by a few years. Maybe that was caused by a discussion at the same time to extend line 5 through Rua de Sto.Amaro to let join it with line 1 at the north end of Brito Capelo. The conclusion was that it was too dangerous to have trams in the very narrow Rua Sto.Amaro.
- 1917 Rio Tinto - São Pedro da Cova. At first only coal, from 1918 also (limited) passenger trams. During the first years, at least until 1922, passengers with special passes were allowed to travel on the zorras.

- 4 October 1926 Sta. Eulália - Gondomar Prelada
- 1927 Gondomar Prelada – Quintã.
- 16 June 1934 Palácio - Rua Júlio Dinís – Boavista.
- 1 January 1935 Gondomar Prelada - Souto (850 m). On 1 October 1939 Prelada - Quintã (400 m) was abandoned. It's not known if there was any service between Prelada and Quintã in the period 1935-1939.
- 3 July 1941 Castelo da Leça da Palmeira - Praia da Leça da Palmeira. In 1938 the line was relocated in Leça with removal of the 1887 bridge across the Leça because of construction of Doca no.1.
- 14 December 1941 Circulação das Antas (line 15) Line 9/ already used Rua Silva Tapada and was suppressed on this day. Shortly after that the indication 9/ was given to existing short workings until Areosa. It is not known when line 9/ started using Rua Silva Tapada, but the track was probably there since about 1916/7 to connect with Monte Aventino where zorras could load coal brought by the ropeway.
- 27 May 1945 Praça da Galiza – Lordelo.
- 1947 Arco do Prado – Coimbrões.
- 1947 Fonte da Moura – Pereiró.
- 1949 Boavista - Ramada Alta - Constituição. Tracks were here already in the 1910's, but not used and lifted again.

Designs for a new network

In 1907 the VE made a plan for the metre gauge network they had to build. Lines to Vila Nova de Gaia were not included in this design:

1. Boavista - Rua Cedofeita (opposite direction Rosário) - Praça D.Pedro - Batalha - Rua Heroismo - Campanhã (4900 m)

Car 256 with trailer 24 at Boavista depot in 1962. *W C Janssen/Online Transport Archive*

The reconstruction plate that appears on the bulkhead of car 218 indicating that the car was totally rebuilt in 2008. *K. Lomas King*

77, 253 and 312 seen at Boavista depot on 10 May 1964. *L. Folkard*

Car 253 seen climbing away from São Pedro da Cova in 1964.

Courtesy of the National Tramway Museum, photographer D Trevor Rowe

A night shot of Boavista depot taken on 15 May 1964. *L. Folkard*

2. Hospital dos Alienados (nowadays Hospital Conde de Ferreira) - Rua Costa Cabral - Rua Sta.Catarina - Rua Formosa - Rua Sá da Bandeira (opposite direction Rua D.Pedro) - Praça D.Pedro - Infante (5920 m)
3. Praça D.Pedro - Infante - Marginal - Leça da Palmeira (12,425 m)
4. Praça Carlos Alberto - Rua Oliveiras - Rua Conceição - Trindade - Rua Thomas - S.Lazaro - Batalha - Praça D.Pedro - Praça Carlos Alberto (3242 m)
5. Boavista - Av.Boavista - Cast.Queijo - Matosinhos (4972 m)
6. Praça D.Pedro - Praça Carlos Alberto - Rua Oliveiras - Rua Martires da Liberdade - Lapa - Campo Lindo - Rua Delfim Maia - Amial - Circunvalação (5080 m)
7. Praça D.Pedro - Rua D.Pedro (opposite direction Sá da Bandeira) - Rua Formosa - Rua Ildefonso - Bonfim - S.Roque - Circunvalação (5080 m)
8. Massarelos - Rua D.Pedro V - Rua Torrinha - Rua Cedofeita - Praça Carlos Alberto - Rua Restauração - Massarelos (3728 m)
9. Praça Carlos Alberto - Rua Rosário (opposite direction Cedofeita) - Carvalhosa - Monte dos Burgos - Circunvalação (4352 m)
10. Ouro - Rua Condominhas - Rua Campo Alegre - Rua Piedade - Rua Torrinha - Praça Carlos Alberto (3994 m)
11. Boavista - Campo de Regeneração (nowadays Praça da República) - Rua Gonçalo Cristovão - Rua Sta. Catarina -

Marques de Pombal - Rua Constituição - Rua Serpa Pinto - Rua Vallas - Boavista - (6245 m)
12. Praça D.Pedro - Palácio (1460 m)
13. Obviously the VE intended Praça Carlos Alberto to be a second hub next to Praça D.Pedro.

In 1909 a plan was presented by CCFP to CMP to make a network fulfilling the obligations of the concession. Numbers were allocated to all lines. These lines and numbers were different from the ones introduced in 1912. However the 1909 line numbers, although not used "on the street", were used in the correspondence with the municipality in the period after this plan was presented. The lines were:

1. Leça da Palmeira - Matosinhos - C.do Queijo - Foz - Massarelos - Infante - Praça - M. de Pombal – Hospital dos Alienados (nowadays Hospital Conde de Ferreira) - Circumvalação (Areosa). This was the combination of the lines 1 and 9 of the 1912 design network.
2. Leça da Palmeira - Matosinhos - C.do Queijo – Boavista - Cedofeita (opposite direction Rosário) - Carmo - Praça - Batalha - S.Lazaro - Campanhã. This was the combination of the lines 5 and 12 of the 1912 design network.
3. S. Mamede - Circunvalação - Paranhos - Campo de Regeneração (nowadays Praça da República) - Carmo - Praça - Batalha - Ponte - Gaya - St. Ovidio. This was the

161 is just leaving Praça on 16 May 1964 passing 253 which is bound for Antas (via Rua Formosa) on route 15E. It is a sunny day, so windows and sunblinds are suitably adjusted. *L. Folkard*

combination of the lines 7/8 and 13/14 of the 1912 design network.

4.	Massarelos - Restauração - Carmo - Praça - Formoza - Bonfim - S.Roque - Circunvalação (São Roque). This was the combination of part of line 2 and line 10 of the 1912 design network.

5.	Circunvalação (Monte dos Burgos) - Carvalhido – Carvalhosa, then a rather complicated two way figure-eight. Rua B.Vista - G.Cristovão - S.Catharina - Batalha - S.Lazaro - Heroismo - Campanhã, second route via Cedofeita - Carmo - Praça - Formoza - Bonfim – Pinto Bessa - Campanhã; and then back from Campanhã - Pinto Bessa - Bomfim - Formoza - Sá da Bandeira - Praça - Carmo - Rosário - Carvalhosa, second route via Heroismo - S.Lazaro - Batalha - S.Catarina - G.Cristovão – Rua Boa Vista – Carvalhosa - and both routes together Carvalhido - Circumvalação (Monte dos Burgos). (both routes crossed at Formosa/S. Catarina) This was a combination of the lines 6, 17, 11 and 12 of the 1912 design network.

6.	Praça - Carmo - Jardim do Carregal - Rua Triumfo (now D.Manuel II) - Palacio / Opposite direction: Palácio - Rua Triunfo - Rozario - Breyner - Cedofeita - Carmo - Praça. This was line 3 of the 1912 network.

7.	Ouro (Largo Calém) - Condominhas - Ant.Cardoso - Boavista - Cedofeita (opposite direction Rosário) - Carmo - Praça. This was the never opened line 4 of the 1912 design network.

8.	Massarelos – Rua D. Pedro V - Torrinha - Cedofeita

(opposite direction Rosário) - Carmo - Restauração - Massarelos. This was a combination of parts of the lines 2 and 16 of the 1912 design network.

9.	Carmo - Conceição - Trindade - F.Thomaz - Bragança - S.Lazaro - Batalha - Praça - Carmo. This was line 19 of the 1912 network.

10.	Boavista - Serpa Pinto - Constituação – Rua Alegria – D. Bragança - S.Lazaro - Batalha - Praça - Carmo - Rosário (opposite direction Cedofeita) - Carvalhosa - Boavista. This was line 20 of the 1912 network, although not yet complete realised at that time.

In the early years only few tram services terminated at Praça Dom Pedro (since 1910 Praça da Liberdade, often just called Praça). This practice was still retained in the 1909 design. In 1912 the network was reorganised at the same time as the introduction of line numbers. Now almost all services terminated at Praça da Liberdade and the track layout was changed to a complex two-way circle to accommodate this. In 1930 the layout was changed to a one way circle with passing tracks at the east and west sides; until 1928 Portugal drove on the left.

Tramlines

In 1912 route numbers were introduced. A lot of changes occurred over the years. The lines as existing or foreseen in 1912, their precursors and their developments in later years are given in the next overview. Time-table information is for work-days, day-time. In the evenings and on Sundays it could be different.

Car 155 climbing out of San Pedro da Cova in October 1965 not long before trams were withdrawn from the route. *W R Stillman/Online Transport Archive*

1 - Linha Marginal
Praça da Liberdade - Infante - Alfândega - Massarelos - Foz (Castelo) - Castelo do Queijo - Matosinhos (Brito Capelo) - Leça da Palmeira.

The section between Infante and Leça was the original line of CCAPFM of which the first part opened in 1872. The line was electrified in 1896 between Infante - Massarelos and Arrábida - Estação do Ouro and in 1897 in several stages to Leça da Palmeira (Rua do Arnado). In 1898 the line was extended from Rua do Arnado to the Castelo da Leça.

The section between Praça D.Pedro (Liberdade) and Ingleses (Infante) was opened in 1881 by CCFP. This part was electrified in 1898 but only became part of the Marginal line in 1902.

Around 1940 some route changes occurred because of construction of the first dock at Leixões. In 1941 the line was also extended in Leça da Palmeira from the Castelo to Praia da Leça near Rua António Nobre. Extension of the docks caused the line to be cut back to Matosinhos on 4 March 1960. From the same day the trams of line 1 continued during daytime to form line 15/ at Praça da Liberdade. On 3 October 1960 this connection with line 15/ was replaced by through working with the lines 9 and 9/. With the closure of the line 9 group on 17 September 1967, line 1 was cut back at the inner end to Infante during daytime, followed on 17 November 1968 with the complete closure of the part between Praça and Infante.

On 11 September 1993 the section between Castelo do Queijo and Matosinhos was closed and on 10 September 1994 the remaining part of the line. From 1950 until 1967 there existed a supporting line numbered 1E between Praça and Massarelos.

Time tables for 1914, 1926, 1942 and the 1950s show a 15 minute frequency. In 1926 eight tram+trailer combinations were needed, but later this was seven. In 1935 seven Belgas or carros bogies were needed. In 1964, after trailer operation on line 1 had ceased, it was every 7.5 minutes In June 1973 it was a 15 minute frequency again, but 10 minutes during peak hours.

2 & 18(2nd) - Linha da Restauração
Praça da Liberdade - Carmo - Massarelos - Foz (Castelo) - Castelo do Queijo.

The section from Cordoaria to Massarelos was opened in March 1873 by CCAPFM. Through services to Foz via the Marginal existed for several years but in other years the services were limited to the Ouro depot or Massarelos.

This line was electrified between Cordoaria and Arrábida in 1895. With the electrification of the Marginal the trams of this line continued up to Castelo do Queijo. In 1914 the line was extended to Fonte da Moura - Boavista - Carvalhosa - Carmo - Praça da Liberdade; introducing the line-number 2/ for the new part. The route numbers were changed somewhere after Carmo when no confusion about the direction was possible anymore. It's not clear of the route number change was done always at the same spot. Originally it might have been done at C.do Queijo for both directions. In the final years the change from 2 or 18 to 2/ or 18// was done at Massarelos and the opposite change at Avenida Gomes da Costa.

Tramcar 361 waits at the Gondomar terminus for departure time. Photo taken around 1965. *P Eaton*

Car 209 is on its way to Monte dos Burgos at the crossing Rua da Nossa Senhora da Fátima with the Rua de Oliveira Monteiro with the tracks of line 20/21. *J. Jordan*

On 16 June 1934 the route via Palácio de Crystal - Rua Julio Dinis was opened and this resulted in a new line 18+18/ as a variation of line 2+2/ with also a line 18E running via Palácio de Crystal - Boavista - Castelo do Queijo until Foz. For this line 18E the loop around the Castelo da Foz was constructed. Probably line 18E was suspended with the closure of the colonial exhibition in the Palácio de Crystal of 1934.

From the late 1940's until 3 March 1960 route 2 cars terminated at Carmo in both directions while route 18 cars terminated at Praça. By then route 2/ and route 18/ cars ran via Carvalhosa and route 2// and route 18// cars ran via Palácio. From 3 March 1960 all cars to/from Restauração terminated at Praça and all cars to/from Boavista terminated at Carmo until 21.00 and at Praça after 21.00. The route numbers 2// and 18/ disappeared because of this change. The line number 2 was not used anymore since the closure of the Carvalhosa route on 24 July 1978. On 6 February 1980 the line number 18// disappeared. Services were reduced to Viriato - Massarelos - Foz (Castelo) - Castelo do Queijo - Fonte da Moura - Boavista; from 1991 extended from Viriato to Carmo again. After 10 September 1994 line 18 was the last remaining route of the classic Porto tram. On 11 June 1996 a bus service with the line number 24 was introduced, following almost exactly the route 18. But the tram

survived, although frequencies decreased from every 15 minutes (with seven cars) to every 35 minutes (with three cars). This service was the beginning of the heritage network.

Time-tables for 1914 and 1926 show a 15 minute frequency. A 1935 13 trams were needed, eight with 28 seats and five with 23 seats. In the 1950s there was a 20 minute frequency on both the lines 2 and 18, giving together 10 minute headways. In 1964 it was 30 minute on both lines, together 15 minutes. In 1972 it was 36 minutes and 18 minutes respectively.

3 - Linha do Palácio
Praça da Liberdade - Carmo - Jardim do Carregal - Rua do Triumfo (nowadays Manuel II) Palácio de Crystal. Opposite direction Rua do Triumfo - Rosário - Breyner - Cedofeita - Carmo.

The line to Palácio was opened by CCFP and was originally part of a circular line. With the electrification on 17 September 1903 it became a point to point route. The line disappeared after opening of the route via Rua Julio Diniz and closure of the 1934 Colonial exhibition in Palácio de Crystal. It returned for a short period in June and July 1936.

The 1914 time-table showed a 30 minute frequency between 12.00 and 18.00 without service in the morning and evening. Also in 1926 there was a 30 minute frequency, but now from about 08.30 until midnight. One tram-car was enough for this service.

3(2nd) - Linha do Lordelo
Praça da Liberdade - Carmo - Palácio - Lordelo.

This line was opened on 27 May 1945 and closed on 31 December 1958. Time-table in the 1950s was every 10 minutes with a 20 minute travel time.

Subsequently the existing line 4/ was renumbered 3.

4 & 3(3rd) - Linha do Bessa
Praça da Liberdade - Carmo - Carvalhosa - Boavista - António Cardoso - Campo Alegre - Condominhas - Rua d'Ouro.

This line 4 was never opened, the rail being lifted during World War I for the line to São Pedro da Cova, but the number was used from July 1927 for an existing line, which often was only indicated by its destinations, but sometimes had the indication B: Alfândega / Infante - Praça da Liberdade - Carmo - Carvalhosa - Boavista - Bessa.

Car 131 at Carmo. *K. Lomas King*

The line B, Alfândega - Bessa time-table for 1926 gave a 15 minute frequency between about 08.30 and 19.00 without early morning and evening service. There were four trams necessary. In 1935 these were indicated as "carros de bancos".

It is likely that the line was then extended to Fonte da Moura. Later the line was cut back to Praça da Liberdade. On 21 December 1947 line 4 was extended to Pereiró. The parallel line 4/ via Palácio was renumbered into line 3 on 3 March 1960. Line 4 disappeared with the closure of the Carvalhosa route on 24 July 1978. This last line 3 was cut back to Boavista on 6 February 1980 and suspended on 30 April 1984.

The time-tables for 1964 give 15 minute headways for both the lines 3 and 4, together every 7.5 minutes. In 1972 this was 18 minutes and 9 minutes.

184 at Ponte da Pedra, 25 July 1972. *Photographer unknown*

5 & 19(2nd) - Linha da Boavista
Praça da Liberdade - Carmo - Carvalhosa - Boavista - Bessa - Fonte da Moura.

This electric tramline shared the tracks with the steam tramline between Boavista and Bessa since 1904. In November 1914 (or February 1915?) the line was extended to: Castelo do Queijo and Matosinhos (Roberto Ivens/ Rua Sto.Amaro) as successor of the steam tramline. CCFP wanted double track in Rua Brito Capello to let both lines 1 and 5 use this street. However until 1930 the municipality of Matosinhos did not allow double track and the trams of line 5 kept using Rua Roberto Ivens until 1 July 1930. From this date line 5 followed in Matosinhos the route of line 1 to Leça da Palmeira. On 16 June 1934 the route via Palácio de Crystal - Rua Julio Dinis was opened and this resulted in a new line 19 as a variation of line 5. At the end of 1947 line 5 was cut back to Carmo while line 19 kept its terminus on Praça da Liberdade. The lines 5 & 19 were cut back to Matosinhos on 3 March 1960. Line 5 disappeared with the closure of the Carvalhosa route on 24 July 1978. Line 19 was cut back to Boavista on 6 February 1980. It closed on 11 September 1993.

The 1926 time-tables give for line 5 a 15 minute frequency with the use of 6 motorcar + trailer combinations. A 1935 overview gives six trams with 32 seats for line 5 and six with 23 seats for line 19. In 1942 both the lines 5 and 19 had a 30 minute frequency, giving together a 15 minute headway. In the 1950s this was respectively 20 minutes and 10 minutes with a 50 minute travel time. In 1972 it was 36 minutes and 18 minutes.

6 - Linha do Monte dos Burgos
Praça da Liberdade - Carmo - Carvalhosa - Carvalhido - Monte dos Burgos

138 seen at Monte dos Burgos terminus in 1972. This terminus was located on the circular road. *Michael Russell*

Cars 144 and 216 reach the summit of Rua das Carmelitas as they arrive at Carmo from Praça in 1972. *Michael Russell*

This line was opened between Praça da Liberdade and Carvalhido in 1909, extended to Monte dos Burgos in 1911. It remained single track and passing loops beyond Carvalhido until 1957.

From the late 1940s until 1960 short workings between Praça and Carvalhido had the number 6/ while line 6 was limited between Carmo and Monte dos Burgos. From 3 March 1960 line 6 was connected with the lines 15 and 15E until the closure of the group 15 lines on 23 July 1967. Line 6 was cut back to Carvalhido on 11 October 1976 and closed on 9 May 1977.

The 1914 time-table gave a 30 minute frequency. In 1926 and 1942 this was 12 minutes, for which four tram-cars were needed. A 1935 overview gives one tram with 32 seats and three with 28 seats. In the 1950s on both the lines 6 and 6/ was a 10 minute frequency. In 1964 the frequency was 7.5 minutes and in 1973 10 minutes.

7 - Linha de São Mamede
Praça da Liberdade - Carmo - Rua dos Mártires da Liberdade - Praça da Republica - Arco d'Agua - Amial - São Mamede - Ponte da Pedra.
This line was opened from Praça da Liberdade to São Mamede in 1910 and extended to Ponte da Pedra on 10 March 1912. From 14 November 1948 the Rua dos Mártires da Liberdade became one way in the direction of Praça da Republica and the trams in the direction of Carmo went through Rua da Boavista and Rua de Cedofeita. Short workings to Amial had the route number 7/. The line 7// had the route: Batalha - Gonçalo Cristóvão - Praça da Republica - Arco d'Agua - Amial - São Mamede. In the late 1940's line 7 terminated at Carmo, while the line 7/ trams continued to Praça da Liberdade. The part from Arco d'Agua to Ponte da Pedra closed on 6 October 1975, the remainder on 20 June 1977.

The 1914 time-table had a 40 minute frequency. In 1926 there was a 20 minute service over the whole line, for which four cars were needed, and another 20 minute service as far as S.Mamede for which another three cars were necessary. A 1935 overview gives as necessary four "carros bogies", three trams with 32 seats and three with 28 seats. In the 1950s the headways on both the lines 7 and 7/ were 12 minutes, while in line 7// it was 18 minutes. In 1964 this was 15' (peak hours 10') for both the lines 7/ and

7// and 15' for line 7//. In 1973 line 7 had 15 minutes, line 7/ 10 minutes and line 7// 15 minute services.

8 - Linha de Paranhos
Praça da Liberdade - Carmo - Rua dos Mártires da Liberdade - Praça da Republica - Paranhos (Campo Lindo).
The line to Paranhos was opened as a mule tramline by 1892 (probably in 1886) with the outward route via Rua Sta. Catarina and the inward route via Rua de Cedofeita. This line was electrified in 1904. From 14 November 1948 the Rua dos Mártires da Liberdade became one way in the direction of Praça da Republica and the trams in the direction of Carmo went through Rua da Boavista and Rua de Cedofeita. Line 8 was closed on 29 November 1976.

The 1914 time-table gave a 10 minute service, except when there was a line 7 tram. In that case the headway for line 8 was 20 minute. On the combined lines 7+8 there was a 10 minute frequency. In 1926 line 8 had a 20 minute service with two trams, in the 1950s it was 13 minutes and in 1973 20 minutes with 10 minutes during peak hours. A 1935 overview gives four trams for line 8: two each with 28 and 23 seats.

9 - Linha de Costa Cabral
Praça da Liberdade - Praça Marquês de Pombal - Rua Costa Cabral - Areosa - Águas Santas.
The first part of this line was opened as a mule tramline in 1875. It was electrified from Praça to Hospital Conde de Ferreira in 1899. Line 9 was extended to Areosa on 28 May 1910, to Águas Santas on 22 January 1911 and to Ermesinde on 8 February 1916. In 1925 the line was diverted following a new road passing under the Minho/Douro railway. Until 1941 short workings terminating in Rua de Silva Tapada had the indication 9/. These short workings disappeared with the opening of line 15 and the indication 9/ was soon used for existing short workings until Areosa. STCP added a route 9// Batalha - Ermesinde.

The basic setup in the STCP period was:
9 Bolhão - Areosa - Ermesinde, from 3 October 1960 Praça - Bolhão - Areosa - Ermesinde with during daytime through working with line 1;
9/ Praça - Bolhão - Areosa, from 3 October 1960 during daytime through working with line 1;
9// Batalha - Bolhão - Areosa - Ermesinde with through working with line 14/ until the closure of the latter on 1 January 1959. From 3 October 1960 line 9// was cut back to Bolhão.

On the lines 9 and 9/ mainly bogie cars were used, on line 9// 2-axle cars with trailers. The group of lines 9 closed on 17 September 1967.

The frequency in 1914 was 40 minutes over the whole line and 20 minutes until Alienados (Hospital Conde Ferreira). During peak hours the latter was 5 minutes. In 1926 there were three services: Praça - Ermesinde with 20 minutes using five trams, Praça - Areosa with 10 minutes also using five trams and Praça - Silva Tapada 20 minutes using two trams. A 1935 overview gives that there are nine bogie trams, five trams with 28 seats and two trams with 23 seats used. In the 1950s there was the line 9 (Bolhão - Ermesinde) with 20 minutes, 9/ (Praça - Areosa) with 10 minutes and 9// (Batalha - Ermesinde) with 30 minutes. In 1964 all three lines (9 Praça / Sá da Bandeira - Ermesinde, 9/ Praça / Sá da Bandeira - Areosa, 9// Bolhão - Ermesinde) had 15 minute headways.

10 - Linha de S.Roque
Praça da Liberdade - Bonfim - São Roque.

The first part of this line was opened as a mule tramline to Bomfim in 1894, and extended to Corujeira in 1898. It was electrified to São Roque in 1902 and extended to Venda Nova in 1912. In 1918 line 10A, later indicated as 10/, was opened using the freight line to São Pedro da Cova branching off line 10 in Rio Tinto. This route was mainly intended for the transport of coal from the mines and the number of passenger trams was for many years limited, passengers being also carried on the coal cars. In a letter to the mine company of 6 July 1922 CCFP specified the number of passengers allowed to travel on a coal car: two passengers plus an employee of the mine plus one of the three directors or three engineers of the mine, making a total of four passengers. The CM of Gondomar permitted the use of passenger trailers behind zorras, but it's not clear if this was ever done in regular service, or that the CM of Porto did not allow it (possibly) or that it wasn't practical (possibly too). On 4 October 1926 the branch line from Santa Eulalia to Bouça Cova (Prelada) was opened. This branch was extended in 1927 to Gondomar Quintã and from Prelada to Souto on 1 January 1935. A 1935 overview gives five trams with 28 seats for line 10 and eight trams for line 10A (S.Pedro da Cova and Gondomar) being two with 28 seats and six with 23 seats. This line was in the documents sometimes called 10B, but also 10A, which made the research a bit confusing. With STCP it was line 10//. The short branch Prelada - Quintã was already closed by that time. The section between Circunvalação and São Caetano, where the lines 10/ and 10// branch off from line 10, received double track in 1932.

In the late 1940s the city centre terminus was moved from Praça da Liberdade to Bolhão introducing a supplementary line 10E: Praça da Liberdade - São Roque. The group of lines 10 closed on 1 January 1967.

The 1914 time-table showed three services, Praça - Venda Nova, Praça - Circunvalação and Praça - Corujeira, all with a 60 minute frequency making a 20 minute headway on the most busy part. In March 1926 there were four services: Praça - Venda Nova 15' with six tramcars, Praça - Circunvalação 30 minute with two trams, Praça - S.Pedro da Cova, 120 minute with one tram and Praça - Sta.Eulália with three trams giving together with the S.Pedro da Cova service 30 minute headway. On the shared part of all services was a 7.5 minute frequency. In 1944 there was a 10 minute frequency on the lines 10 and 10// giving together a 10 minute on the shared route with on line 10/ to S.Pedro da Cova 15 return trips with irregular intervals. In the 1950s the frequencies were 15 minutes for the lines 10 and 10//, 50' minutes for line 10/, all terminating at Bolhão and 10 minutes on line 10E Praça - São Roque. In 1964 it was 10 and 10//: 15 minutes, 10E 15 minutes with 10 minutes during peak hours and 10/ 50 minutes with 10 minutes during peak hours.

11 & 12 - Linhas de Campanhã
11 - Praça da Liberdade - Bonfim - Campanhã.
This line opened in 1911.

12 - Praça da Liberdade - Batalha - São Lazaro - Campanhã.

The predecessor of line 12 was opened as a mule tramline in 1876 and part of the mainline of CCFP from Campanhã to Foz. It was electrified in 1900. These two lines had a remarkably complex history. For some years around 1914 the lines 11 & 12 were combined as a circular service. The clockwise route was allocated the number 11, the anti-clockwise route the number 12. Later both services were made independent from each other again. Probably in 1926 or 1927, line 11 was converted into a circular route in both directions via Bonfim and São Lazaro. At that time line 12 gained a loop terminus at Campanhã using Rua Garrett (now Rua Padre António Vieira), but apparently within a few years this circular line 11 was reversed to the original route via Bonfim only. In 1951 line 11 was changed in the outbound direction to Praça - Batalha - Bonfim - Campanhã while line 12 was cut back to Batalha - São Lazaro - Campanhã. Line 12 closed on 3 March 1960. On the same day line 11 was converted again into a circular line, but only counter-clockwise: Praça - Batalha - São Lazaro - Campanhã - Bonfim - Bolhão - Praça. Line 11 was closed on 10 June 1964.

In 1914 both circular lines had 15 minute headways. The time-table for 1926 gave 20 minutes with two trams on line 11 and 7.5 minutes with four on line 12. In the 1950s it was 12 minutes on line 11 and 7 minutes on line 12. A 1935 overview gives two "carros bancos" for line 11 and four 23 seat trams for line 12.

13, 14 & 15 - Linhas de Vila Nova de Gaia
13 - Praça de Almeida Garrett – Rua Loureiro - Ponte D.Luiz I (upper deck) - Santo Ovidio. This line opened 28 October 1905 and closed 3 May 1959.

Lines 13 and 14 were diverted in 1912/13 to run from Praça da e Liberdade via Batalha.

14 - Praça de Almeida Garrett - Rua Loureiro - Ponte D.Luiz I (upper deck) - Devesas.

This line opened 28 October 1905 and was in 1913 extended to Arco do Prado. In 1947 the line was extended to Coimbrões. Since the late 1940s there was a supporting line 14/ Batalha - Coimbrões, which was connected with line 9// at Batalha. Line 14 was closed on 1 January 1959.

The 1914 time-table gave 45 minute frequency on both lines with 22.5 minutes on the shared section. In 1926 it was 20 minutes and 10 minutes, in 1942 10 minutes and 5 minutes and in 1945 12 minutes and 6 minutes. In the 1950s there was on both line 13 and 14 a 10 minute service with additionally a 30 minute service on line 14/. A 1935 overview gives for both lines each four trams with 23 seats.

A tranquil scene as 220 waits at Paranhos terminus in 1972. *Michael Russell*

15 (1st) - Infante - Ribeira - Ponte D.Luiz I (lower deck) – Vila Nova de Gaia. This line was never opened, the rail being used for the line to São Pedro da Cova.

15 (2nd) - Linha de Circulação das Antas
This circular line was opened on 14 December 1941 and had three variations:

15 - Praça da Liberdade - Batalha (until 1951 Bolhão) - Bonfim - Antas - Rua Costa Cabral - Praça Marquês de Pombal - Praça da Liberdade; (anti-clockwise)
15/ - Praça da Liberdade - Praça Marquês de Pombal - Rua Costa Cabral - Antas - Bonfim - Bolhão - Praça da Liberdade; (clockwise)
15E - Praça da Liberdade - Bolhão - Bonfim - Antas - Rua Costa Cabral - Praça Marquês de Pombal - Praça da Liberdade. (anti-clockwise) Line 15E was opened in 1951.
All three lines were operated one way only. The group of lines 15 closed on 23 July 1967.

In the 1950s these lines had 15 minute headways. In 1964 it was 15 minutes for both the lines 15 and 15E, making together 7.5 minutes, and 7.5 minutes on line 15/.

16 (1st) - Linha de Circulação de D.Pedro
Massarelos - Rua D.Pedro V - Piedade - Torrinha - Carmo - Praça da Liberdade - Infante - Alfândega - Massarelos.
This line opened in 1909. Probably in 1934 the route via Rua D.Pedro V was closed and line 16 received a new route: Torrinha - Carmo - Praça da Liberdade - Batalha. It was closed on 13 April 1945. Later the existing line 17/ was renumbered to 16.

The 1914 time-table showed 60 minute frequency in both directions, apparently one tram alternating. In 1926 two trams circulated during daytime each one way every 30 minutes. In the evening it was just one tram every 30 minutes shuttling between Praça and Massarelos via Rua Torrinha. A 1935 overview gives four "carros bancos". Apparently at that time the frequency was better than in 1926.

17 & 16 (2nd) - Linha de Santa Catarina
Batalha - Gonçalo Cristóvão - Praça da Republica - Boavista.
This line had its origins with the circular urban mule tramline opened in 1875. It was completely electrified by 1904 and became for many years the only line not reaching the hub of the network at Praça da Liberdade. In later years this line was extended in stages to Bessa - Fonte da Moura - Castelo do Queijo - Foz and Leixões. Trams operating the Batalha - Leixões route had the line number 17/, since the 1950s this was line 16. The section Batalha - Boavista was closed on 31 October 1977, the remainder in 1978.

In 1914 the frequency was 30 minutes (Batalha - Boavista). In 1926 there were two services, Batalha - Foz every 20 minutes with five trams and three trams that strengthened Batalha - Fonte da Moura to every 10 minutes. A 1935 overview gives twelve tramcars with 28 seats necessary. In the 1950s there was a 15 minute service on line 16 (Batalha - Matosinhos) and 20 minutes on line 17 (Batalha - Foz) with extra services Batalha - Gomes da Costa. In 1964 on both lines was a 15 minute frequency giving 7.5 minute headways on the shared route. In

1972 this was 18 minutes and 9 minutes respectively.

18 (1st) - Linha da Maquina
Praça da Liberdade - Carmo - Carvalhosa - Boavista - Bessa - Fonte da Moura - Foz (Cadouços) - Castelo do Queijo - Matosinhos (Roberto Ivens)
This was the original trunk line of CCFP. Electric trams to and from Campanhã, Carmo and Praça interchanged trailers at Boavista until closure of the steam tram route via Cadouços on 9 November 1914.

The 1914 time-table gave 20 minute to 30 minute headways, depending the time of the day.

18 (2nd) Refer to routes 2 and 18 (2nd)
Since 1934 the line number 18 was used again for a variation of line 2.

19 (1st) - Linha de Circulação da Trindade / Pequena Circulação
Praça da Liberdade - Carmo - Rua das Oliveiras - Conceição - Trindade - Fernando Tomás - São Lazaro - Batalha - Praça da Liberdade.
This line was opened in 1909 and closed in 1917, the rail being used to build the route to São Pedro Da Cova. In November 1912 the time-table of this line gave a 30 minute' frequency from 08.00 to 22.00 in both directions. The 1914 time-table showed 60 minute frequency in both directions, apparently one tram alternating.

19 (2nd)
From 1934 the line number 19 was used again for a variation of line 5.

20 - Linha de Grande Circulação
Praça da Liberdade - Batalha - São Lazaro - Rua Alegria - Constituição - Boavista - Carvalhosa - Carmo - Praça da Liberdade.
At first this line was operated only on the eastern half from Praça da Liberdade via Rua da Alegria to Constituição. The tracks between Constituição and Boavista were installed but unused and probably lifted in 1917. About 1919/20 the connection was made between Rua de Constituição and Rua Antero Quental and line 20 became a circular line with the route Praça da Liberdade - Batalha - São Lazaro - Rua Alegria - Constituição - Praça da Republica - Rua dos Martires da Liberdade - Carmo - Praça da Liberdade. On 31 December 1949 the line was extended to Praça da Liberdade - Batalha - São Lazaro - Rua Alegria - Constituição - Boavista - Palácio - Carmo - Praça da Liberdade. This was finally the line as originally projected in the 1912 schedule, except for going via Palácio instead of Carvalhosa. A line 21 was introduced, at first presumably as enforcement of line 20 on the Northern and Eastern part as in the archives an announcement was found that line 21 was extended to Castelo do Queijo on 19 August 1951. On 22 December 1951 line 21 was changed to become equal to line 20 but following the Carvalhosa route. The trams running clockwise showed the numbers 20/ and 21/. The lines 20 and 21 were closed on 20 July 1968.

In 1914, line 20 was not yet circular, the frequency was 30 minutes. In 1926 it was 20 minutes with two cars for each direction. A 1935 overview gives six trams with 23 seats. In the 1950s it was 15 minutes on both lines 20 and 21 in both directions giving 7.5 minute headways on the shared route. The 1964 time-tables showed 10 minutes on both the lines 20 and 21 (together 5 minutes) and 12 minutes (together 6 minutes) for the trams in the opposite direction of the lines 20/ and 21/.

A - Infante - Praça da Liberdade - Praça Marquês de Pombal - Rua Costa Cabral - Areosa
B - Bessa - Boavista - Carvalhosa - Carmo - Praça da Liberdade - Batalha - São Lazaro - Campanhã
C - Boavista - Carvalhosa - Carmo

Lines A, B and C were suspended by November 1914. A replacement shuttle service with one tramcar was introduced on Infante - Praça da Liberdade and services on lines 5 and 12 were strengthened. It seems later a new line was opened, which first received the indication B and later 4: Bessa - Boavista - Carvalhosa - Carmo - Praça da Liberdade - Infante - Alfândega.

Pre-1946 closed lines

Although the network was at its largest extent in the 1950s, a number of routes were closed before take-over by STCP. A summary:

1. Line 2 and others. Tunnel. Already in the mule tram period, but also used by the electric trams, was a "tunnel" through buildings between Jardim do Carregal and Rua do Rosário. Probably late 1925 or 1926 this tunnel was replaced by the later route. It was sold to the neighbours in 1927.
2. Line 5 Matosinhos. Rua Sousa Aroso - Rua Roberto Ivens closed in 1930. Rua Brito Capelo received double track over its whole length and line 5 joined there line 1 to Leça da Palmeira.
3. Line 9 Ermesinde. The original route via Rua Portocarreiro with a level crossing of the Minho/Douro railway was changed in 1925 to operate via Rua Rodrigues de Freitas

passing under the railway. With this the line in Rua Portocarreiro was abandoned.
4. Lines 9, 10 and 11. The original mule tram route from Praça to Bolhão was in both directions via Rua D. Pedro (later Elias Garcia) - Cancela Velha - Formosa. Later the trams in down direction went through Rua Sá da Bandeira. Around 1916 to 1920 the area north of Praça was demolished for the construction of the Av. dos Aliados. With that Rua Elias Garcia and Cancela Velha were also demolished and the trams went both directions via Sá da Bandeira.
5. Line 10// Gondomar. The original 1927 route went from Sta. Eulália via Prelada to Quintã. In 1935 a new route from Prelada to Souto was added. The route between Prelada and Quintã was officially suspended in 1939, but it is unknown if there was regular service during the (whole) period 1935-1939.
6. Line 13/14 Porto. Except for the route via Batalha there was between Praça and the Ponte Luís I also a route via Rua Loureiro and Rua Chã. Prior to and with the 1906 concession it was expected that the lines to Vila Nova de Gaia would be routed via a new direct road. However the area between Praça Almeida Garrett and the Sé was still very densely built. The necessary demolition of this area was only done in the 1940s. It seems trams initially ran from a stub terminus on Praça de Almeida Garrett, at the front of São Bento station, via the very narrow but busy Rua Loureiro and Rua Chã, being the nearest feasible alternative. With the revision of the layout at Praça de Liberdade in 1911 to accommodate the network reorganisation of 1 January 1912, also the layout of Praça Almeida Garrett was changed with retaining the possibility to reverse direction for trams coming out of Rua Loureiro. A map dated 10 January 1913 shows the layout of Praça Almeida Garrett with line 13 going through Rua Loureiro and line 14 up Rua 31 de Janeiro to Batalha. However an overview of the

Constructora tram CCFP 104 with a freight trailer, perhaps 83, in front of the São Bento railway station on 27 October 1943. *Photographer Unknown*

tram lines made around 1913 says that all trams of line 13 were running via Batalha, but line 14 trams could use both routes. Probably in 1914 the Loureiro route was abandoned and all trams to/from Vila Nova de Gaia went via Batalha. The unused rails were probably lifted for reuse on the 1917 line to São Pedro da Cova; in 1937 CMP was demanding reinstatement. Even in 1950 STCP made a not realised plan to reinstall a track in Rua Chã and Rua Loureiro. The idea was that the trams would ride from the bridge via Loureiro to Praça Almeida Garrett and then turn to the right into Rua Sto. António (31 de Janeiro) to go via Batalha to Vila Nova de Gaia again. This was a solution to make it possible to change Rua Sto. António into a one way street. When the new direct road to Ponte D. Luis I was finally built, plans for tram replacement by trolley-buses were already being formulated.

7. Line 13/14 Vila Nova de Gaia. When the tramline over the Ponte Luís I was opened, at the south side of the bridge there was the hill of Serra do Pilar and the Av. da República did not yet exist at this point. After excavating this part of the hill, the trams could then follow the straight route. Until then they had to circumvent what is now Jardim do Morro via Rua Rocha Leão. It is not known when the trams left the route around the hill and started using the straight route, but it was between 1907 and 1914.

8. Line 16 Circulação de Massarelos. This circular line went via Rua D. Pedro V and Rua da Piedade until 1934. After that it was a point to point line Rua da Torrinha - Batalha with abandoning of the tracks in Rua D. Pedro V and Rua da Piedade.

9. Line 18 Steam tram. With the closure of the steam tram

in 1914, the route Fonta da Moura – Foz Cadouços - Rua de Gondarém - Castelo do Queijo - Rua Roberto Ivens (Rua Sousa Aroso) also disappeared. A few short sections were later used by freight trams, and a permanent way yard existed on the former line at Fonte da Moura. The section in Rua Roberto Ivens North of Rua Sousa Aroso was electrified and used by line 5 until 1930.

10. Line 19 Circulação da Trindade. With the closure of this line in 1917 disappeared the route Ruas das Oliveiras - Conceição - Trindade - Fernando Tomás - D. João IV.

Built but never used

A strange Porto tradition of installing tram tracks but never using them.

Early times: On the upper deck of the Ponte D. Luís I was since the opening of the bridge in 1887 a track (probably 900 mm gauge) for mule trams owned by the State (Ministério das Obras Publicas) and to be used by tram companies having concession for lines south of Porto. The track was never used and on request of CCFP lifted in 1903.

Installed around 1914 and lifted around 1917 to provide rails for the route to S. Pedro da Cova:
• From Avenida da Boavista via Ruas António Cardoso - Campo Alegre - Condominhas to Ouro (Largo António Calém). This route was to be used by the never opened line 4.
• From Infante - Rua do Infante D. Henrique - Rua de São João - Praça da Ribeira - Cais da Ribeira to the lower deck of the Ponte Luís I. The plan was for a line to Vila Nova de Gaia via the lower deck, but that was never realised.

Car 116 is in Rua do Amial in 1973. *A. G. Murray*

Car 117 climbs up Rua das Carmelitas to Carmo in 1973 passing the celebrated bookshop Lello. *A. G. Murray*

- Boavista - Rua de Nossa Senhora da Fátima - Largo Ramada Alta - Rua Serpa Pinto - Rua da Constituição (Rua Antero Quental). This route was to be used by the circular line 20. However this line was made into a shorter circular line via Praça da República instead of Boavista using the tracks of the lines 7 and 8. Reinstalled after WWII and used by the lines 20 and 21.

Modern times

- There are new but never used tracks at Castelo do Queijo, Circunvalação at the Ocean and along the sea-front in Matosinhos (Av. Norton de Matos) until Av. da República. There were also new tracks on Av. do Castelo do Queijo between Castelo do Queijo and Circunvalação, but these were lifted before ever being used. These tracks were meant for the never realised re-extension of tramline 1 to Matosinhos.
- Batalha line 22 was originally intended to run via Rua de Entreparedes and Praça dos Poveiros, but revised after work began to make a shorter route via Rua Santa Catarina. On Praça da Batalha is a triangle with two unused legs and a spur ending at the exit of Rua de Entreparedes.

Maybe there were more, but no records of these have been found.

Business as usual 1: extra fares and special revenues luggage

services – If it can be moved, we can handle it, for a price of course, one could say of CCFP. Non passenger services were profitable, like luggage transport for overseas ship travellers to the Leixões Harbour. Fish, vegetables and other goods that could bother or dirty the passengers had to be carried in trailers. Fish came from a purpose built spur in Matosinhos harbour to the fish market, where trawlers unloaded and fisherman sold the catch to fishmongers. This service was operated with a tram and specially modified fish trailers with a tray to hold the brine from the fish. Although today many details of this operation are not known, photos show several locations, other than the fish market, with a tram and its sardine trailer unloading in the streets. In Cordoaria fish market a spur existed for around 60 years, in 1892 as a mule operated one, as a siding on the as yet unelectrified section of the circular route from Carmo (nowadays route 18 & 22 terminus) to Porta do Olival and then around the Jardim de Cordoaria to the Hospital de Santo António, and later, as an electric one, but now branching from top of Rua da Restauração, corner of the Hospital St António.

Business as usual 2: advertising, selling power and public lighting – Moving trams could carry moving ad panels, giving greater visibility to the products, CCFP soon realised this, and sold virtually every bit of space to display the ads. Either they were large outside panels, bulkhead doors with enamels or monitor roof tinted windows; advertising was a lucrative business. Today enamels and monitor roof tinted windows are no longer used, but dash panels, roof panels and body panels are still largely used. After all money is money. And advertisers had the cost of paying for the material, as well as renting tram space. Small industries were modernising and replacing stationary steam engines by electric motors, so CCFP sold power whenever factories needed it. Public lighting utilities were normally gas operated, but soon CCFP was in the business of providing electric power.

Business as usual 3: 3rd party revenues services and mail – In Porto post office trams were never used, as far as is known; but trams carried mail boxes for many years. Services seemed to begin in 1914 and ended in 1966, a fee being paid by CTT, the postal services company, for each mail box carried on the trams. An employee of CTT would open the box and remove the letters. Although it is not confirmed, some retired employees mentioned that a closed box car went to the old post office at Cancela Velha, near where later the new town hall was constructed, to collect letters and parcels, delivering them at Post office headquarters in Batalha. The Cancela Velha route was abandoned, and Rua Cancela Velha demolished about 1918, before the construction works of the town hall started.

Business as usual 4: non-passenger revenues, sidings and coal – Zorras are often called coal-cars; this is not accurate. The word zorra means an all-purpose freight vehicle. If moving coal and dumping ashes was a primary and most common use, it wasn't the only one, neither were they built with that purpose. CCFP's early fleet (1876) had animal traction zorras, 20 years before coal was needed. Transporting sand from sidings built near the Douro River, stone from CCFP's quarry at Arrábida, etc. were also common tasks. Coal was another merchandise people could buy for many years in Massarelos.

Local industry used to hire zorras to replace the lack of lorries. They brought goods and sent general merchandise using the existing railway connecting points, at railway stations like Campanhã, S. Bento and Boavista, and at the Leixões harbour. Even coal for operating plants was bought from the Empresa das Minas de S. Pedro da Cova, and moved using CCFP freight services. Porto customs, at Alfândega near Infante, also had extended rail facilities with both a CCFP and Portuguese railways siding.

Sidings were sometimes a couple of metres long, and tracks and aerial lines could be customer property, CCFP owned, or a combination of both. However connecting points with main tracks were often CCFP property, so when the contract ended, they were lifted, leaving the remaining client tracks in situ. Sidings could also be complex, including several branches and reaching warehouses and have for instance weighbridges or other technical facilities.

Building temporary sidings for removing debris when new streets or large public buildings were in construction was also a common business practice. Examples are the opening of Av. dos Aliados and new Town hall construction, literally demolishing and flattening the entire area; the new building of the College of Engineering at the Porto University in Rua dos Bragas, and new Av. do Gama (nowadays Av. Marechal Gomes da Costa) from Avenida da Boavista in the direction of Praça do Império. Some of these tracks were not ballasted, as they were lifted according

to the work's progress, saving endless man hours.

In Matosinhos the highest concentration of freight branches of the entire network could be found. A postcard published by the Vinicola company shows a half-closed towing box car, with the designation Rebocador Eléctrico, advertising the fact it was no longer animal pulled but electrical, and moving zorra trailers loaded with wine barrels.

Very little data has survived about most of these freight branches. The following overview is based on documents and correspondence but is probably not complete. The years refer to those identified in the documents, the exact years of existence are mostly not known.

Mines, coal transfer and coal merchants

1. São Pedro da Cova, mines. Opened in 1917, originally the whole line from Rio Tinto was used by coal trams only. There existed passes for travelling on the coal trams. In 1918 most of the line was also used by the trams of line 10/, but from the terminus of this line the coal line to the mines continued until the closure of the Massarelos power plant in the 1950's. In São Pedro da Cova was a mineral narrow (probably 90 cm) gauge railway (Ribeiro da Covilhã)
2. Rua Silva Tapada – Monte Aventino. Coal was also transported from the mines in São Pedro da Cova to Monte Aventino by a rope way, from where it was transported by tram to Massarelos or other clients. Monte Aventino was storage and distribution centre, where coal was also picked up by clients with ox cars and later lorries. Still existed on 1 June 1946.
3. Rio Tinto - São Caetano on the line to São Pedro da Cova close to the connection with the line to Venda Nova, Delfim da Silva. Coal merchant (1920, 1931, Still existed on 1 June 1946.)
4. Santa Eulalia, Manoel de Sousa Matos. Coal merchant (1920, Still existed on 1 June 1946.)
5. From Rua do Bomfim via Campo 24 de Agosto to Rua Santos Pouzada, a service line for unloading coal and other merchandise. (1919)
6. Central do Ouro, CMP, ca. 70-80 ton coal/day from Monte Aventino. (1919)

Factories and other private companies

7. Branch line from Boavista via Rua de 5 de Outubro to the factory of A Constructora (not electric)
8. Branch line Coats & Clark Vila Nova de Gaia - March 1908, 700 m from S. Ovídio. March 1939 lifted.
9. Avenida da Boavista, G & G Graham & Companhia, next to Rua Azevedo Coutinho (1912, 1917 100m, This was a major customer for coal from S. Pedro da Cova. Still existed on 1 June 1946.)
10. Boavista, Empresa de Combustiveis, Lda. Rua de Agramonte (1930, Still existed on 1 June 1946.)
11. Boavista, Companhia Manufactura de Artefactos de Malha, Avenida da França (1930)
12. Restauração, Fabrica de Moagem do Corpo Santo. (1910, 1914 135 m)
13. Rua da Piedade into Fabrica União Fabril Portuense (1910)
14. Branch line from Rua da Rainha (S. Mamede) via Travessa de Valle Formosa and Rua de S. Diniz to the Fabrica Portuguesa da Refinação, 305 m (1911, 1914 298 m)
15. From Rua Formosa into Alexandre Braga, de Empresa Geral de Transportes (1924)

16. Matosinhos. Real Companhia de Vinicola do Norte de Portugal. (1898 or 1899) Probably lifted in 1941 when it had not been used for many years.
17. Matosinhos, from Rua Brito Capelo into the Fabrica de Conservas on the corner with Avenida de Meneres (1899)
18. Matosinhos, from Rua Brito Capelo (Rua Castelo do Queijo) into the Armazems Bento José Pereira da Cunha just to the north-east of the crossing with the railway. (1899, Still existed on 1 June 1946.)
19. Matosinhos, from Rua Brito Capelo via Avenida Meneres to the Fabrica de Brandão Gomes & Ca. Also a connection from the line in Rua Roberto Ivens to this branch line.
20. Rua do Ouro, Fábrica da Gaz. (1904)

Markets
21. From Rua Formosa into Sa da Bandeira next to Mercado do Bolhão. (1892), no tramline yet in this part of Sa da Bandeira.
22. Siding at the fish-market at Cordoaria (not electric)
23. Branch line to Praça do Peixe / Cordoaria (1929).

Railway stations and port facilities
24. Estação São Bento, Rua da Madeira. (1913)
25. Estação de Campanhã. Still existed on 1 June 1946.
26. Boavista, Estação PPF. (1909)
27. Miragaia, Alfândega.
28. Cais das Pedras / Monchique. Cais do Paixão (1914, 152 m)
29. Leixões, breakwater north from Castelo da Leça.
30. Leixões, breakwater south from Rua Brito Capelo via Rua Godinho. There was an industrial line branching off Rua Brito Capelo into Av. da República and then following Av. Serpa Pinto to the end. Here there was a branch into Rua S. Sebastião. Another branch went through Rua Godinho to the mole. It is likely that the Serpa Pinto spur was lifted in 1935 and the spur to the mole was directly connected by a new spur branching off Brito Capelo into Rua Godinho. Still existed on 1 June 1946.
31a. Short branch of the CCAPM line to the river side at a location called "Insua do Ouro". Made in November 1884 and meant to transport materials for the construction of the Porto de Leixões.
31b. Prado station of the NG São Gens - Leça line for constructing the moles of Leixões had side spurs for both CCAPFM out of Rua Juncal de Cima (later Brito Capelo) and CCFP out of Rua Juncal de Baixa (later Roberto Ivens). 1884.

Sand and stone lines
32. Siding for loading of sand at A Cantareira.
33. Line for sand at Castelo da Foz (existed 1930-1935)
34. Castelo do Queijo connection branched off the main line in front of the sub-station and at the other end with the old tracks of the steam tram to Circunvalação.
35. Pedreira (quarry) da Arrábida (1933, Still existed on 1 June 1946.)

Construction and demolition sites
36. Rua da Piedade into Julio Diniz for construction of Julio Diniz (1930)
38. Siding off line 12 at Rua Barão de Nova Cintra, 478 m transporting material for Quinta das Oliveras (1930, letter to Serviços Municipeis Aguas e Saneamento)
39. From Avenida da Boavista to Rua Seralves siding for transport of spoil destined for the construction of Avenida do Gama (now Avenida Gomes da Costa) (1925)
40. From Praça da Liberdade into Rua Laranjal for removal of debris of the demolition of buildings in the future Avenida dos Aliados area. (1917)
41. Branch line from Rua de Cedofeita into Rua dos Bragas for the transport of construction materials for the new Faculdade de Engenharia (1927)
42. Siding for the transport of spoil from Avenida da Boavista to Rua António Aroso.

Dump sites for ash from Massarelos
43. Boavista / Julio Diniz corner Western side (1930)
44. Matosinhos, Circunvalação into Avenida Norton de Matos (1920)
45. From Avenida do Brasil via Rua do Padrão to Rua de Gondarem connecting with the old steam tram line (1917, 1918)
46. Rua Dom Pedro V. José da Silva Maia & Ca. (1927)
47. Siding on line 16 for landfill in connection with the construction of Escola Industrial Infante D. Henrique. (1928)
48. Castelo do Queijo (1929)
49. On route 1, Conde de Campo Belo (1931)
49a. Alto da Serraes Fânzeres, Still existed on 1 June 1946.

CCFP sites (except depots)
50. Massarelos power plant. Still existed on 1 June 1946.
51. From Avenida da Boavista via the old steam tram line to the permanent way yard just after the Rua Serralves. (1938?)
52. From Rua de Ameal via Rua da Telheira to substation no.2 at Largo da Telheira. Removed 1941.
53. From line to São Pedro da Cova to substation St. Eulalia. (1928, Still existed on 1 June 1946.)
54. Rua do Castelo do Queijo into substation no.1. (1910, Still existed on 1 June 1946.)

Diverse and unknown
55. From Estrada Real no.3 (Amial / Vale Formoso?) via Rua Delfim Maia to Igreja de Paranhos.
56. Rua da Piedade, parallel to track of line 16 (1925)

Gradients
Porto, especially the central area, is hilly. The trams had to deal with the slopes too. An overview of the most important gradients:

Location	Incline	Length (m)
Rua dos Clerigos	1 in 10 (10.2%)	110
Rua 31 de Janeiro (Sto. António)	1 in 10 (10.1%)	280
Infante (Eastern side of the Jardim)	1 in 10 (10%)	94
Lapa	1 in 12 (7.8%)	170
Rua da Restauração	1 in 13 (7.6%)	590
	1 in 25 (4%)	164
Rua das Carmelitas	1 in 14 (7%)	72
Rua Mousinho da Silveira	1 in 16 (6%)	306
	1 in 25 (4%)	64
	1 in 31 (3.2%)	102
	1 in 33 (3%)	86
Rua Senhora da Luz	1 in 21 (4.8%)	65

Destination indications

The trams' destination was at first indicated with small plates on the roofs above the platforms. With the UEC and Brill cars came blinds as they were used with many other systems. Also the older cars received these destination blinds. However except those of the UEC cars, these couldn't be illuminated.

When in 1909 trams were ordered from Brill and UEC, CCFP specified the names for the blinds:

Fim	Campanhã	Palacio
Leça	P.D.Pedro	Massarelos
Matozinhos	B.Vista	Carmo
Carreiros	Bessa-	Circulação
Fóz	Matozinhos	Batalha
Inf.D.Henrique	Bessa	Devezas
P.D.Pedro	P.D.Pedro	Sto.Ovidio
M.de Pombal	S.Roque	Reservado
C.Cabral	Paranhos	Theatro
B.Vista-Povoa	P.D.Pedro	Fim

When a few years later a third batch of trams was ordered from Brill, the film counted more destinations and some names had changed:

Theatro	Bomfim	Bessa
Aguas Santas	P.da Liberdade	Bessa-
Ariosa-	Massarelos	Matozinhos
Circumcão	Ponte da Pedra	Campanhã
C.Cabral	S.Mamede	P.da Liberdade
M.Pombal	Paranhos	M.Burgos-
Inf.D.Henrique	Batalha	Circumcão
Alfândega	P.da Republica	P.E.Libertador
Foz	Carmo	Biblioteca
Carreiros	Palacio	Circ.D.Pedro V
C.do Queijo	Ouro	Circ.Trinidade
Leixões	Boavista	Grande
Venda Nova	Boavista-Povoa	Circulação
S.Roque-	Matozinhos	Ribeira
Circumcão	Devezas	Lordello
Corojeira	S.Ovidio	Reservado

On 31 August 1912 Brill was requested, if possible, to change Carreiros into Ava Brazil, as this name was changed.

In 1926 CCFP changed to a rather primitive system with stencil plates in front of light boxes. The same system was used for the line numbers since about 1946.

List of available destinations stencils

There were not only stencils for the regular termini, but also for the cross-overs that could be used for short workings in case of delays. A tram leaving the depot had to take all stencils with it that were on the service it was going to work. Each stencil had a unique location and where two different routes had the same name referring to a specific place, only one of the routes would use such a stencil; on the other hand the same place could have two different stencils, according to route number. In the first case location Prelada existed both in line 6 and 10\\, but the destination was used only in line 10\\; 2nd case is Passeio Alegre, where line 2 and 18 cars used PASSEIO ALEGRE and sometimes

the crossover, and line 17 used FOZ and the loop. Every rule has an exception and Porto trams qualify for it. The following is a list of available stencils; it's not a complete one, and should be read with caution, as the spelling of location names changed from time to time (AMIAL replacing AMEAL, and ERMESINDE replacing ERMEZINDE are examples).

A. AROSO	GONDOMAR
A. SANTAS	ILHÉU
ALFANDEGA	INFANTE
ALTO DA SERRA	LARGO CALÉM
AMIAL	LEÇA
ANTAS	LEIXÕES
ARCA ÁGUA	LEIXÕES-PALACIO
AREOSA	LORDELO
Av. BRAZIL	M. de POMBAL
AV.GOMES DA COSTA	MASSARELOS
BARROCAS	MATOSINHOS
BATALHA	MTE. DOS BURGOS
BESSA	P. DE PEDRA
BOAVISTA	PALACIO
BOLHÃO	PARANHOS
BONFIM	PASSEIO ALEGRE
C.DO QUEIJO	PEREIRÓ
C.24 DE AGOSTO	PR DA REPUBLICA
CAMPANHÃ	PRAÇA
CARMO	PRAÇA-CARMO
CARVALHIDO	PRAÇA CARVALHOSA
CARVALHOSA	PRAÇA CORDOARIA
CIRCUNVALAÇÃO	PRAÇA-INF.TE
COIMBRÕES	PRAÇA PALACIO
CONSTITUIÇÃO	PRELADA
CORIM	RAMADA ALTA
CORUJEIRA	REPOUSO
COVILHÃ	RESERVADO
DEVEZAS	RIO TINTO
DOCA MERCADO	S. CAETANO
ERMESINDE	S. MAMEDE
F. DA MOURA PALACIO	S. ROQUE
F.DA MOURA	S.P. DA COVA
FOZ	S.TO OVIDIO
FOZ MARGINAL	SÁ DA BANDEIRA
FOZ CARVALHOSA	SILVA TAPADA
FOZ-MOLHE	SOUTELO
FOZ PALACIO	STA EULALIA
G. CRISTOVÃO	VIRIATO
GODINHO	

List of "via" and other hung boards

Hung painted plate boards were not illuminated, and at night time were only visible a couple of metres away from the tram. They were dual side painted, and had as well as the "Via" indication sometimes the "Operario" indication. On "Operario" services special cheap tickets for workmen were valid. These were white, with letters in black. In old CCFP days, it appears they had reverse colours, i.e. black background and white letters. Other indications (hung boards) like "Giesta", "Infante" or "Gaz" were also available, but their use or purpose is not known today. The "Directo" had a red background with a "Directo-Areosa" on the other side, with white letters. "Directo" was used on

Car 171 turns from Rua de Oliveira Monteiro into Rua da Boavista at Carvalhosa in 1973. *A. G. Murray*

outbound trams on which tickets were only valid beyond a certain stop. In this way people who didn't want to travel that far were encouraged to use short working trams leaving space to those who wanted to go further. After 1967 when cars of line 1 were cut back to Infante a new indicator was introduced: "Correspondência Infante-Sá da Bandeira carreira 149", informing passengers their ticket was for both the tram and the bus trip, or vice-versa. The bus (or the trolleybus) was a permanent replacement and extension for the tram while line 1 existed until 1994.

Via Carvalhosa
Operario
Via Palácio
Via C. Cabral
Via S. Lázaro
Via R. Formosa
Gaz
Infante
Directo
Directo-Areosa
Via Marginal
Via P. Cordoaria
Giesta
Correspondência Infante-Sá da Bandeira carreira 149

Rolling Stock

Porto car builders and suppliers – early mule trams were

ordered from several European manufacturers with The Starbuck Car and Wagon Company as most important supplier.

But after the early years CCFP's own workshops began constructing cars, literally copies of the ones acquired from abroad. From the turn of the century (19th to the 20th) until WWI A Constructôra, a substantial engineering company at Carcereira, which shared some directors with CCFP, was building passenger bodies, freight trailers and some freight cars for electrical operation and Companhia Aliança electric freight cars bodies. The trucks, mainly Brill, and electric equipments for these cars were bought abroad. For the extension of the services required by the concession, most cars were ordered from J.G.Brill of Philadelphia, USA, but also five from UEC-United Electric Car Company of Preston, UK. Since WWI all trams were built in the operator's workshops, even in STCP days, with the exception of ten cars ordered in 1928 with Familleureux in Belgium, for which insurance money after the Boavista fire was used.

How many trams, or even how many types of trams existed in the earlier, pre-STCP period was until recently unknown. Documents in archives have been consulted but some ambiguities remain. Fleet-lists, especially with car-type indication, appeared to be scarce. Based on over 20,000 individual reports about tram cars, often about accidents, it could be reconstructed which fleet numbers were occupied over the years. With old photos the relevant types could often be detected. With fleet-lists for some years of the 1910's the reconstructed fleet-lists for the period until 1924 were confirmed, at least for the electric passenger trams. The period 1925-1946 is confusing on an individual car level. But it has been possible to get a reasonably reliable, although not perfect, picture of how many trams and of what types existed over the years. The largest mysteries about the types are with the freight trams. It is not always clear what the exact type was. There are also indications about the existence of two zorras on bogies, but positive evidence has not (yet) been found.

Mule cars & old type trailers
CCAPFM
Little is known about the tram-cars of CCAPFM. The closed cars, at least the first delivered, were of the type with the semi-circular etched glass windows above the main windows like trailer 8. The first two cars, which were delivered in September 1871, were of the type of trailer 9, now at Crich. In October 1871 an open car arrived with seven transverse benches. A chance mention in "Tramways. Their Construction and Working" by D. Kinnear Clarke, indicates that the first three cars of CCAPFM were built by Metropolitan of Saltley (Birmingham).

Museum car no.8, which it is believed came from CCAPFM probably to be used on the Restauração line, has evidence that it was built by Starbuck. No more is known about the builders of CCAPFM cars. In March 1872, when services started there were probably six cars available. In May 1872 another six cars arrived. The "O Primeiro de Janeiro" of 3 June 1879 notified that after the fireworks of the Senhor de Matosinhos festivities the company ran a convoy of 30 mule tram-cars. When merging with CCFP, CCAPFM had a total of 31 tram cars, but it is unknown how many there were of the several different types.

CCFP until 1892
In 1874 CCFP acquired from Starbuck 28 tram-cars for passengers:
- 2 double deck cars with seats for 20 inside
- 10 closed cars with seats for 20
- 12 closed cars with seats for 12
- 2 open cars with transverse seats
- 1 open car with longitudinal seats facing inside (Risca-ao-meio)
- 1 open car with seats in the middle (probably a knife-board arrangement)

The 20-seat closed cars were of the 7-window type. Later the capacity was reduced to 18 seats, but that will not have been because of a modification of the cars, but because of calculating 47 cm per passenger instead of 42.5 cm.
In 1875 three more cars were acquired from Starbuck. In 1875/6 two open and two closed cars came from the Societé Metallurgique Charbonnièr Belges and in 1876 two other open cars from Leon et Eugene Deletrez (Paris). In October 1877 a car was presented in the Palácio de Crystal built by the workshops of CCFP. By 1878 the number of cars was increased to 43. But the origins and types of the new cars are unknown. The cars 44 & 45 were newly constructed in 1879. In 1882 the number of passenger cars was raised from 45 to 51 by acquisition and reconstruction of 6 cars of the Larmanjat system, which operated during a short period from Lisbon to Sintra and Torres Vedras. These received the numbers 46-51. In 1883 the workshops constructed car 52. in 1885 two new cars were

constructed, a closed car received the number 53 and the other new car the number 32, evidently replacing an older car as the size of the fleet was extended by only one car. For 1885 the annual report gives a specification of the different types of cars:
- 20 large closed cars
- 9 medium sized closed cars
- 13 small closed cars
- 1 large double deck car
- 10 open cars

The workshops continued building new cars. In 1887 the series 54-57, in 1889 large closed car 58, in 1890 59 & 60, and finally in 1891 61 & 62.

Steam locomotives
In the beginning of 1878 CCFP tested locomotives from Winterthur and Henschel for use on the line Boavista - Foz (Cadouços). The Winterthur was returned, it had been at an exhibition in Paris in 1887 before coming to Porto. After Porto it was tried in Lisboa too, but wasn't bought by the CCFL either. The Henschel locomotive (works number 964), the first tram locomotive made by Henschel, was acquired together with three other Henschels (works numbers 972, 994 & 995) when CCFP received permission to use locomotives. These locomotives were of the same type as delivered later to companies in Italy, Germany and Holland. Henschel 964 was also tried in Lisboa without resulting in an order there.
In May 1878 CCAPFM tested a Henschel locomotive (probably works number 972, later delivered to CCFP) between Rua dos Inglezes (Infante) and Matosinhos. On 27 June 1878 both companies obtained permission for the use of steam locomotives. For CCFP this permission was valid between Boavista and Foz (Cadouços), for CCAPFM between the Ouro depot and Matosinhos, but the latter never used this permission.
Because of the extension of the line in 1882 from Foz to Matosinhos CCFP ordered another two locomotives from Henschel (works numbers 1383 & 1384) and also bought a Merryweather (possibly works number 95 built in 1880, which means it might have been 2nd hand) in 1882. Maybe it was used first by a sub-contractor for the construction of this line. The fleet numbers were 1-6 for the Henschels and 7 for the Merryweather.
In 1886 four more steam tram locomotives were acquired, this time second hand from France. The origins of these four locomotives are not (yet) known, but when acquired they were located in Cette (since 1928 Sête), a town that never had steam trams. Possibly they were engines delivered by Merryweather to the Tramways du Sud in Paris in 1876 and perhaps used in Béziers (Compagnie Régionale des Tramways du Midi) from 1879 until 1882 and/or in Cette for major port construction works, which were carried out in the period 1882-1888. In that case their weight and power were less than the Merryweather locomotive acquired in 1882. It took the workshops almost two years to get all four locomotives ready for service. Three of these locomotives, with the numbers 8-10, were sold for scrap to the Companhia Alliança in 1900, the fourth in 1903.
In 1900 locomotives 3 & 7 received new boilers, followed by the numbers 1, 2 & 4 in 1901 and 5 & 6 in 1903. All the new boilers were supplied by Companhia Alliança. In 1914 the steam traction ended. On 20 November 1916 the locomotives were offered for sale but not sold until 1919. Note that they were not sold to the line running between Lixa and Penafiel, whose metre

gauge locomotives were constructed between 1912 and 1915.

CCFP since 1893

The merging of the fleets of both companies must have resulted in renumbering, but no information has been found about how this was done. Also no reliable data exists about the different types. The total fleet must have comprised 93 tramcars, of which 31 came from CCAPFM and the others from CCFP. The assumption is there were 10-15 open cars mainly of the cross-bench type but also a few of the "Risca-ao-meio" type (facing longitudinal seats). The closed cars were most likely all of the latter type.

Trailers in the Electric Era

Of the cars which were converted into electric, the identity of only seven of them is known. In 1901/2 the mule cars / trailers 59, 60, 62, 85, 55, 54 & 61 became the electric tram cars 18, 28, 29, 30, 31, 32 & 35. The electric cars 18, 30, 32 and 35, (mule cars / trailers 59, 85, 54 & 61) were of the 8-window original CCAPFM type. Electric tram 29 (62) was of the 7-window CCFP type. For cars 28 and 31 the type is unknown.

When mule car operations were ended, there was no more need for small light cars which were by preference used on the lines with steep inclines, at least when the traffic demand was not very high. CCFP enlarged the trailers, in many cases probably by extending the platforms. Of some cars the body might have been enlarged. As part of the program probably also the bottom frames were reinforced to withstand better the forces when used as trailer. Museum car 8 clearly had these changes.

In total 29 or 30 mule cars / trailers were converted into 25 or 26 closed and 3 open motorcars. In 1908 one was converted back. Most of the unconverted mule cars were retained as trailers until the arrival of the bogie trailers and the closure of the steam tramline. By 1899 there were five series of fleet numbers, all beginning at 1. This must have been confusing. In 1904 the trailers were renumbered into the 100-series. Before 1904 probably they still had the old numbers. In 1905 five trailers (136, 151, 153, 167 & 169) were withdrawn from service. In 1907 the remaining cars were renumbered into the series 1-59. The same year large trailer No.27 was scrapped and two years later 31 had also disappeared leaving 57 cars in 1909:

- 5 closed trailers with seating for 20 on 2 benches: 1, 15, 16, 19 & 30
- 24 closed trailers with seating for 18 on 2 benches: 2-14, 17-18, 20-26, 28, 29
- 3 closed trailers with seating for 14 on 2 benches: 33, 36 & 37
- 6 closed trailers with seating for 12 on 2 benches: 32, 34-35, 38, 47 & 48
- 10 closed trailers with seating for 10 on 2 benches: 39-46, 49 & 50
- 2 open trailers with seating for 16 on 2 benches: 51-52
- 1 open trailer with seating for 18 on 2 benches: 53
- 1 open trailer with seating for 28 on 7 benches: 54
- 3 open trailers with seating for 24 on 6 benches: 55-57
- 2 open trailers with seating for 20 on 5 benches: 58-59

At Boavista depot at exactly midnight on 3/4 June 1973, cars 287, 284, 315 and 313 are ready to work extra journeys for sardine workers. *Michael Russell*

269 is receiving attention to one of its bogies at Boavista depot in 1973.

Michael Russell

A movie shows the trailers 10, 41, 23 & 32 behind a steam locomotive. Car 10 has an eight-bay body with semi-circular topped windows similar to "104", car 32 is also similar to "104" but with 6 windows only and a flat roof without clerestory, car 23 had a body with 4 windows and open sections between the saloon and the platforms, car 41 was similar to STCP trailer 13.

In October 1911 trailers 16, 37 & 49 were withdrawn, in August 1912 followed by no.29. In 1912 two trailers with the numbers 4 & 5 were renumbered into 16 & 27, apparently to free their original numbers for the new bogie trailers. This probably also happened with other cars. Six more cars were withdrawn in 1913 and three in 1914. Also in 1914 twenty trailers were sold to the CCF Penafiel á Lixa e Entre-os-Rios: 13-15, 23, 26, 30, 36, 39, 40, 45 and 50-59. In 1915 a car was withdrawn. CCFP had only 23 old type trailers left in 1916, which were all of closed types:

- 1 with seating for 20: 19
- 13 with seating for 18: 8-12, 16-18, 21, 22, 25, 27 & 28
- 1 with seating for 14: 33
- 3 with seating for 12: 32, 38 & 47
- 5 with seating for 10: 41-44 & 46

In 1918 a car was withdrawn and five more in the period 1920-1925. In the Boavista fire of 1928 four trailers were lost. Apparently in 1930 there was a renumbering: all trailers were grouped into the series 1-20 without gaps. That means at that moment there were 7 bogie trailers and 13 old cars. Two of the old cars were withdrawn in 1931 and one more had disappeared by 1934. In 1940 there were five large and four small cars remaining. From 1941 until the takeover by STCP in 1946 there were eight left.

STCP

It seems that STCP kept only three large (9, 10, 11) and two small (8, 13) former mule trailers in service. Probably six or seven old electric cars were converted (or possibly converted back) into trailers. According to the annual reports there were in 1947-1949 25 trailers, of which five in storage. Over the years the number of trailers gradually went down. Around 1966 they

were all phased out. Trailer 8 was one of the first museum cars. Trailer 9 was sold to the Crich museum in the UK. Trailer 22 was electrified again later to represent the oldest type of electric tram in the museum.

Trailer Usage

Before the advent of electric traction, CCFP used mule cars as trailers on the steam tram service between Matosinhos and Boavista. Porto bound trains were divided at Boavista, and hauled by mules to Carmo, Praça, and Campanhã. After electric traction reached Boavista by 1901 these arrangements continued using the new technology. The Marginal had become the "main line" by 1902 with through running from Praça. Between Infante and Leça the eléctricos hauled up to three old mule cars until they were replaced by the bogie trailers around 1910. Since then the use of one bogie trailer was the standard on line 1 between Leça and Infante except possibly for several years during in World War 1, although sometimes a 2-axle trailer was used. Trailers had always been detached at Infante, due to the ten percent gradient at the bottom of Rua Mousinho Silvera (East side of the Jardim do Infante). Line 1 trailers were normally based at Massarelos.

Line 5 used former mule cars as trailers and later also Fumista trailers 18-20, which were berthed between peaks at Boavista on a siding beside the depot. Trams going to Praça detached trailers at Carmo, and attached on the return journey, the trailers resting on the curve by the churches at Carmo, where a slight gradient permitted hand shunting. Trailer operation on the lines 1 and 5 ceased on 3 March 1960 when these lines were cut back to Matosinhos. Between 1960 and 1966 summer holiday weekend trailer operation continued between Boavista and Castelo do Queijo, and possibly Foz, due to heavy beach traffic.

STCP installed turning circles at Ermesinde and Coimbrões in 1947 for through services with trailer operation on the lines 9// and 14/: Ermesinde - Batalha - Coimbrões. It is unclear exactly when through trailer operation began, but it appears to have been sometime between 1947 and 1950; it may have been that there were insufficient trailers available at first as STCP converted a few of the oldest electric trams to trailers in the late 1940's. When trolleybuses replaced trams on route 14 in 1959 the trailer operation was reduced to run Ermesinde to Batalha, then from 4 October 1960 cut back to Bolhão. On the combined 9//+14/ route only old narrow trailers could be used because of the bridge. When tram services across the bridge were suspended, the Fumista trailers 18-20 were also used on line 9//. Trailer operation ceased on 31 December 1966.

From 1904 trailer operation required the use of brakesmen by law. This was relaxed from 1908 if continuous air brakes were used, resulting in an order for 25 motor car sets for the Brill bogie and the 24 Constructora cars and 58 sets for trailers doubtless for use on line 1. This requirement was suspended from circa 1920 due to post-war difficulties, and re-equipment began in 1938, continuing until 1956, but some brakesmen were still required until the end of trailer operation.

Electric cars and new build trailers

In this section an overview is given for the different types of electric tramcars. All 4-wheel cars had Brill-21E trucks unless mentioned otherwise. All bogie cars had maximum traction type trucks unless mentioned otherwise. In 1922/23 at least 14 cars received Siemens D58 motors: 149, 151-154, 161, 165, 166, 261, 285 and 296-299. Maybe two more cars received these motors.

In 1923/24 the GE270 55hp motors were introduced, which were used on all new and reconstructed cars since then until STCP took over. The painting schedule of the earliest years is not known, but photos prove there were different schemes. After 1903 the trams were painted green and white, in 1933 this was changed to yellow and white. STCP changed the colours to ochre with cream after take-over in 1946.

After the 1920s open platforms were given windscreens when trams received a major overhaul. In October 1912 a test was done with fenders on 4-wheel Brill car 275, but this did not yet result in equipping the fleet with these. Providence fenders were only mounted on the cars in 1925/6. After about 1946 these were gradually replaced by gate-and-tray lifeguards. Replacement of the destination blinds by stencil boxes, which could be illuminated, started in 1926. Replacement of the line number indications from metal plates to stencil boxes was started around 1946. Change of seat cover from rattan to artificial leather started in 1950.

Type names

Over the years many names were used to indicate the type of a car. The use of some names disappeared, while others were only used in later times. Several different names could be applied to the same type of car at the same time. During the first years of the electric trams those which were made out of mule cars / steam tram trailers were called "Transformados". The remaining mule cars / trailers were at that time called "Carros ordinários", "Carros simples" or "Carros antigos". In later years the latter

became "atrelados antigos" to distinguish them from the new build trailers. "Carros antigos" became now the indication for all pre-1904 electric trams, but later they were called "Carris-18", "Carris pequenos" or "Carris abertos". The latter was used when they were the last ones with open platforms. Of course before "carros abertos" were open cars, but those had disappeared by that time. The 6-window cars built by Constructora were always called "Constructora", however in the 1930s they also belonged with the older trams to the group "Carros de bancos", a group that was by STCP called "Risca-ao-meio". The "Carros Ingleses" were for a long time also called "UEC" and in their early years "Mountain & Gibson" cars. Until the second half of the 1920s the cars built by Brill were just "Brill", but they became "Brill-23" when it became necessary to distinguish them from the larger types. The four 7-window cars made by Constructora in 1907/9 were ambiguous, sometimes "Constructora", but later often "Brill-23". The 8-window Brill type was "Brill-32" but became "Italianos" when STCP installed Italian made motors. "Brill-40" was sometimes used for Brill-bogie trams, but also "CFP-40 (plataforma) fechada". "Familleureux" was used together with "Belgas". "CFP-40 (plataforma) aberta" and "CFP-28" or "Carris-28" were used for the trams also called "Fumista bogie" and "Fumista", but the very first car called "Fumista" was the open cross-bench mule tram that arrived in October 1871 with CCAPFM.

Renumbering

CCFP was rather fond of renumbering. However there is very

Car 77 propelling rails on a pair of "Joaninhas" towards Trindade at Praça da República in 1973. Note the attendant riding at the front with a gong.
Michael Russell

Car 53 at Castelo do Queijo in 1976, approaching the substation. *Michael Russell*

little information about the way this was done. Records from the workshops or depot are failing for CCFP period. Fleet-lists had mainly to be reconstructed from reports, often accidents, about individual cars with the help of old photos to determine the type of car. Only for the 1910's are several fleet-lists found which specify the car types by fleet number. This makes a reconstruction of the history difficult. In most cases without knowing the reasons or way of renumbering at least the following occurred:

CCFP had numbered the passenger mule trams / trailers and the freight trams both starting with 1. When the fleet of CCAPFM was merged with the one of CCFP this principle was retained, but it's unknown how the cars were renumbered to make new series 1-93 for the passenger trams and 1-24 for the freight trams. When the first electric trams were introduced, they received their own series starting with 1. A few years later this was repeated with the introduction of the first two electric freight trams. In 1901 there were electric trams (carros eléctricos) with the numbers 1-30, electric freight trams (rebocadors eléctricos) 1-2, mule trams / trailers (carros antigos) with the numbers 1-93 (with gaps) and freight trailers (zorras) with the numbers 1-41. Probably this was confusing, for renumbering was done giving all vehicles a unique number. The electric freight trams apparently received the numbers 81-82 and four new ones the numbers 83-86. The trailers were numbered above 100 with the freight trailers getting the highest numbers. The exact ranges are unknown, but photos show freight trailers with numbers in the 190 and 200 series. When in 1904 the unique bogie car was acquired from Brill, it received the number 90.

There is photo evidence that about 1901-1903 several cars were renumbered.

In 1905 the fleet of electric freight trams was renumbered again, now starting from 220.

In 1907 (February-June) the complete fleet of electric trams was yet again renumbered by adding 100. At the same time also the trailers were renumbered, apparently becoming 1-59 for the passenger trailers and 60-98 for the freight trailers. Until the outbreak of WWI the fleet grew significantly in size, but, at least by CCFP standards, little renumbering was done in the period until 1924:

1-59 (with gaps) trailers
60-98 (with gaps) freight trailers
101-170 (with gaps) pre-1907 trams
171-190 Brill
191 Brill-bogie (until 1913 it had the number 190)
199-201 7-window Constructora
220-248 electric freight trams
250 7-window Constructora
251-295 Brill
296-300 UEC (from delivery until 1913 these cars had the numbers 202-206)

In 1924 most of the fleet was renumbered. There is no information about the way this was done. The reasons are unclear, but about this time many cars received new electric equipment. Maybe the (new) car number indicated the type of electric equipment rather than the type of car. Although it's unknown for many numbers, which type of car it was, it's known

which numbers were occupied before new build trams started to arrive in 1926:
1-59 (with gaps) trailers.
60-98 (with gaps) freight trailers.
100-136 Probably these the trams with the highest numbers had Siemens D58 motors.
151-186 Number 150 appeared in 1928.
201-249 Probably all these trams had GE270 motors.
301-328 electric freight trams.

Between 1924 and 1946 many cars received a new number again, but in most cases the reason why it was done is unknown; probably a reconstruction or new electric equipment. Also new cars were built. As a result gaps were created in both 100 series, but the 200 series continuously grew without leaving a gap. Probably because of this the electric freight trams were in 1929 renumbered again, now to 50-78. About the same time the remaining trailers were numbered into a group 1-20, apparently by giving the cars with the highest numbers the lowest available, and the freight trailers from 79 up. Since 1935 the numbers from 199 down were filled, which resulted that in 1946 the following numbers were in use:
102-110 (five pre-1904 cars and four 6-window Constructora)
120-132 (with gaps, total seven cars: five Brill-23 and two 6-window Constructora)
155-158, 160-161 (all six were pre-1904 cars)
165-320 (without gaps, no pre-1904 cars, one 6-window Constructora, all other types)

After take over by STCP the fleet was renumbered yet again. This happened probably in 1947. The renumbering data of STCP has survived. As of STCP the type of car for each number is known, it gives also the overview for the types of cars and their numbers during the final period of CCFP.

Transformado types
The first electric tramcars of Porto were converted from existing mule cars / steam tram trailers of the 7-window type. In the first years electric equipments and trucks for these transformations were acquired from several suppliers. Apparently CCFP wanted to compare and choose the best for later acquisitions. The first three cars with the numbers 1-3 were put in service in 1895 and had equipment of Thomson-Houston. In the first years electric equipment was also bought from Walker, Siemens & Halske and Schuckert & Co. The trucks came from the same companies but in October 1897 three trucks were ordered from BSI to be equipped with Walker motors. As construction of new cars started in this period, it is not clear for which cars these were used.

The first Brill-21E trucks were ordered in 1898. These trucks became the standard in Porto for all 4-wheel cars. For electric equipment CCFP in later years limited their orders to Siemens-Schuckert and General Electric. Probably the oldest motors of Thomson-Houston and Walker were already replaced by 1905.

On 31 December 1902 there were 24 transformados numbered 1-14, 18-19, 28-32 and 35-37. As there existed different types of mule cars / steam tram trailers, also transformados sub types existed. All transformados seating arrangement was on longitudinal benches and except for one the capacity was 18 on benches of 4.25 m. It is unknown exactly how many cars there were of each sub type. Of the "7-window" type there were at least 11 and the "8-window" type at least 6

transformados. Of the other transformados no photos have yet been found to determine the type.

One unique car with the number 67 was made in 1904 out of two small type mule cars. This last car had a capacity of 22 on two benches of 5.35 m, but no photos are known. This car had disappeared from the fleet by 1917.

In 1905 a last transformado was put in service with the same measurements as the others, but no photo is known. Maybe this car, which had the fleet number 70, was not a transformado but the former car 17, which had disappeared from the fleet by 1904. In 1908 one electric tram was converted into a trailer again. It is likely that this was car 2.

New cars of the 7-window transformado type.
In 1898/9 the workshops of CCFP constructed three new cars, which without doubt were also fitted with imported trucks and electric equipment. These cars were supposed to have the numbers 10-12 but were renumbered to 15-17. It is not clear if they ever had the numbers 10-12. The bodies were copies of the 7-window transformado type.

The local firm A Constructora delivered four new cars in 1900 getting the numbers 20-23. These were also copies of the 7-window transformado type. It is probable that 20-22 had Schuckert & Co equipments. 23 had equipment of Siemens & Halske.

German cars
In 1898 two complete cars were ordered from two different suppliers in Germany. Without doubt CCFP wanted to compare both.

Siemens & Halske delivered a tram with four large windows and seating for only 16 on two benches of 4.06 m. The car received the number 14, but about 1901 it became the number 26. The car was also unique in having two front lights. It was still on the fleet list of 1916 but had disappeared by 1919.

Schuckert & Co. delivered a 5-window type of car of clearly German origin as known from constructors like Herbrand and Falkenried. This car had seating for 18 on longitudinal benches of 4.30 m. Originally it had the number 13 but about 1901 it was renumbered 27. It had disappeared from the fleet by 1916.

A Constructora cars with 5 windows
The local firm A Constructora delivered six cars with bodies which were inspired by the 5-window Schuckert car. The history of these cars is rather confusing. The first two were delivered in 1901 and had Siemens & Halske equipments and the numbers 24-25. In 1902 the next two were delivered with Schuckert & Co equipment and the numbers 33-34. However car 34 had a runaway accident in Rua de Santo António in November of the same year and doubts arose about the braking capacity. The fleet numbers 33-34 disappeared by 1903. In 1903 the last two cars of the type were made by A Constructora, getting the numbers 38-39 and Walker equipment. In 1904 the cars with numbers 55-56 appeared and are considered to have been 33-34 after getting new equipment. It is assumed that car 24 swopped numbers with 8-window transformado 35 around 1904. The cars seated 18 on two benches of 4.31 m and had larger platforms than the transformados and the Schuckert car. It seems that on at least some of these tramcars the 5-window arrangement was changed to six windows without changing the size of the body.

Car 252 is at Passeio Alegre, Marginal in 1978. Note the ladies by the river doing their laundry. *Michael Russell*

Car 271 is on the very narrow section of road at Cais das Pedras in 1978. *Michael Russell*

Car 137 is seen from above as it descends Rua da Restauracao in 1980. *Michael Russell*

Carros Antigos

No distinction was made in later years by CCFP between the several types of pre-1904 cars. As they became the smallest and oldest cars, after arrival of the Brills they were mainly used on less important services and the first to be phased out. Of the 39 cars available in 1909 there were probably 19 remaining by 1925. The 1928 Boavista fire losses for these trams are estimated to be seven cars, probably two being of the 5-window Constructora type. In the period 1934-1940 there were still 12. STCP took over 11 but kept only four of these cars on the fleet and renumbered them 101-104. 101 was a transformado type with 7 windows, 102-103 were 5-window Constructora cars but with the window arrangement changed into 6 and 104 was an 8-window transformado. This last car still had open platforms. All four were scrapped in 1959. A replica of 104, built in 1997, is now part of the museum collection. The other seven cars, or at least some of them, extended their life as trailers. One of them, 7-window trailer 22, is reconverted to an electric museum-tram.

Open cars

In 1903 the workshops converted an open mule car to an electric car with the number 40 and truck and electric equipment from Walker. According to Guido de Monterey this was an open car with seats for 28, which was adapted by the workshops for service on the longest lines. It made a successful test run on 21 February 1904. The fate of this car is a mystery as it was not reported anymore after 1904.

In 1904 A Constructora converted two further open mule cars into electric cars. They had transverse benches of 1.80 m. The official capacity was 28, with 4 on a row. It seems the seats on the benches on the platforms were not counted officially as photos prove they had seven benches inside of which five reversible, and also a bench on each platform. The cars received the numbers 41-42, after 1907 141-142. In 1924 they were renumbered into 100 & 115. They were used on all routes except to Vila Nova de Gaia. As they had many seats and little standing capacity compared to other tramcars, they were often used on longer routes like line 9 and later also line 10. Probably both were lost in the Boavista fire of 1928. A replica of 100 was made in 1995.

Brill-bogie of 1904 (STCP 249)

An enquiry was made in 1904 of J G Brill Philadelphia, seeking to buy a four-motor cross-bench bogie car of the patent Narraganset type for use on the western line, which was described as "practically level… but in the last 1,000 metres is …the inclination of 10%". After some correspondence an order was placed instead for a cross-bench car "of the Marseilles type" on maximum traction bogies. Presumably this gave better clearance between bogies and footboards on the sharp curve at Infante, but less adhesion on the gradient. Within a week of confirmation the order was changed to a Brill Patented Grooveless Post semi-convertible car on Brill 27G bogies with four Siemens motors, which overcame the challenges at Infante. The car had seats for 32 in two-plus-one configuration and open platforms.

The original number of the car was 90, since 1907 it was 190 and about 1913 it became 191. From 1918 until 1924 the car was used as a trailer on the Marginal. In 1924 it received two GE 270 motors and the number 226. Probably at this time also windscreens were provided. From this period the car was only used on line 9. With STCP it had the number 249. In 1965 the equal wheel bogies with 55 hp motors were replaced by maximum-traction bogies also with 55 hp motors, but in 1967 it received 68 hp motors from car 257 when this car was scrapped. STCP used the car on several routes, in later years most often on line 16(2nd). In 1970 the car was sold to the Shade Gap trolley museum in Pennsylvania USA.

A Constructora cars with 6 windows (STCP 105-111)

In 1904-1906 A Constructora delivered 24 cars with six windows on both sides and seats for 22 on 2 benches of 5.22 m in length. The trucks came from Brill, the electric equipment from Siemens. These were the last Risca ao Meio type cars. They were numbered 43-54, 57-66 and 68-69. Car 46 was used in 1907 for reconstruction into a 7-window car. After arrival of larger types of trams their status diminished. In later years they were mainly used on less important routes like 11 and 16 (1st). A number probably perished in the Boavista depot fire of 1928. Others were scrapped to supply parts for new cars. In February 1940 there were still nine in the fleet. A total of seven survived into STCP days and were numbered 105-111. One survives as museum tram 163 (ex-107), another as works tram no.48 (ex-111). The other five were scrapped in 1959.

A Constructora cars with 7 windows (STCP 112-114)

In 1907 A Constructora built two new cars with seven windows at each side and seating for 23 on transverse benches. These cars received the numbers 199-200. In the same period the workshops of CCFP converted car 46 as well into a similar 7-window car. This car received the number 201 and seems to have had shorter platforms. For many years these three cars had a second trolley pole to be used on the Ponte D.Luís I. Finally in 1909 A Constructora built a more or less similar body. This was done without having a confirmed order from CCFP, probably as an attempt to get the order for the new cars the latter was intending to buy. A Constructora failed to get this order but eventually CCFP acquired the body and used it for electric tramcar 250. This car was the last to receive open platforms. All later acquisitions involved cars with windscreens. Probably it was a little different from the other three and a short version of bogie trailer 1.

In later years these cars were often incorrectly considered to be cars of the Brill-23 type. Three of them came to STCP who numbered them 112-114. It is not clear what happened to the fourth. No.112 was scrapped in 1978, no.113 is in storage since 1987 and no.114 was sold to the USA in 1980.

Bogie trailers (STCP 1-7)

The body of trailer 1 was, together with the small body which was used for 250, built in 1909 by A Constructora without having received a firm order. After some discussion CCFP acquired both bodies, used the large one for trailer 1 and ordered from A Constructora six more bodies of this type together with type 23D bogies from Brill. The trailers 2-7 came into service in 1911/12 and were equipped with air brakes. They sat 32 on transverse 2+1 seats and spent all of their lives working on line 1 between Infante and Leixões, at first towed by 4-wheel cars, later behind the bogie cars.

In 1918 there was a proposal to convert these trailers into motor cars. This plan was never realised. Their original open platforms were given windscreens around 1948. In 1960 they were withdrawn from daily working. During the months July and August (weekends only) until 1966, they were used for leisure traffic on line 1 to the beaches. The bogie trailers were unique as the only pre-1920 cars which were never renumbered and the only pre-STCP class of which no car was ever renumbered. 1 belongs to the museum collection, the others were scrapped.

Carros Ingleses (STCP 115-119)

In 1909 CCFP ordered five cars with 7 drop windows each side from UEC of Preston in England. They sat 23 on transverse 2+1 seats. Together with the 4-wheel Brills these were the first tramcars with windscreens. The original numbering was 202-206, but in 1913 was changed to 296-300. The electric equipment came from Siemens-Schuckert. 296-299 had new Siemens D58 motors in 1923. The original trucks were Mountain and Gibson 21EM, a Brill 21E copy. In a later period CCFP changed the original platforms into the common Brill type. At first they were used, together with the Brills, on the most important routes. After arrival of the larger types of trams in the late 1920s they changed to less important lines. All came to STCP and were renumbered to 115-119. One of them is now museum car 247 (ex-118). Of the others two were withdrawn in 1967, one in 1974 and the last in 1977.

Brill-23 (STCP 120-146)

In 1909 CCFP ordered 20 cars from Brill with seven windows at each side and 2+1 seating for 23. These four-wheel cars were built to the Brill Patented Grooveless Post semi-convertible design. Together with the Ingleses these were the first tramcars with windscreens.

In 1910 a second batch of 25 cars of this type was acquired. The saloons of these cars were 33 cm longer, a feature which can be recognised by comparing the width of the corner pillars.

In 1912 a third batch of again 20 cars was ordered from Brill. The saloons of these cars were 7½ cm longer than those of the second batch. The Brills became a great success. In 1920 CCFP placed a fourth order, this time for wider cars. Lack of finance caused this order to be delayed and finally cancelled in March 1923.

Brill order	Year	In service	Numbers	El. eq.	Length/ width saloons
16717	1909	1910	251-270	GE80	556/220 cm
17302	1910	1911/2	271-295	Siemens	589/220 cm
18349	1912	1913/4	171-190	Siemens D53	597/220 cm
21180	1920	cancelled	30 cars	-	597/243 cm

Brill delivered the trams "knocked down" and in the white. The first was common practice of Brill with overseas deliveries to save on transport costs. The 2.15 m wheelbase trucks were delivered with axles but without wheels. The wheels and electric equipments were ordered by CCFP separately. As a result the workshops in Porto had to assemble the components, install the electric equipments and wheels and paint the trams. This explains the lapse of time between order and putting in service of the trams.

At first the 65 Brills were, together with the Carros Ingleses, used on the most important routes. After arrival of the larger types of trams in the late 1920s they were changed to less important lines. By 1926 CCFP started with the widening of this type of car into the Brill-28, a type of reconstruction that apparently later was more rebuilding with use of parts of the old car, and from 1938 the Brill-28 Plataforma Salão model. Gradually new electric equipment was installed. In February 1940 there were still 38 Brill-23's. This type of reconstruction was done until 1946. When STCP took over 27 cars still remained of the Brill-23 type, partly even with old electric equipment.

Six Brill-23's were probably scrapped around 1950. Of the others ten were modernised in the 1950s. Of the unmodernised cars eight were withdrawn in 1967, two in 1980 of which one (122) went to the USA. The last unmodernised car (144) was used until 1986, stored for some years and then scrapped as source for parts of replica 104. Of the ten modernised cars, one was withdrawn in 1967, one in 1968, two in 1989 and two in 1990. In 1995 129 was scrapped to supply parts for replica 100, 134 is stored since that year, 143 is available for rental services and 131 for normal service after a major overhaul.

Brill-28 (STCP 150-199)

To get cars with more capacity CCFP started to widen Brill-23 cars. This facilitated having the seats in a 2+2 arrangement instead of the old 2+1. The seating capacity became 28. Not only were old Brill-23 cars reconstructed, new cars of the Brill-28 type

were also constructed. In total there probably existed 50 Brill-28 cars, 28 new build and 22 reconstructed out of Brill-23 cars or built with the use of parts of older trams. It is assumed that 167-168, 172-177 and 180-199 were newly constructed, while the others were the product of reconstructed older cars. In the late 1950s the seat arrangement was changed from 2+2 to 2+1, reducing the number of seats from 28 to 23 and increasing the standing capacity. As this became the most numerous type of the fleet, they were used on all lines, except for the lines to Vila Nova de Gaia. STCP 160 seems to have been further modernised by STCP effectively becoming a Brill-28 Plataforma Salão. STCP 169

Car 275 is seen near Álfandega in 1980. *Michael Russell*

A busy scene as cars 199 and 193 and 113 are returning to Boavista depot after the morning peak in 1986. *Michael Russell*

An unidentified two-axle car is seen with bogie car 274 at Foz in 1986. *Michael Russell*

(currently 217) seems to have been widened and given extended platforms by CCFP. It has the short saloon of a Brill-23 car from 1909.

One car, 151 which was probably the first widened from Brill-23 to Brill-28, was scrapped around 1950. Eight cars were withdrawn in 1967 of which one (172) went to the USA. Further withdrawals were three cars in 1972, four in 1973, one in 1974, two in 1978, two in 1980, three in 1981, one in 1982, five in 1983, three in 1984, four in 1986, three in 1986 and the last six in 1989.

Brill-28 Plataforma Salão (STCP 200-223)

In 1938 (re)construction of the Brill-28 ceased in favour of reconstruction into the Brill-28 Plataforma Salão model. In many aspects similar to the Brill-28, these cars had longer platforms increasing the standing capacity. Probably none of these cars was totally new but all were reconstructions out of Brill-23 cars or constructions with the use of parts of older cars. With CCFP the numbers of these cars were scattered in a large range mixed with cars of other types. Moreover in most cases the car numbers had already been used for a number of years before the Brill-28PS trams were constructed. In a few photos taken in this period Brill-23 cars can be seen having numbers later used by Brill-28PS cars. In total there were 24 of these cars. Probably 18 of these trams were made out of Brill-23 cars, and six with the use of parts of older trams; five of the Constructora 6-window type and one Carro Antigo. The use of the Brill-28 Plataforma Salão trams was equal to those of the Brill-28 type. In the late 1950s the seat arrangement was changed from 2+2 to 2+1, reducing the number of seats from 28 to 23 and increasing the standing capacity. Four were withdrawn in 1967, one in 1978, two in 1989, one in 1990, two in 1991, three in 1993, two in

1994 and one in 2010. When the tram operations of STCP were diminished, these were the most numerous of the last cars in use.

Brill-32 (STCP 250-261)

In the years 1926-1928 the workshops constructed 12 cars similar to the Brill-28 type but with an extended saloon with eight windows at each side instead of seven. Seating was 2+2 for 32. These cars were numbered 250-253, 256-259 and 284-287 by CCFP. The last four were renumbered by STCP into 254, 255, 260 & 261. They were used on all lines except to Vila Nova de Gaia. During the second half of the 1930s they were used most of the time on lines 5 and 7. The original GE270 motors were later replaced by more powerful Italian CGE motors from the Belgas and bogie Fumistas, which gave them the nickname Italianos. In the late 1950s the seat arrangement was changed from 2+2 to 2+1, reducing the number of seats from 32 to 26 and increasing the standing capacity. Two cars were withdrawn in 1967, six in 1974, one each in 1977 and 1978 and the last two in 1981. Two cars were sold to the USA (Yakima, Washington) and two to Buenos Aires (Argentina). Car 250 is now part of the museum collection.

Brill-bogie (STCP 270-277)

In 1926 the workshops of CCFP constructed two new bogie cars inspired by the existing car of 1904. They had the numbers 200 & 201. These cars were wider than the original, which made a seating arrangement of 2+2 possible. Total capacity was 40 seats. In 1928 six more cars were constructed. These were a few centimetres wider and longer and numbered 268-273. When introduced they were mainly working on line 1. In 1929 they were also in service on lines 2, 7 and 9. In the early 1930s their service was equally divided over lines 1, 5 and 9, which shifted for the last 12 years of CCFP to mainly line 9 with some services on line 1. STCP used the Brill-bogies mainly on lines 1 and 9. With the renumbering done by STCP 268-269 became 274-275 and 200-201 became 276-277. From 1950 these cars received 89 hp BTH-114 motors. In the late 1950s the seat arrangement was changed from 2+2 to 2+1, reducing the number of seats from 40 to 32 and increasing the standing capacity. They kept working on line 1 until the closure of that line in 1994. All eight cars survive although 273 has left Porto for Crich. Car 274 belongs to the museum collection, 275 is available for hire and the other cars are in storage.

Belga (STCP 280-289)

After the 1928 Boavista fire ten new tramcars were acquired from Familleureux in Belgium. Delivered late in 1928, most entered service in the beginning of 1929. The bodies with large windows and without clerestory roofs are completely different from the American style, which was normally used in Portugal. For the rest the Belgas were in many aspects equal to the Brill-bogie cars. They too had seating for 40. CCFP numbered the Belgas as 274-283. The first six were renumbered by STCP into 284-289. When introduced these cars worked mainly on lines 1 and 2 with some services on the lines 5, 6 and 7. After 1930 line 1 had most of the Belgas, with additional services on line 5 during the first half of the 1930s and on line 9 in later years. STCP used them on the same lines as the other bogie tramcars. In 1947 CGE 68hp motors were fitted and then from 1949 these cars received 89 hp BTH motors. STCP also equipped them with retractable steps. In the late 1950s the seat arrangement was changed from 2+2 to 2+1, reducing the number of seats from 40 to 32 and increasing the standing capacity. One car was withdrawn in 1967, two in 1974 and one in 1982. The others kept working on line 1 until the closure of that line in 1994. Some performed limited service until 1998, 283 was used until 2004 and 287 until 2006. Car 288 belongs to the museum collection, 287 was used during the season for several years as information and ticket office of Carristur but is now available for service whereas 280, 283 284 and 285 are in storage.

Fumista (STCP 300-315)

Probably inspired by the Belgas the workshops constructed

Interior of the Massarelos power plant around 1920. Several items of vintage electrical plant survive in the power house. *Siemens Corporate Archive*

in 1929 and 1930 sixteen 4-wheel cars with the same type of modern roofs as the Belgas. The very deep windows however were in the more American style, although not of the semi-convertible construction as the Brill cars. Instead of that the windows could be removed during the summer period. When open it was allowed to smoke in the cars, which gave them their nickname, although the practice of converting the cars twice a year was abandoned in later years, and the windows were replaced by a type that could be opened by the passengers. Seating was for 28. During CCFP period they served mainly on line 2, with much working on lines 6, 7, 8 and 9 in the mid-1930s. They were also used on other lines except to Vila Nova de Gaia. Originally the cars had two trolley-poles with the resistance at the centre of the roof. Later the configuration was changed into one trolley-pole and resistances on the roofs of the platforms. CCFP numbers were 296-311. STCP renumbered the first four 312-315. In the late 1950s the seat arrangement was changed from 2+2 to 2+1, reducing the number of seats from 28 to 22 and increasing the standing capacity. Three Fumistas were withdrawn in 1967, two in 1972, three in 1974, five in 1976, one each in 1977 and 1978. The last one, 315, was withdrawn in 1984 and is now part of the museum collection.

Fumista-bogie (STCP 266-269)

In December 1930 the workshops started the construction of four cars, which were a bogie version of the Fumistas. Like the other bogie cars seating was for 40. When introduced they were divided over lines 1, 5 and 9 with some services on line 7. In the second half of the 1930s they normally worked on line 9 with some services on line 1. STCP used them on the same lines as the other bogie tramcars. With CCFP these cars had the numbers 312-315. STCP renumbered them 266-269, and fitted CGE 68hp motors in 1947. From 1950 these were again remotored with 89 hp BTH-114 motors. In the late 1950s the seat arrangement was changed from 2+2 to 2+1, reducing the number of seats from 40 to 32 and increasing the standing capacity. The cars 267, withdrawn 1990, and 269, withdrawn 1991, belong to the museum collection, the other two cars, withdrawn in 1980 and 1981, were sold to the USA.

Fumista trailers (STCP 18-20)

In 1934 the workshops made three trailers of the Fumista type with seating for 20. Trailer use with CCFP was limited to lines 1 and 5. STCP used these trailers also on line 9// after through running to Vila Nova de Gaia was abandoned by suppressing line 14/. They were numbered 18-20 and in service until 1966. Trailer 18 is part of the museum collection.

Cars modernised by STCP

In the late 1940's and 1950's STCP had plans to modernise the antique fleet. However this was not realised except for some reconstructions and the building of the Pipis and a prototype for a new generation of trams. One of the modifications done was extending the platforms of existing cars. This was a slow process and with later reconstructions the cars also received platform doors, folding steps and in the case of car 124 even a more modern roof without clerestory. Ten of these cars were Brill-23s, but also a Brill-28, three Ingleses and one each of the 6- & 7-window A Constructora cars received extended platforms. Of the unmodified Brill-23 cars none is left in Porto, but several of the modernised trams still exist.

The Tramways of
PORTO

Boavista Depot
As reconstructed
after the 1928 Fire

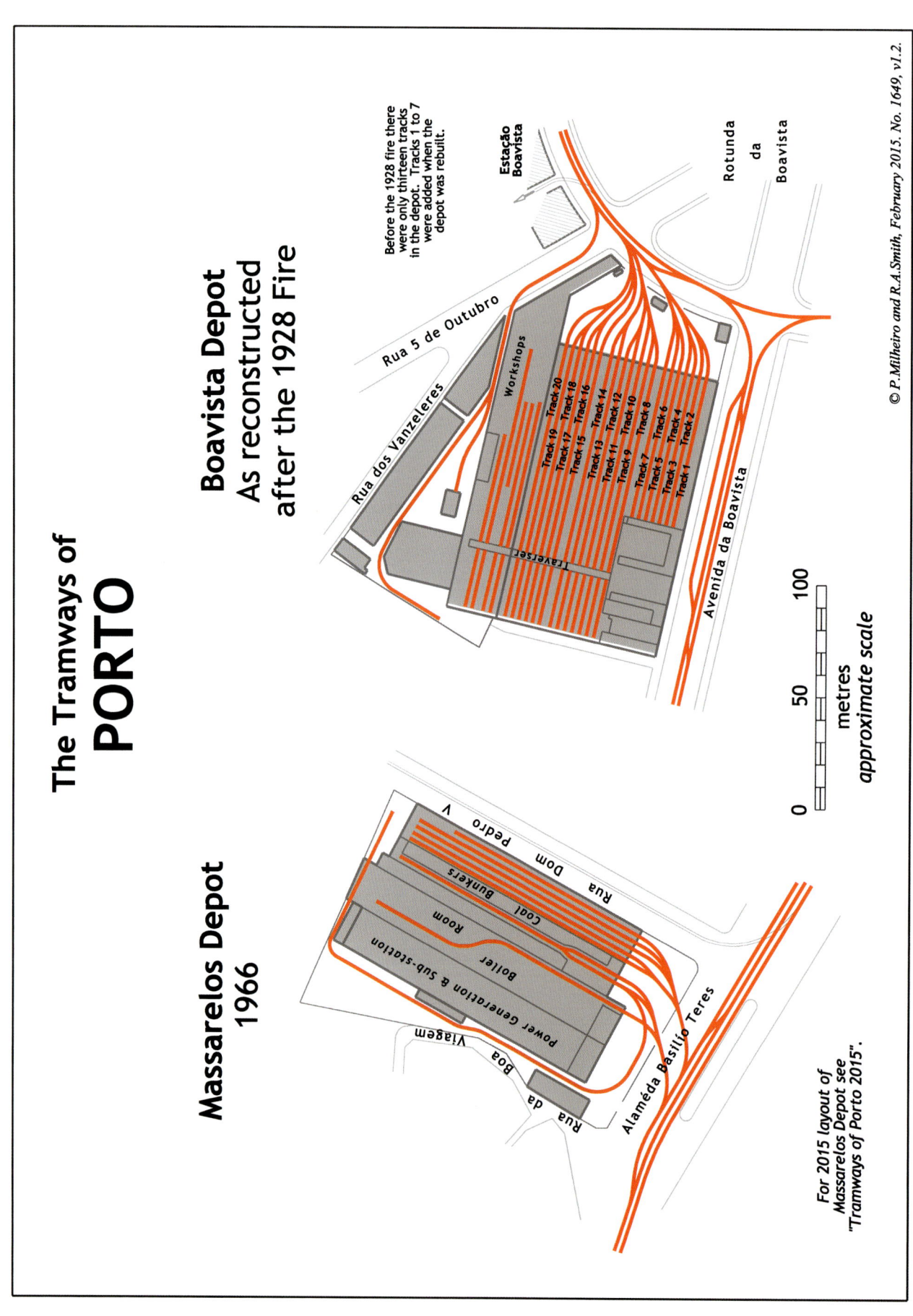

Before the 1928 fire there were only thirteen tracks in the depot. Tracks 1 to 7 were added when the depot was rebuilt.

Estação Boavista

Rotunda da Boavista

Rua 5 de Outubro

Rua dos Vanzeleres

Workshops

Traverser

Track 19 Track 20
Track 17 Track 18
Track 15 Track 16
Track 13 Track 14
Track 11 Track 12
Track 9 Track 10
Track 7 Track 8
Track 5 Track 6
Track 3 Track 4
Track 1 Track 2

Avenida da Boavista

Massarelos Depot
1966

Rua Dom Pedro V

Cool Bunkers

Room

Boiler

Power Generation & Sub-station

Boa Viagem

Rua da

Alaméda Basilío Teres

For 2015 layout of Massarelos Depot see "Tramways of Porto 2015".

0   50   100
metres
approximate scale

© P.Milheiro and R.A.Smith, February 2015. No. 1649, v1.2.

Modern bogie car plans

After 1930 with the construction of the four double truck Fumistas, no further bogie cars were built. Although STCP needed larger trams and in spite of projects for extra-long bidirectional trams, PCC look-a-like type, resembling Madrid Fiat Ferroviaria built modern units, they remained on paper. Porto bogie trams would have had a capacity for 100 passengers, and would have been 13.65 m long, and probably have had Commonwealth type trucks with four motors.

Pipis (STCP 351-373)
Pipizinhos (STCP 400-405)

STCP wanted to modernise the old fleet. As available funds were insufficient to acquire new trams from abroad, new constructions were limited to simple cars with antique types of trucks and electric equipment. A tram set consisting of motor car + trailer was designed. These cars were all built for unidirectional operation, having doors on the right side only, because STCP after 1946 was installing reversing loops at most of the termini; but as manufacturing costs had to be kept low, and the time and man-power available were scarce, they were not as robust as the older cars and lacked comfort with hardboard exterior panels, no upholstery, no clerestory roof being unpleasant during summer, no bulkheads or lowered platforms, and thus needing an extra step, making climbing aboard difficult and slow. The technical solution was neither modern nor innovative: Brill 21E trucks with 20 years old GE 270 motors, and the standard G.E. B-54 controllers.

However after finishing the first trailer it was decided to construct motor cars only. The trailers under construction were finished as motor cars getting electric equipment from Transformados. In total five of these small Pipizinhos with seating for only 15 were constructed in 1947. The trailer, which was finished and put into service with the number 28, was converted into motorcar 405 in 1949. The small Pipizinhos had the numbers 400-405. 24 of the large Pipis, which were intended to be motor cars, were built during the years 1947-52. These cars had seating for 28 and were numbered 350-373. The use of the Pipis was limited to the lines without stub-termini. Neither could they go to Vila Nova de Gaia. As the Pipis were less flexible being single ended, they were among the first to be withdrawn in the 1960s. Car 405 was scrapped in 1955 after a sad runaway accident in Rua da Restauração causing two fatalities. All the other small cars and 17 of the large cars were withdrawn in 1967 and another one in 1968. Five more were withdrawn in 1972 and only 373 made it to 1974. The cars 373 and 400, the latter converted to a trailer with the number 25, belong to the museum collection.

Car 500

This was the most modern tramcar in Porto until the appearance of the Eurotram of the Metro do Porto. The design of the car was very similar to that of trams in Madrid and Torino. It was a prototype for the intended new generation of trams including double-ended modern bogie cars, but these new cars were never built. Car 500, built in 1951, is single-ended and has a Brill 79E 4-wheel truck and CGE 68hp motors. It was the only Porto tramcar with seats for driver and conductor and folding doors closed while driving. It was withdrawn from service in 1972 becoming one of the first museum cars.

Car 212 seen at night at Carmo in 1988. *Michael Russell*

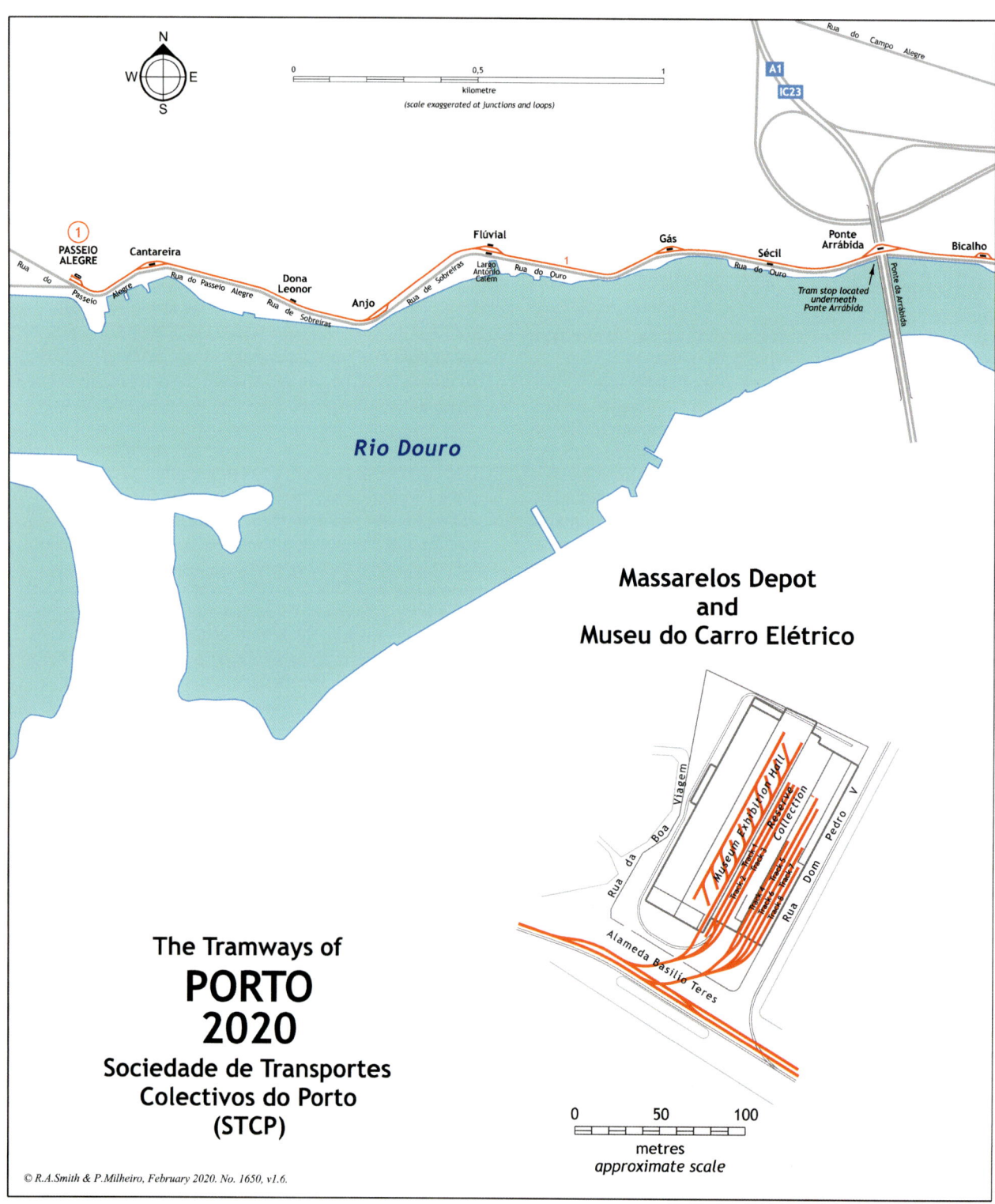

N W E S

0   0,5   1
kilometre
(scale exaggerated at junctions and loops)

A1
IC23

Rua do Campo Alegre

1
PASSEIO ALEGRE

Cantareira

Dona Leonor

Rua do Passeio Alegre

Rua de Sobreiras

Anjo

Largo António Calém

Rua do Ouro

Flúvial

Gás

Sécil

Rua do Ouro

Ponte Arrábida

Bicalho

Ponte da Arrábida

Tram stop located underneath Ponte Arrábida

Rua do Passeio Alegre

Rio Douro

Massarelos Depot
and
Museu do Carro Elétrico

Rua da Boa Viagem

Museum Exhibition Hall

Reserve Collection

Track 1 · Track 2 · Track 3 · Track 4 · Track 5

Rua Dom Pedro V

Alameda Basilio Teres

0   50   100
metres
approximate scale

The Tramways of
PORTO
2020
Sociedade de Transportes
Colectivos do Porto
(STCP)

© R.A.Smith & P.Milheiro, February 2020. No. 1650, v1.6.

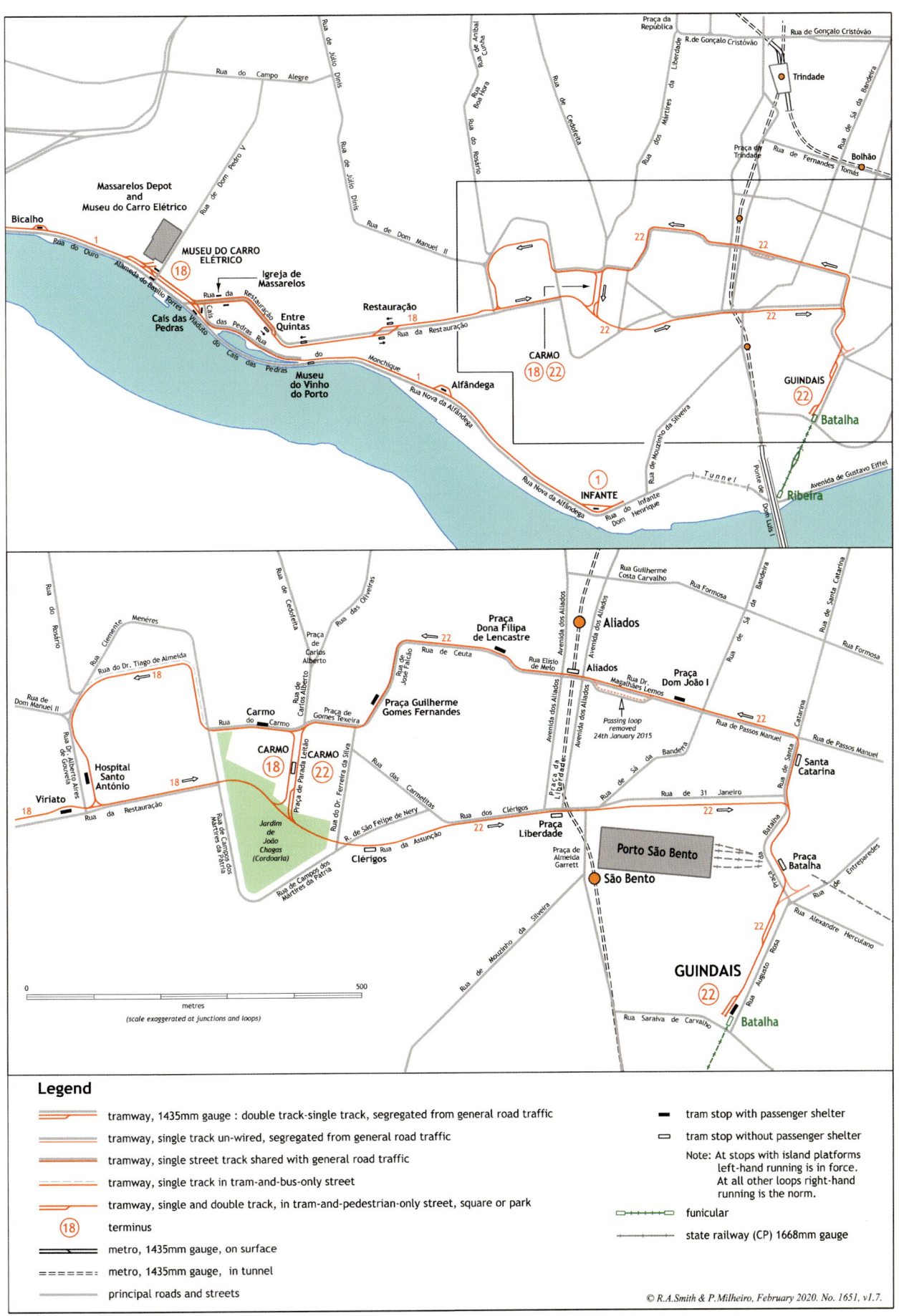

Car 272 seen at Passeio Alegre on a sunny day in 1991. *Michael Russell*

Car 77 at Foz with small trailer and what appears to be a rotary converter in 1995. *Michael Russell*

An unidentified two-axle car seen in front of the Igrejas dos Carmelitas and do Carmo. It is often difficult to take this shot due to the proliferation of pedestrians and traffic and challenging lighting; in this view these challenges have been overcome by the photographer. *Commercial postcard*

Car 218 at Cais das Pedras in 2004. *Michael Russell*

Cars 220 and 218 at the junction at Massarelos in 2005. *Michael Russell*

Car 104 passing under Ponte Arrabida during the annual parade in May 2006. *Michael Russell*

Electric freight and works cars

In 1899 five sets of equipment were ordered from Walker of which two for 2 x 50 hp and three for 2 x 25 hp. As 2 x 25 hp was the common type for passenger trams, it is supposed that the stronger equipment was meant for freight cars. On 12 January 1902 a frontal collision was reported between Rebocador 1 and electric tram 7 on the curve of Cais das Pedras. The fact that the freight tram-car had a fleet number suggests that there were at least two of them. The number also suggests that at that moment the fleet number 1 was in use for five different vehicles: steam locomotive, electric tram, mule tram-car / trailer, electric freight tram and a freight tram trailer. This must have been a little confusing and CCFP had reason to renumber some. It is likely that the electric freight trams received the numbers 81-82.

In 1903 three more electric freight trams were constructed. They were indicated as wagons eléctricos and received the numbers 83-85. In 1904 86 was added to the fleet. According to the annual reports of 1905-07 there were 3 wagons eléctricos cobertos and 2 wagons eléctricos abertos (open wagons). It seems one of the first electric freight cars had only a short life. In 1905 the zorras were renumbered into the 220-series.

On 30 November 1911 an emery grinder was ordered from The London Emery Works. Neither in the correspondence or the bookkeeping anything more is found about this, but maybe the equipment was used for a grinder car.

For the transport of coal from the mines in São Pedro da Cova to the new power plant in Massarelos as well as for the removal of the ashes a large fleet of new open electric freight trams was required.

With help of the inventories of 1913-1916 & 1919 the fleet of that period is summarised in the table in the fleet list. The constructors of the electric equipment are from the correspondence. The constructors of the bodies were only mentioned in 1913/14. In 1916 the capacity was not given.

In 1908 CCFP discussed with the Coats & Clark company about the transport of goods to and from their site in Vila Nova de Gaia. From 1907 until 1939 there was a branch line of 700 m from Sto. Ovídio to Coats & Clark. CCFP gave the alternatives of the use of 5 t zorras on a Brill 21E truck with 2x Siemens D54 motors or 10 t zorras on bogies with 4x Siemens D54. In 1922 12 motors type D58 were ordered from Siemens for bogie cars. At that moment CCFP had only the 1904 Brill bogie motorcar, but in use as a trailer, and seven trailer cars with bogies not suitable for motors. Probably these D58 motors were in fact used for 4-wheel cars; the Brill car received two years later GE270 motors. It is not known for certain if the zorras 225-226 with 10 t capacity had bogies, but it seems likely. In that case, they were probably the only bogie cars which crossed the Ponte Dom Luis I and had double trolley poles to be used on the bridge.

In 1924 the freight trams were renumbered in the 300-series. On 31 December 1925 there were 28 zorras. Two zorra motors were lost in the Boavista fire of 1928. In 1929 three new zorras were constructed. The same year the fleet was renumbered in the series 50-79. On 9 September 1931 there were 27 zorras. In 1932 a motorcar for overhead wiring repair was constructed. In 1939 two new zorras were constructed. On 19 February 1940 there were 2 10-tons and 26 6-tons zorras. Both in 1943 and 1946 one new zorra was constructed. STCP constructed in 1949 another 2 new zorras, again 2 in 1950 as replacements for 2 others and finally in 1951 a zorra with a metal body. In the 1950s STCP had still 31 freight and works cars. With the closure

of the Massarelos power plant, the competition of road vehicles for commercial freight transport and the virtual disappearance of the trams, also most of the zorras and vagões disappeared. Overhead wire repair car 49 and the zorras 58 & 66 are part of the museum collection, grinder car 48, break-down assistance car 76 and zorra 77 are still available but see no use anymore and are now considered to be museum trams.

Mule- and trailer zorras

Of the four major groups of trams, electric passenger cars, mule cars / trailers for passengers, electric freight cars and mule cars / trailers for freight, the history of the last is the most obscure. Both mule tram operators had freight cars, but of CCAPFM it is only known there were six freight cars when this company merged with CCFP in 1892. In 1886 the CCAPFM reported to have transported 10,839 tonnes of material to be used for constructing the Porto de Leixões. It is likely this transport was from the branch at the riverside constructed in November 1884, to Prado at the narrow gauge railway-line from where it was brought by NG trains to the construction sites near Senhor do Padrão (South mole) and Castelo da Leça (North mole) Probably part of this transport included the components of the Titan cranes of Leixões that were ordered from Fives-Lille in France in 1886 and assembled in Leixões the same year.

In 1874 CCFP acquired from Starbuck six mixed cars for the transport of passengers and luggage, of which two were small, and four freight cars with 4 t capacity of which two were open and two closed. It is not clear if the mixed cars had accommodation for passengers or were just meant for the transport of luggage in combination with passenger trams. In 1876 CCFP specified there were six luggage cars and four freight cars. In later years new cars were constructed and probably also existing cars reconstructed from one type to another. By 1882 the total number of cars was twelve of which probably three were used for luggage and nine for freight. By 1892 CCFP had about 18 cars for freight and luggage.

In 1899 A Constructora delivered 13 open freight cars, 6 with high and 7 with low sides. The same company delivered six more cars in 1901, probably raising the fleet to its largest extent of 41 vehicles. These new freight trailers were probably used together with the two electric rebocadores built at the same time for the transport of wine between the new site of the Menezes company and Alfândega. By 1905 there were 38 freight trailers, one of them was scrapped in 1907. By 1919 there were probably still about 36 freight and works trailers. Type / use are specified in the table in the fleet list.

In later years the number of freight trailers decreased, but also some new were constructed. However little data is known. In the Boavista fire of 1928 one zorra atrelada was lost. In 1933 three new zorra trailers were constructed. In 1938/40 there were 8 zorras atreladas for 4 tons and 1 for 5 tons. In 1943 a new freight trailer was made, probably the last time this happened. Their last use was mainly for the transport of fish from Leixões to the city centre attached to a motor car on which the peixeiras (fishwives) travelled. Freight trailer no.80 is now part of the museum collection.

No historical data is known of the small trucks, in Porto nicknamed Joaninhas. These tiny vehicles were until the 1990s used by the permanent way department. It happened frequently that two were used to carry lengths of rail. These combinations were sometimes pulled, but often pushed by a zorra. In the latter

Car 213 at Aliados in 2011 with the city hall in the background.
Michael Russell

case on the first Joaninha an employee of STCP was seated continuously beating with a hammer on a large bell to warn the other traffic. Joaninhas were also used to carry a welding power supply for track repair. In 2002 one was used, fitted with scaffolding, to mount the overhead wires above the new tracks at Cordoaria. One of the most remarkable jobs was when one was hauled by museum trolleybus 1 participating in the 1997 tram parade. The trolleybus ran on its own power with one trolley-pole on the overhead of the tram and the other connected by a cable with the Joaninha.

Power supply

In 1895 the necessary electricity was generated by the first power plant of CCFP situated at Arrábida just west of the location of the current bridge with this name. In 1895 there were two 150 hp generators installed by Thomson-Houston but the capacity was soon increased. At Arrábida were also the first depot and workshops for the electric trams. In 1912 the new, much larger power plant at Massarelos was opened. Arrábida was closed by 1915. Between 1919 and 1928 there was also a small power plant in São Pedro da Cova, but this was replaced in 1928 by a sub-station in Sta. Eulália. Other sub-stations were located at Massarelos, Telheira, Castelo do Queijo, Contumil and Corpo da Guarda (near São Bento station) In 1922 the capacity of Massarelos was increased. In 1930 new more powerful units replaced part of the old machines and another new unit acquired in 1936 made it possible to abandon all pre-1930 machines.

On 1 September 1929 the Companhia Hidro-Electrica do Varoza started the supply of electricity via overhead cables between Ameal (substation Varoza) and Sta. Eulália. Two years later a connection was made between the Varoza sub-station and Telheira.

The amount of electricity bought from Varoza, later U.E.P (União Electrica Portuguesa) was limited. In 1933 the Massarelos power plant generated 20,891,370 kWh and only 17,900 kWh was acquired from the U.E.P.

The necessary coal for Massarelos came from the mines in São Pedro da Cova. The mine had three shafts. A film shows that the coal from one shaft was transported by a narrow gauge railway to the site of one of the other shafts. From one shaft the coal was transported by a rope-way to Monte Aventino and from there by zorras to Massarelos. This rope-way, in use from 1916 until 1970 and scrapped in 1972/3, was the mode of transport for

the coal to the Rio Tinto railway station and the Monte Aventino coal depot. From Monte Aventino the coal was distributed to the local Porto based customers. This was done partly by tram and for this reason there was a freight branch. From 1917 another shaft of the mine had a connection with the tram network and the coal for the power plant as well as for other customers was directly brought by tram from this mine shaft to its destination.

In 1933 almost 44,400 tonnes of coal was used. With a load of 6000 kg this means about 7400 zorra trips from São Pedro da Cova or Monte Aventino to Massarelos or on average just over 20 trips each day. As the coal of São Pedro da Cova was of poor quality, about 11-12 zorras must have left Massarelos each day for removal of the ashes to the dump places. To monitor the amount of coal, and probably also ash, in 1911 a weigh bridge pattern no.524 with 20 tons capacity was ordered from Henry Pooley & son Ltd, John Bright Street, Birmingham. In the 1950s the Massarelos power plant was phased out.

Depots

The most important CCAPFM depot was Ouro at Largo Calem. CCFP had its major depot at Boavista. During the mule tram period a second depot was located at Bolhão. When electric operations started the depot and workshops were at Arrábida. In 1901 Boavista was connected to the electric network and became depot for the line Boavista - Campanhã. In 1903 the workshops were moved to Boavista and this site also became the major depot for the electric trams. Massarelos was depot for the zorras involved in the coal transport and, maybe only in later years, the trailers operating on the Marginal.

On 30 November 1874 the stable of Boavista was destroyed by a storm. Five men were injured and 60 mules were killed or severely injured. Services were suspended for several days. Boavista suffered twice from fire. The first time was on 3 November 1876 and the second time on 27 February 1928. No facts are known about damage caused by the first fire so this was presumably limited. With the second fire the paint shop and the western part of the depot was destroyed together with 21 motorcars, 4 trailers, 2 zorras and 1 zorra trailer. Which cars exactly were destroyed is not known, but probably most were of the older types. Two days after the 1928 fire the trams of the lines 1 and 16 were moved to Massarelos to return to Boavista only two years later.

218 poses in front of a splendid art deco façade at Carmo in 2017.
K. Lomas King

General view of the main hall of the tram museum in 2007. *B. King*

131 is at the Batalha terminus on 19 June 2008. *L. Folkard*

For many years Boavista had insufficient capacity to stable all trams inside. During the night there was a row of trams on the second track on the Rotunda da Boavista in front of the depot.

When in 1908 CCFP bought the VE it gained a large parcel of land at Massarelos, built a new depot and in 1912-1915 a new large power station. This is the only depot in existence today.

Not realised plans for new depots

During the twenties, there were plans for constructing a new depot on the north side of the sub-station at Castelo do Queijo, with 12 tracks with pits, and capacity for 60 trams, but this was never done. CCFP also owned another large parcel of land in Rotunda da Boavista, not far away from the original depot, which became STCP property after the takeover. However it was sold for real estate development.

Parking the trams and trailers, and conducting light repairs and daily maintenance was a nightmare, as cars from the entire network, with the exception of the coal cars and the trailers of line 1 allocated at Massarelos had to leave Boavista in the morning and return at the end of the day. Plans were then made to construct a new depot at Covelo, near Marquês de Pombal and this would provide easy access to at least lines 9, 15 and 20. This depot was to be constructed in a municipal park, with a capacity for 60 trams, 11 tracks with pits, a reverse loop, and a permanent way yard area. But for legal reasons, the use of land could not be assigned to any other purpose except a public park, so the project was abandoned. STCP never found an alternative place, but as Providence life fenders were removed, and replaced by the gate and tray type, almost all trams had platform beams shortened around 15 cms, saving 30 cms per car linear parking space; it is not clear if this is related or not, but no other reason for cutting and shortening most of the trams has been found.

In 1988 Massarelos replaced Boavista as a running depot, but due to road works the trams returned to Boavista three times: March - July 1991, June - November 1998 and February - May 1999. In August 1999 Boavista was finally completely evacuated and demolished. The Casa da Música (Music Hall) was built on the site.

Staff

A tram company is more than rails, trams and overhead wires. It is also the people working to drive the trams, collect the money, doing maintenance etc. Porto trams had most of the time a crew of two: driver and conductor. But as there was for many years an almost continuous stream of trams going up from Praça to Batalha and Carmo, many people, especially in the afternoon used the trams just for these short but steep stretches. To be sure that all passengers paid, auxiliary conductors were used on these short sections during the afternoon. So then the small, 2-axle trams had a crew of three. Trailers had often a staff of two too. On all trailers had to be a brakesman, except if the tram+trailer combination was equipped with air-brakes, devices that were not installed on all vehicles. Zorras had a crew of two too. Except for the driver there was an auxiliary who had to take care for changing the points and turning the trolley-pole. With passenger trams that was the responsibility of the conductor, although often a passenger took the points-bar and mucked in. Only freight trailers could do with only a brakesman.

Incidents

Of course the Porto trams suffered accidents. Safety was not yet on the level as nowadays. The number of accidents that occurred was so high that few people now can imagine, but It wasn't much different than with other contemporary tram companies or with factories, construction works, etc. To current standards, a company suffering that much (serious) accidents as was common 100 years ago, would be immediately closed by the authorities. Fatalities were not rare. In the period that ambulances did not yet exist, trams were used to bring people involved in an accident to the hospital. It is a sad part of the history and it is impossible to recall the thousands of accidents that happened with the Porto trams since its beginning in the 1870s. But there were some serious accidents to be remembered.

On 13 October 1911 there was a head on collision between two steam trams on the single track section between Rua de Gondarém and Castelo do Queijo. The weather was bad and the trams were running late. A tram coming from Matosinhos and hauled by locomotive 7 (the Merryweather) didn't wait in the passing loop at Castelo do Queijo for the crossing with the one in the opposite direction, but continued with maximum speed. In the curve where the line turned into Rua de Gondarém and vegetation obstructed the view, it smashed into the oncoming tram hauled by locomotive 3 (a Henschel). Driver and fireman of locomotive 7 were killed. The driver and firemen of the other locomotive, three conductors and several passengers were injured, some of them seriously. Both locomotives were heavily damaged and because of that the service on the line had to be reduced from every 20 minutes to every 30 minutes until both locomotives were repaired.

Two months later, on 10 December 1911 a real disaster occurred. Carro Inglês 203 was on the way from Leça to Praça with two trailers but had a technical failure. The tram was replaced at Massarelos by 150, a Constructora tram. Apparently the driver was in a hurry because of the delay and probably drove too fast while heavy rainfall had caused the grooves of the rails to be filled with mud. The tram derailed in a sharp curve at Monchique and the motorcar went with the first trailer into the river. The brakesman of the second trailer managed to keep his car on the quay. 14 people were killed and 31 injured. Tram 150 and the remains of the trailer were recovered from the water. The electric tram was repaired, but the trailer scrapped.

On 18 September 1955, at about 11.10, Pipizinho 405 was on its way up in Rua da Restauração when near Rua Jorge Viterbo Fereira the wheels slipped and the tram slid down with increasing speed in spite of attempts of the crew to stop it. At the curve the tram derailed and crashed. A young mother and her four week old baby died. Fifteen other passengers and two STCP staff were injured. The wreckage of the tram was scrapped.

Not all incidents were sad though. Twice (at least) a baby was born in a Porto tram.

On 12 September 1916, at 04.30 at Carmo a baby was born in tram 276, a Brill, on its way from Boavista to commence service on line 9. On 13 May 1936, at 07.40 on Praça Carlos Alberto a baby was born in tram 311, a Fumista, which at that moment was in service on line 6 and on its way to Praça.

Heritage Tram Operations

In 1996 the operation of the only remaining tram-line 18 was changed. The frequency decreased from every 15 minutes (with seven cars) to every 35 minutes (with three cars). Also the working times were reduced: 09.00 – 19.00. A positive step was the re-introduction of working of the tram on Sundays. In this way the last of the classic tramlines was converted to the first line with heritage characteristics. In 1998 the river bank between Infante and Massarelos was reconstructed, including the installation of new tracks and overhead wires. Trams had disappeared on this route in 1994 with the closure of the last part of line 1, but now returned early in 1999.

Due to construction at many locations (tunnel near the Hospital de Santo António, Rua da Restauração, Castelo do Queijo) services were limited further. Line 18 was closed between Massarelos and Carmo for this in July 1999 and on 29 May 2000 with the route Boavista - Massarelos closure the last old tracks went out of public service. Only Infante - Massarelos remained, newly rebuilt. This enabled STCP to state that the tramway had never closed. On 15 February 2001 line 18 was reintroduced between Massarelos and Viriato using new tracks. On 11 March 2002 line 1E was extended to Foz (Cantareira), using new tracks on the reconstructed riverside. Line 18 was extended in December 2005 from Viriato to Carmo again. On the 21 September 2007 a new line with the number 22 was opened in the centre. Early in 2009 line 1 was extended in Foz from Cantareira to Passeio Alegre.

The Centenary of the tramway in Porto in 1995 had been preceded by a major contraction of the system to the extent that there had been some doubt that the tramway would survive until the actual ceremony. However a visit to Masserelos in 1997 revealed that they still had a maintenance staff of 35. This didn't look like a tramway that was about to close. When the majority of route 18 was abandoned in 2000, leaving only the short (but newly rebuilt) section between Massarelos and Infante, the management were insistent that the tramway would expand again.

It was when Porto was chosen as the City of Culture in 2001 that the tramway really regained its foothold in the city centre, although it was to be some years before the investment was to be realised. It appeared that cash for the rebuilding of the tramway infrastructure came directly from the budget of the "Porto 2001 City of Culture" event.

The first fruit of the project was the rebuilding of route 18 which climbed Rua da Restauração, and which brought the tramway back up the hill into town at Viriato, albeit a fair distance from the historic centre.

By this time, conductors on the trams had been dispensed with, the universal S.T.C.P. "Andante" ticket and its equipment having taken their place. Tram passengers now entered at the front and passed their pre-paid blue Andante tickets across a reader affixed on the platform within sight and hearing of the operator, a valid ticket emitting a bleep and triggering a green light. An invalid ticket showed red and alerted the operator with a different sound.

However the cars were not fitted with track brakes, and the steeply-graded 18 route passed in the shadow of a line of sycamore trees, which could cause difficult rail conditions at certain times of the year. In addition to cars losing adhesion, unable to pull away from the traffic lights in the middle of the street, they had been known to run back down the grade with wheels slipping, not being able to drop sand behind them. As it was planned to operate trams on streets with even steeper grades, a technical solution had to be found. Initially, a spare motorman would ride on the car acting as emergency brakesman on the rear platform by activating the sander there.

The Faculty of Engineering of Porto University designed an entirely new braking system whereby the cars were equipped with a static converter mounted on the roof of the car producing 24 volts DC from the line current. This then powered a pair of magnetic track brakes, one above each rail, mounted on a custom-built frame fixed to the truck. A new console was provided in each cab, activated by a key, and the track brake applied by depressing a button, with a foot pedal as an alternative. The action of these powerful magnets not only brought the car to a swift standstill, irrespective of gradient or rail conditions, but had the distinct advantage of holding the car stationary, as opposed to conventional track brakes, which lose their grip as soon as the wheels stop turning.

The new braking system naturally produced greater forces on the car body, and attention turned to the condition of the cars themselves. The remaining serviceable trams were all faithful copies of the 1900 Brill semi-convertible design, some being original Brill cars. These have many substantial window pillars, designed so that the window frames can slide up into the car ceiling, and extend right up into the roof, producing a particularly heavy and robust construction. The cars weigh around 13 tons, almost as much as some British double deckers. Over the years, timber depreciation in the bodies had been routinely patched up; however this was costly and time-consuming. A new plan was needed.

It was decided to refurbish individual cars by building brand new bodies for them to the original design (albeit with minor alterations), mounted on new steel underframes. The cars to be so treated would be stripped of their electrical equipment, seats, interior fittings, window frames and ornate timber facings for eventual fitting in the new vehicle. The clerestory roof would also be removed and retained, as would the truck and lifeguards. The remains were scrapped. The new bodies were constructed entirely in house at Massarelos using the existing workforce and joinery equipment, and using patterns which had existed since the cars were built. As such, they should not be considered replicas, but a continuation of the tradition of turning out car bodies of the original series.

The trucks, however, were reconditioned by EMEF (the main railway engineering contractor) at Custoias, with new axles, broader wheel tyres and new axle boxes complete with roller bearings. In some cases, new pinions and gears were also fitted. When the trams were assembled, with their original interiors and trimmings, and put on the road, the travelling public were completely unaware that they were travelling in new cars.

As part of the Porto 2001 City of Culture, rails were laid through the Parque da Cordoaria, including pointwork for a city centre loop, together with a further isolated section, including a passing loop, in Rua de Passos Manuel. After some years of stagnation due to road repairs, the 18 was eventually extended past the Hospital Santo António, linked up with the existing track in the park, continued round the hospital and back on itself at Rua da Restauração, forming a large loop.

More rails appeared in Praça da Batalha leading down to the newly reopened funicular at Guindais. However the work appeared to stop after the 2001 event came to a close with a lack of cash and political apathy being suggested as the reason. Delays to construction of an underground car park at Carmo were also responsible. Substantial sections of track had been laid, traction poles installed and overhead wire strung, but there were large gaps. It is said that the city fathers took a close look at the situation, and decided that it would actually cost more to rip up the existing track and reinstate the roadway (considered then as a serious option) than to complete the project, so once again tracklaying began in Porto city centre. In the summer of 2007, the final gap in Praça de Liberdade linking Rua dos Clerigos and 31 Janeiro was closed, in conjunction with metro works immediately underneath. On 22 September 2007, the historic city circle line 22 was inaugurated, and immediately proved popular with local residents and tourist alike. In 2009 a special tourist tramline "T" was introduced on the route Infante - Massarelos - Cordoaria - Batalha - Carmo - Massarelos - Infante. However prices were too high and frequency too low to make this a success and the service was suspended in June 2011.

The tram-services as at 2021:

1 - Infante - Alfândega - Massarelos - Foz (Passeio Alegre)
18 - Carmo - Viriato - Massarelos
22 - Batalha - Carmo

The basic frequency was every 30 minutes on all lines during daytime with two trams working on line 1 and one each on the lines 18 and 22. The service on line 1 is every is every 20 minutes using three cars. Start and end of the services depend on the season. Due to the availability of drivers services can sometimes be less or more as announced.

Today there are 5 overhauled trams, and they all went through the same major reconstruction process: 131, 205, 213, 218 and 220. 4 trams are necessary for regular operation and besides the 5 new cars, 216 is also available and fitted with track brakes, the 600V/24V converter, new tires and roller bearings. Rental fleet 2 axle trams (143,191,203) and Brill bogie 275 were not overhauled, but (except 275) received the static converter and the track brakes. 287, a 4 axle Belgian car was also retrofitted with a static converter and track brakes, although it was not used in regular service after the installation of the track brakes. 287 was later overhauled and put in service again in May 2017. In the overhaul process 221 was scrapped due to financial constraints.

Extending the line from Infante to Praça via Rua Mousinho da Silveira was also part of the intended Millennium network, but due to lack of money for rail investments after 2009, STCP decided not to install the tracks, preventing the return of bogie trams to Praça. Current aspirations include reopening line 1 to Matosinhos following the construction of a cruise liner terminal.

From Castelo do Queijo to Rotunda da Boavista and to Matosinhos (Av da Republica) and physically connecting with Metro line A, plans were also abandoned, although rails were installed, with a pair of points in the Metro line. Following construction work at Castelo do Queijo, rails were installed, but not used. History repeats itself.

Museum

Since the 1960s a few trams were earmarked to become part of the collection of a future museum.

When trailer 8 was saved from scrapping in 1964, no one could have guessed the future would offer one of the most fascinating operating Tram Museums in the world, as all vehicles can be used as when they were new and constructed for service. It's not clear why STCP decided in 1964 to keep five window Starbuck built trailer 8 while selling eight window 9 and scrapping 10; probably because of the influence of John Price, on his second visit to Porto to buy trailer number 9 for the National Tramway Museum at Crich.

STCP Museum didn't have premises, but these trams and trailers could be seen and photographed in Massarelos or in

Boavista depot, or even hired (at least 373). By the late 80s with the closure of the Boavista depot for tram operation, and with only a couple of lines in operation, one of the Board members was fond of the idea of using those cars and a few others of the now surplus ones, to preserve for posterity a unique legacy. Initial Museum plans suggested a joint venture with other museums at Alfândega, but insufficient progress led to a plan involving the use of Boavista depot, lines 1 to 7, to display the tram collection, and including some buses. Due to legal problems however, Boavista depot was vacated in August 1999 after the public works decision to build Casa da Musica (The Music Hall) on the site. Massarelos main building, in the former boiler room, since 1960/61 used for the permanent way department, was chosen instead; in the beginning only the first half of the space, and later the entire area when the quantity of cars increased. And on 18 May 1992, Museums day, Porto Tram Museum was officially inaugurated. Until 1997 trams were methodically overhauled, and today's Museum's roster includes 25 vehicles.

Once a year the museum organises a parade with museum trams. This is likely to occur on the first Saturday in May, or failing that, the second Saturday in May.

Tickets

On the Metro do Porto, the buses of STCP and several private operators the intermodal Andante system is used. This system is based on zones. Tickets can be purchased from vending machines on all metro stations and stops. For tourists there are 24 and 72 hour tickets available valid on the whole network where the Andante is valid. These tourist tickets can be bought at counters of several ticket selling points like in the São Bento railway station. However the Andante tickets are not valid on the heritage trams or the funicular. On the trams single trip tickets are sold by the driver as well as 24 hour tickets only valid on the trams, including admission to the museum. Single funicular tickets are sold at the top and bottom stations.

220 is seen in 2017 travelling along Rua do Cais das Pedras passing the Igreja do Corpo Santo de Massarelos (Parish church of Massarelos) which dates from 1394. *K. Lomas King*

213 seen at Carmo on 19th June 2008. *L. Folkard*

The annual parade of historic tramcars from the Porto museum never fails to provide an impressive spectacle, evoking the great days of tramways. Ten of the cars here proceed in stately fashion along Cais das Pedras during the cavalcade of 4 May 2019. *Michael Russell*

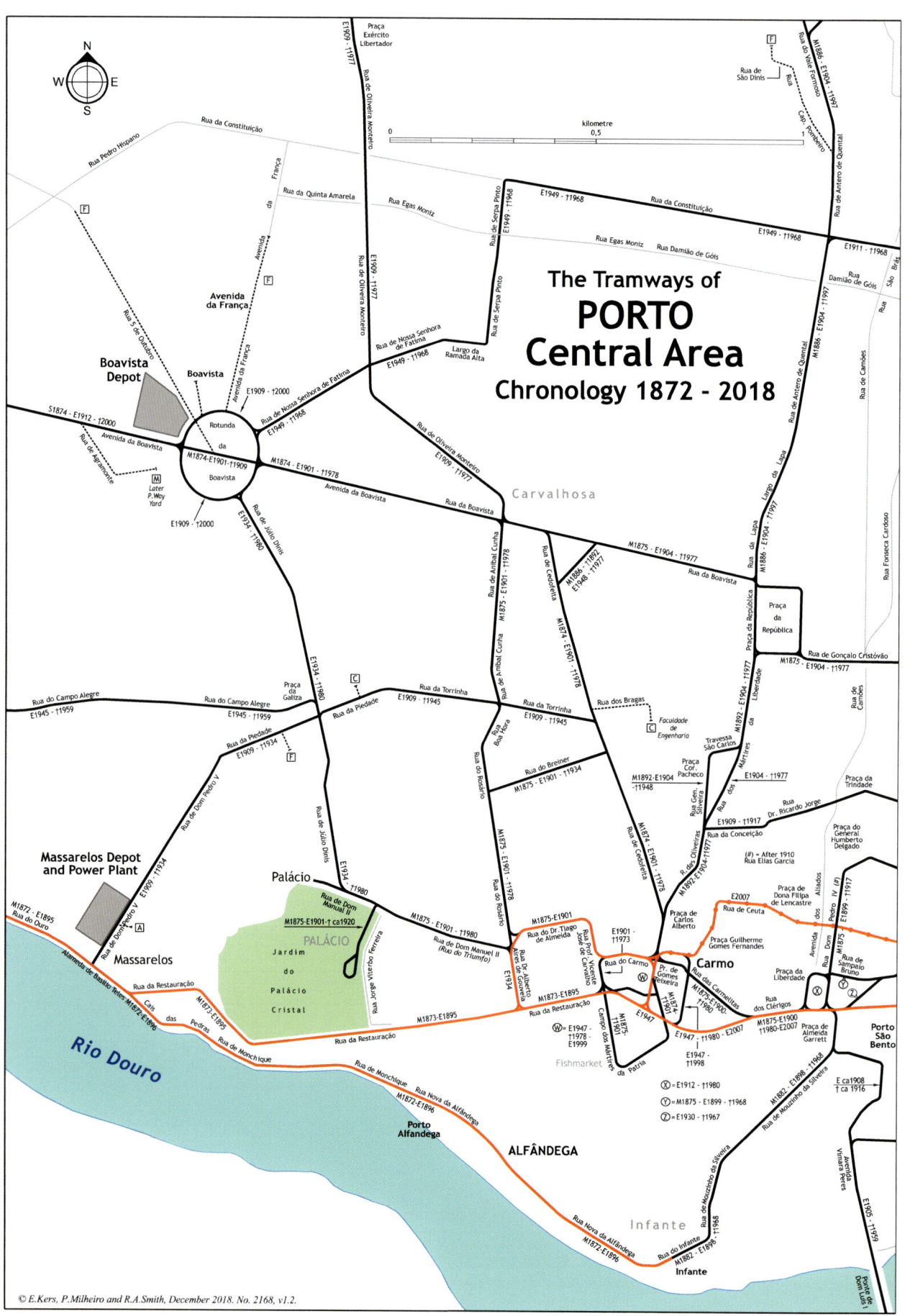

The Tramways of
PORTO
Central Area
Chronology 1872 - 2018

© E.Kers, P.Milheiro and R.A.Smith, December 2018. No. 2168, v1.2.

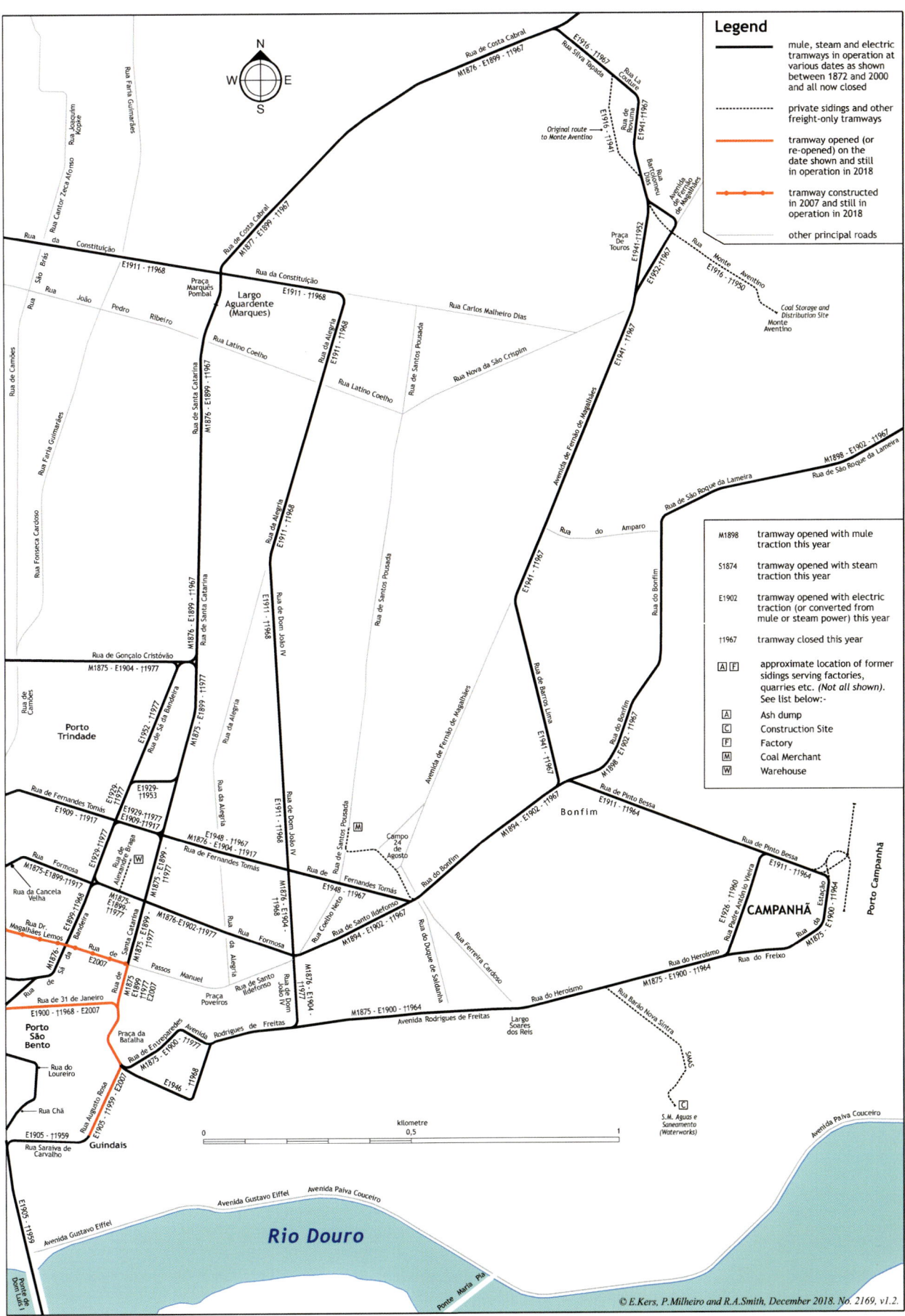

Legend

mule, steam and electric tramways in operation at various dates as shown between 1872 and 2000 and all now closed

private sidings and other freight-only tramways

tramway opened (or re-opened) on the date shown and still in operation in 2018

tramway constructed in 2007 and still in operation in 2018

other principal roads

M1898	tramway opened with mule traction this year
S1874	tramway opened with steam traction this year
E1902	tramway opened with electric traction (or converted from mule or steam power) this year
†1967	tramway closed this year
A F	approximate location of former sidings serving factories, quarries etc. *(Not all shown).* See list below:-
A	Ash dump
C	Construction Site
F	Factory
M	Coal Merchant
W	Warehouse

© E.Kers, P.Milheiro and R.A.Smith, December 2018. No. 2169, v1.2.

The Tramways of
PORTO
Chronology 1872 - 2018

Companhia Carris de Ferro do Porto (CCFP) (and predecessors)
until 30 June 1946
Serviço de Transportes Colectivos do Porto (STCP) from 1 July 1946

N
W E
S

0   0,5   1   2   3
kilometres
scale exaggerated at junctions and loops

For Matosinhos area
see separate map

Note A
Original route prior
to harbour extension
in 1938

Leça da
Palmeira
Praia
da Leça
Castelo
Av. do Dr. A. Guimarães
See Note A
Matosinhos
Rua Roberto Ivens
Rua de Brito
Rua do Godinho
Porto de
Leixões
Doca
Mercado
Prado
Avenida Meneres
Capelo

Praça Cidade
do Salvador

M&S1882 - S1888 - †1914
M1872-E1897-†1993
Castelo
de Queijo

M1872-E1897-†2000
Avenida da Boavista
E1914 - †2000
Avenida de Montevideu
Avenida do Brasil

Pereiró
Campinas
Avenida Dr. Antunes Guimarães
E1947 - †1984

Fonte da
Moura
Avenida da Boavista
M1874 - M&S1878 - S1888 - E1912 - †2000
M1874 - M&S1878-
S1888- †1914
Avenida da Boavista

M&S1882 - S1888
- †1914
M1874 - M&S1878-
S1888- †1914

M1872-E1897-†2000

Foz do
Douro
Cadouços

Rua da
Senhora
da Luz
Esplanada
do Castelo
Foz
Rua do Passeio

E1934-†2000
M1872-E1897-†2000

Lordelo
Rua do Campo Alegre
E1945 - †1959
Alegre

Ouro
M1872-E1897
Alegre
Rua de Sobreiras
Arrábida
M1872 - E1896
Rua do Ouro
Rua do Ouro
M1872-E1895

See
Central Area
Map

Rio Douro

Oceano
Atlântico

Ponte
de
Pedra
Rua Godinho de Faria
E1912 - †1975
Rua de Silva Brinco
E1910 - †1975
São Mamede
de Infesta
Rua de Amial
E1910 - †1975
Amial
Sub-station
No.2
Rua da Telheira
E1910 - †1977
Amial
Rua do Campo Lindo
E1910 - †1997
Paranhos
M1886-
E1904-
†1976

E1957-†1976
Monte dos
Burgos
Rua Monte dos Burgos Carvalhido
E1911 - †1976
Rua do Vale Formoso
Carvalhido

E1905 - †ca1908/1913
General
Torres
Avenida da República
E1905 - †1959
Rua Álvares
Cabral
Rua
Dr. Francisco
Sá Carneiro
Devesas
E1905 - †1959
Rua Conselheiro Veloso
da Cruz
Arco do
Prado
E1913 - †1959
Rua do Barão
do Corvo
E1947-†1959
Coimbrões
Vila
Nova
de
Gaia
Jardim
Soares
Reis
E1905 - †1993
Rua de Dr. Soares dos Reis
Santo Ovídio

© E.Kers and R.A.Smith, December 2018. No. 2166, v1.1.

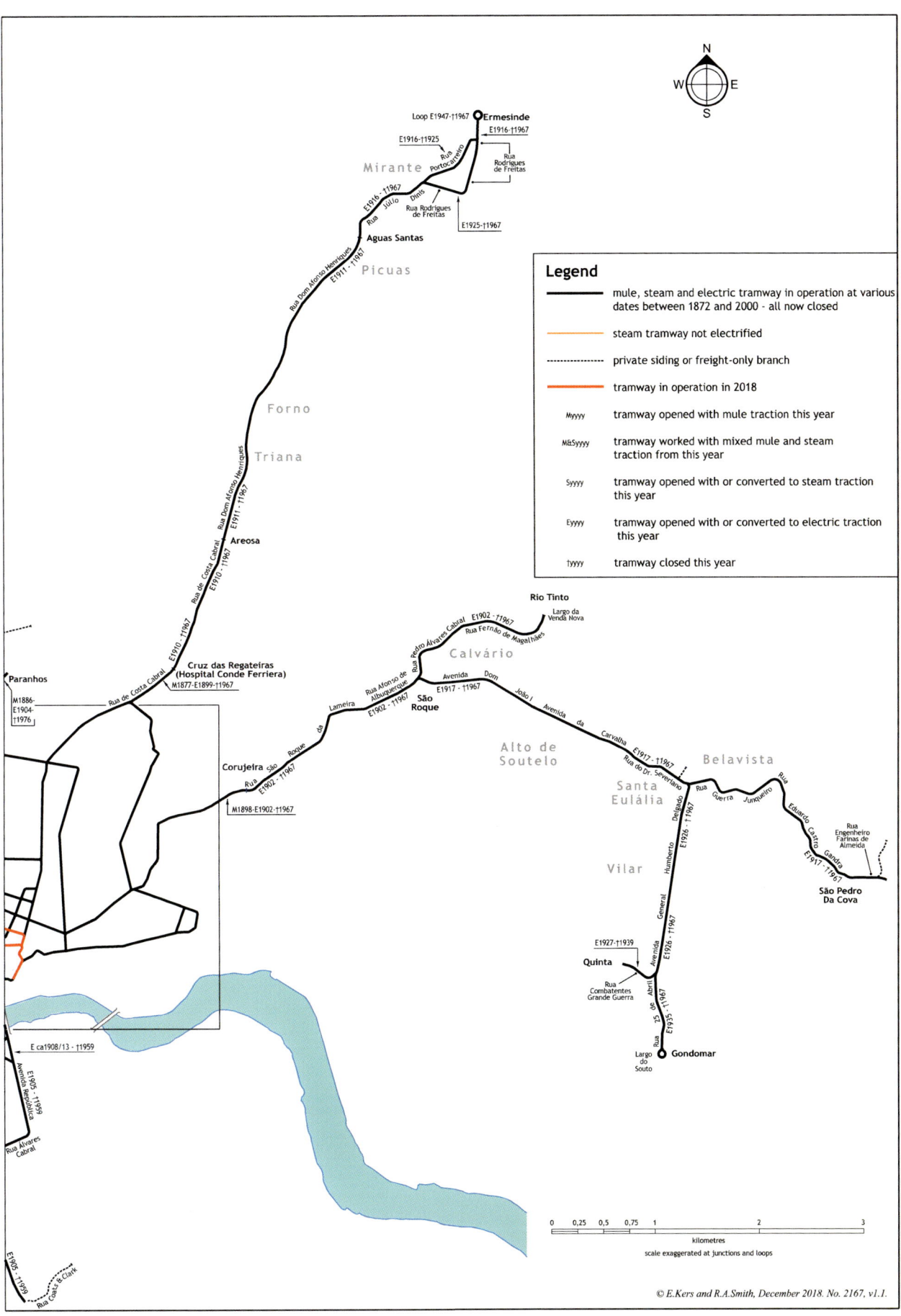

Legend

mule, steam and electric tramway in operation at various dates between 1872 and 2000 - all now closed

steam tramway not electrified

private siding or freight-only branch

tramway in operation in 2018

Myyyy — tramway opened with mule traction this year

M&Syyyy — tramway worked with mixed mule and steam traction from this year

Syyyy — tramway opened with or converted to steam traction this year

Eyyyy — tramway opened with or converted to electric traction this year

†yyyy — tramway closed this year

Loop E1947-†1967   **Ermesinde**

E1916-†1967

E1916-†1925

Rua Portocarreiro

M i r a n t e

Rua Rodrigues de Freitas

E1916 · †1967

Rua Júlio Dinis

Rua Rodrigues de Freitas

E1925-†1967

Aguas Santas

P i c u a s

E1911 · †1967

Rua Dom Afonso Henriques

F o r n o

T r i a n a

Rua Dom Afonso Henriques

E1911 · †1967

Rua de Costa Cabral

E1910 · †1967

Areosa

Paranhos

M1886-
E1904-
†1976

Rua de Costa Cabral   E1910 · †1967

Cruz das Regateiras (Hospital Conde Ferriera)

M1877-E1899-†1967

Rua Afonso de Albuquerque

E1902 · †1967

Lameira

São Roque

Corujeira

Rua São Roque do

E1902 · †1967

M1898-E1902-†1967

Rio Tinto

Largo da Venda Nova

Rua Pedro Álvares Cabral   E1902 · †1967

Rua Fernão de Magalhães

C a l v á r i o

Avenida Dom

E1917 · †1967

João I

Avenida da

Carvalha

E1917 · †1967

A l t o   d e   S o u t e l o

Rua do Dr. Severiano

Belavista

Rua

Guerra

Junqueiro

Rua

S a n t a
E u l á l i a

Humberto

Delgado

E1926 · †1967

Rua
Eduardo
Castro
Gandra

E1917 · †1967

Rua Engenheiro Farinas de Almeida

V i l a r

São Pedro Da Cova

E1927-†1939

Quinta

Rua Combatentes Grande Guerra

Avenida General

E1926 · †1967

Rua 25 de Abril   E1935 · †1967

Largo do Souto   **Gondomar**

E ca1908/13 · †1959

E1905 · †1959

Avenida Republica

Rua Álvares Cabral

E1905 · †1959

Rua Costa & Clark

N

W   E

S

0   0,25   0,5   0,75   1   2   3

kilometres

scale exaggerated at junctions and loops

© E.Kers and R.A.Smith, December 2018. No. 2167, v1.1.

Above: The original builder's plate on car 191 indicating that this car was constructed by CCFP in 1927. *K. Lomas King*

Left: In the time honoured fashion the driver of Porto car 216 changes the stencil on the front of the car before commencing the next journey. *K. Lomas King*

Below: The Porto service cars are aided by modern technology in the form of a display that shows not only the route but the actual position of the car which is helpful as most of the routes are single track and passing loops. *K. Lomas King*

Gallery 4

Mule, Steam and Early Cars

Mule tram built by Starbuck for the CCFP.

Science Museum London - J. R. Stevens

Mule tram built by Starbuck for the CCFP. *Science Museum London - J. R. Stevens*

Left: Trailer 10 around 1950 in front of the Boavista depot. This former mule tram is thought to have originated from the Carril Americano. The platforms were probably extended a little in the first decade of the 20th century after mule tram operations ended. *MCE*

Below: Henschel locomotive no.5 at Cadouços while on its way to Matosinhos. The first carriage probably originated from the CCAPFM and had seats on the roof. The second carriage supposedly came from the Larmanjat company near Lisbon and was reconstructed when bought by the CCFP. The last carriage is an open car with longitudinal benches. The buildings in the photo still exist. The photo was made by Aurelio Paz dos Reis probably around 1902. *CPF.*

Left: This was one of the first three mule trams transformed into electric trams in 1895. Here it is at one of the most iconic locations of Porto, in front of the Igrejas dos Carmelitas and do Carmo. This photo of CCFP 3 was made by Aurelio Paz dos Reis probably around 1902. *CPF.*

The trams 33-34 were built in 1902 by A Constructora and had motors and controllers by Schuckert & co. Number 34 was that year involved in a runaway accident in Rua de Santo António, which prompted questions about the brakes. Both trams disappeared from the fleet, at least with the numbers 33-34, but later returned probably with new trucks, motors, brakes and the numbers 55-56. *CPF*

This works photo is of the tramcar delivered in 1898 by Siemens & Halske to the CCFP. On the photo it has the number 1, but in Porto it received the number 14 and was later renumbered to 26, and then 126. The bow collector was replaced by a trolley-pole. Until the arrival of the Pipis it was the only tram in Porto with two head-lights. Typical for Siemens & Halske at the time were the controllers, which were mounted outside the dashes. The tram appears to have a truck made by BSI. *Siemens Corporate Archive*

This was a former mule tram transformed into an electric tram in 1903. While most of the previous transformed trams probably all originated from the CCFP, this type is believed to originate from the CCAPFM. This location is now called Avenida de Montevideu. The building on the right still exists. The photo was made by Aurelio Paz dos Reis probably in 1903. *CPF.*

Right: In 1904 A Constructora transformed two open mule trams to electric trams. They had the numbers 41-42, from 1907 141-142 and from 1924 or 1925 100 and 115. It is believed that both cars were lost in the Boavista depot fire of February 1928. This photo of open tram 41 was taken at Infante by Aurelio Paz dos Reis around 1905 (1904-1907). *CPF.*

Left: A Transformado with two trailers on the way to Matosinhos. This location is where the current route 1 terminates. The last carriage is believed to be a former Larmanjat car. The photo was made by Aurelio Paz dos Reis probably around 1902. *CPF.*

This was one of the three mules trams transformed in 1896 to an electric tram. It pulls two trailers, the first an open cross-bench and the second probably a former Larmanjat car. This photo of CCFP 4 was taken by Aurelio Paz dos Reis around 1902. *CPF*

PORTO

Matosinhos

By Pedro Milheiro and Ernst Kers

At Castelo da Leça the original bogie car is about to haul two former Sintra Laramanjat trailers converted to standard gauge across the river Leça to Matosinhos. The metre gauge railway bridge can be seen, with the tramway bridge just visible in the far left. *B. Lennox*

Legend

Comparing old maps with the present situation, it's clear the coastline between Castelo do Queijo and Circunvalação has changed. It is assumed that the construction of the Leixões moles and/or the later extension of the northern mole affected the currents and a small part of Portugal was lost to the Atlantic Ocean because of that. As a result when drawn on a current map it seems the trams ran through the water.

Tramlines

CCAPFM opened its mule tramline in 1872. At first the terminus was indicated in the newspapers as Leça, but soon it was Matosinhos. It is thought that it was at the edge of the Leça river at the northern end of what is now Rua Brito Capelo. In those days it was called Rua Juncal da Cima.

In 1882 CCFP opened its tramline through Rua Juncal da Baixa, now known as Rua Roberto Ivens. Its terminus was at the northern end of Rua de Sto. Amaro, a very narrow street close to the river. CCFP built a wooden bridge across the river, and had ideas to extend the line over it. They probably concluded that it wasn't a good idea with steam locomotives as traction. Using mules would have meant extra costs to allocate some animals and staff to take care of them near the bridge.

CCAPFM probably didn't like the fact that it was easier for the residents of Leça to use the trams of CCFP instead of theirs. So they ordered the Willebroeck company to build an iron bridge and extended their line across it. That happened in 1887. At about the same period Willebroeck also built the Ponte Luis I

bridge across the Douro, which still exists. The new terminus of CCAPFM was near the northern end of the wooden bridge of CCFP.

In 1893 CCFP took over CCAPFM and the lines of the latter were the first to be electrified. The electrification was completed to the terminus in December 1897, but the line was extended to Castelo da Leça early in 1898. There are several photos of trams next to Castelo da Leça.

When the steam tram operation finished, CCFP requested double track in Rua Brito Capelo. They said it was impossible to operate two tramlines, 1 and the new line 5, through Brito Capelo where there was single track. However the Câmara Municipal de Matosinhos refused permission. Instead the route in Rua Roberto Ivens was retained, now reached from the other tramline through Rua Sousa Aroso. Only in 1930 was CCFP allowed to install double track over the whole of Rua Brito Capelo and line 5 was shifted to it. The route in Rua Roberto Ivens was abandoned that year. The wooden bridge had already disappeared.

In the late 1930's the excavation of Dock 1 started. This had huge effects on the route of the tramline in that area. The iron bridge disappeared, to be replaced by a culvert to let the water of the river flow to the dock. It is not certain that the tramline was retained exactly on the same place as the old bridge or shifted a few metres to one of the sides. At the north side of the river/dock the route was shifted to the north and in 1941 the line was extended until close to Rua Santos Lessa. Here a turning loop was made, probably to avoid shunting with trailers. It is

Rebocador eléctrico with a train loaded with wine barrels is ready to leave the site of the Companhia Vinicola Portugueza in Matosinhos around 1900. It seems there is only one controller and the driver has to stand in the centre of the cab looking sideways. Probably there were two of these vehicles, soon reconstructed into closed freight cars with platforms and controllers at each end. *Source unknown*

10·C—LEIXÕES—Castello de Leça

One of the five tramcars built by the UEC for Porto next to the Castelo da Leça. Apparently the tram is moving around its trailer which should be standing at the terminus just outside the view at the right. The tram doesn't have a line number yet, which means that the photo was taken in 1910 or 1911. Next to the tram tracks are the tracks of the narrow gauge railway with some freight wagons. *Commercial postcard.*

Mattosinhos—Ponte sobre o Rio Leça

31—Editor—Alberto Ferreira—Batalha—Porto

Electric tram 18 crosses the iron bridge across the river Leça built in 1888 for the CCAPFM. This bridge was in use until 1938 when it was demolished because of the construction of Doca 1 of the Leixões harbour. The current bridge across the docks is more or less on this same location. The photo was taken around 1903 *(1899-1907) Commercial postcard.*

284 is seen arriving at Matosinhos on 29 November 1970 against a backdrop of dockyard cranes and the lifting bridge partially open on the right. *J. Meredith*

possible that the route at the southern end of the former iron bridge was changed a bit too.

Around 1960 the excavation of Dock 2 commenced and the tram lines were cut back to the south side of the docks. Apparently operating tramlines across the new movable bridge wasn't considered feasible. For the new terminus stub tracks were laid next to the fish market. In 1993 the tramline to Matosinhos was closed.

Branches for freight

Apart from the routes used by the passenger trams, there were also tracks only used by freight trams. This part is more enigmatic. There are maps which show situations that existed. But there are also maps which show situations which were only studied, but never realised. And finally there are maps of which it is not known whether it was realised or not. Moreover on most maps no date is given, so one has to estimate that from data shown on the map.

It is fairly certain that everything shown on the map existed. Going from north to south:
- From Castelo da Leça there was a spur in the direction of the northern mole. This spur is also visible on photos. It probably ended at the base of the mole. When the tramline was extended in 1941 this spur had disappeared already.
- There was a spur to the southern mole through Rua Godinho. It is thought that it ended at the base of the mole. Possibly this spur branched originally off the line in Roberto Ivens. But it might also have had an original route via Av. República - Av. Serpa Pinto. There is a map showing a proposal to reroute line 5 through Av. Serpa Pinto which shows that situation.

At the crossing with Av. Menéres were spurs leading to three destinations.
- A spur to the east for the Vinicola (Menéres) company. Probably there was on the site of Menéres a more complex layout, but no drawings have been found. Probably this connection was made in 1898. Before the electrification the tram companies operated freight trams, but for the transport of wine from Menéres two rebocadores (electric towing cars) and a fleet of zorra trailers were constructed.
- A spur at the west side to the new build Fabrica de Conservas. It is thought that this spur was constructed in 1899.
- A spur branching off the one to the Fabrica de Conservas to the Fabrica de Brandão, Gomes e Ca. It is thought that this spur first branched off from the tramline through Rua Roberto Ivens and was later connected as described to the tramline through Rua Brito Capelo.
- Near the street crossing with Rua Sousa Aroso was a spur to the east into the premises of Bento José da Cunha. These were on the north-east corner of Rua Brito Capelo with Rua Sousa Aroso. It is thought that the spur was made in 1899.
- A map dated 1884 shows tracks branching off from both the mule tramline of the CCAPFM and the tramline of CCFP parallel to the tracks of the narrow gauge railway, to interchange goods between the latter and the tramways. The one branching off from Brito Capelo still existed in 1913; the other one had probably gone by that time. Both tracks were gone in the years before 1930.
- A map probably made in the 1940s, shows a spur from Circunvalação along the coast in northern direction. It was meant for the transport of ash from Massarelos to a terrain where an

Car 288 is in Matosinhos not far from the terminus in 1978. The diagonal car parking often caused delay to the trams. *Michael Russell*

Car 271 passes the junction with Rua de Sousa Arosa in 1988. *Michael Russell*

Seven window car 216 is seen in this busy scene at the Matosinhos terminus on 29 November 1970. *J. Meredith*

"Avenida" was to be constructed between Circunvalação and Praia de Matosinhos. It was probably not constructed, but maybe the line existed but the ash disappeared in the ocean. It is not shown on our map.

Circunvalação is the border between Matosinhos and Porto. But between Circunvalação and Castelo do Queijo were also spurs for the transport of sand, using a part of the old steam tramline, and for the transport of ash.

Narrow gauge railway

At the east, in Senhora da Hora the NG railway to Póvoa de Varzim was constructed in 1875. Although the line itself is of course very interesting, the map does not provide any additional information except that the station of Senhora da Hora shifted from its original location, where now MP line A branches off from the trunk route, to the south in 1900.

More can be said about the quarry line. This line was to transport stone from the quarry in São Gens, east of the NG railway to Póvoa, to the coast for the construction of the moles. No date is known for the commencement of construction of

this quarry line, but as construction of the moles started in 1884, it's probably that year. When the moles were finished in the early 1890's, the line became available for commercial freight and public transport. For maintenance stone still had to be transported, so the connection to São Gens was retained.

The NG railway had its own bridge across the river Leça made of iron with the use of stone pillars. The railway went up to both moles. Normal practice (a movie exists showing this) was that the Titan crane lifted a wagon loaded with stone, manoeuvred it above the spot of the stones destination and then the wagon was tipped over to empty it. After that the wagon was put back on the track.

When the excavation of Doca I started, the bridge had to be removed. The line was cut back to Matosinhos. But it is thought that new tracks were laid around the dock to retain the possibility to transport stone to the northern mole as well as to make the transport of goods to and from the new build quays possible. However in 1938 the broad gauge Linha de Leixões was opened, connecting to the national network in Contumil and Ermesinde and strongly reducing the amount of goods transport by the NG railway.

Metro car 019 at Rua Brito Capelo in Matosinhos on 1 May 2014. Note the lowering of the track at the stop to the rear of the car. *C. F. Isgar*

Metro car 003 at Senhor de Matosinhos terminus 2 May 2014. *C. F. Isgar*

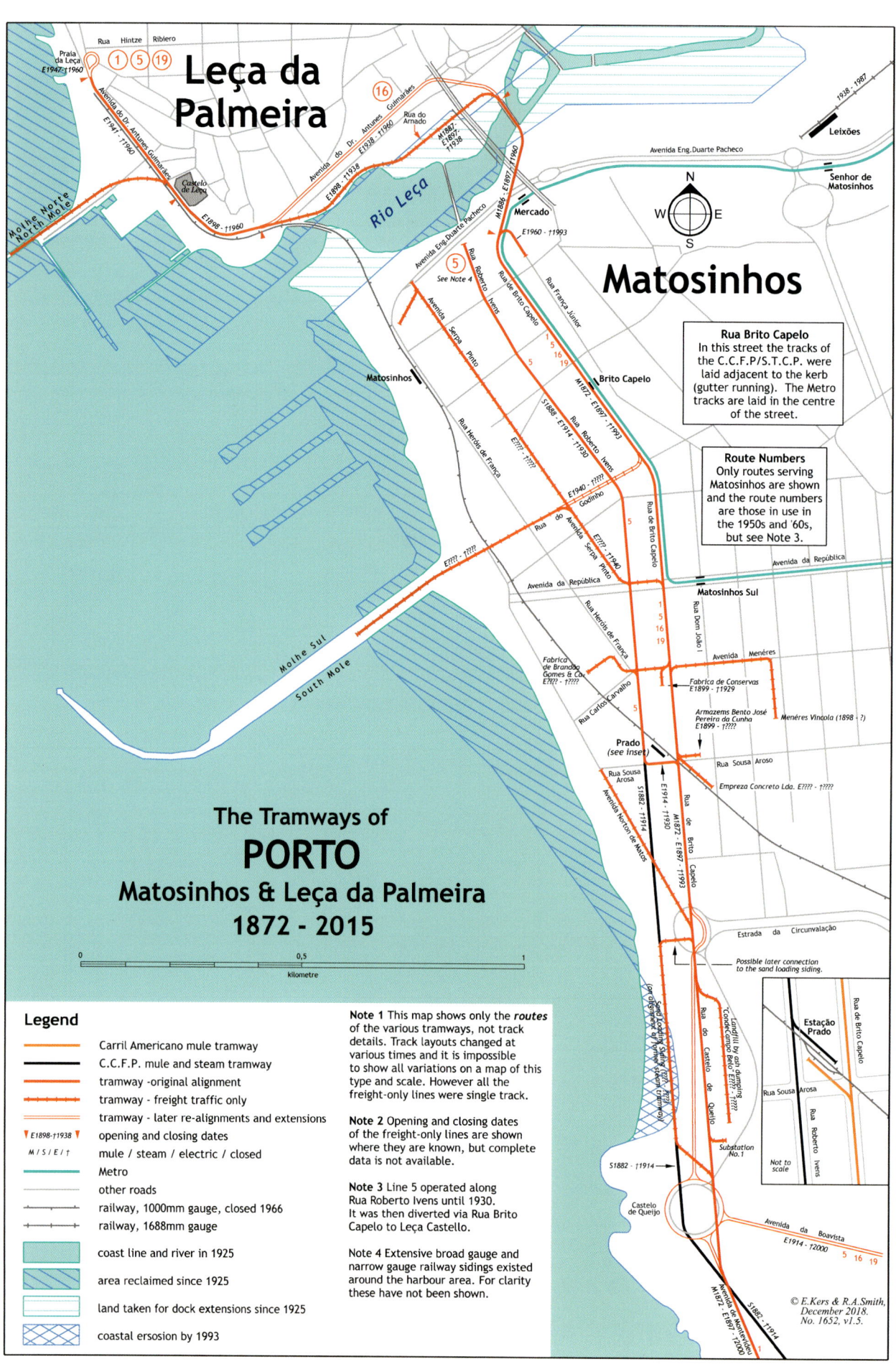

PORTO

Trolleybuses

By Ernst Kers and Pedro Milheiro

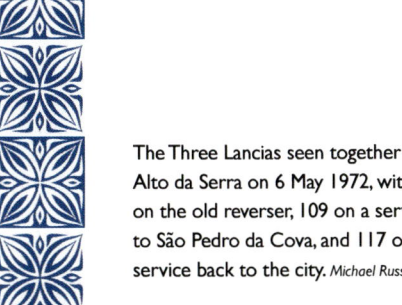

The Three Lancias seen together at Alto da Serra on 6 May 1972, with 150 on the old reverser, 109 on a service to São Pedro da Cova, and 117 on a service back to the city. *Michael Russell*

After taking over the CCFP in 1946, the STCP wanted to modernise the antique tram system by using buses on new and less important lines. The modernisation plans were not realised and as an alternative it was decided in the 1950s to replace the trams by trolleybuses.

This replacement was only partly implemented. In fact the tram outlived its supposed successor.

1959 brought both Vila Nova de Gaia (routes 13 & 14) and Lordelo (route 3) to an end. These lines were replaced by trolleybuses, mainly as it was the easiest way to get to the Carcereira bus (and trolleybus) depot. By the end of 1959, public riding and patronage on these routes were much more demanding than in the previous years. Therefore discontinuing these three tram routes freed resources that could now be transferred to the heavily used routes.

When new buses came into service, a large number of them were assigned to substitute for the trams, while the aerial lines were being replaced by a completely new overhead system, including traction poles. It took 9 months to install new overhead lines for 10-Venda Nova, 11- São Pedro da Cova and 12-Gondomar, and an additional 14 months to have them running to Ermesinde, Travagem and Areosa depot. Gradually the network grew towards the east and north-east. The system was at its maximum extent in the seventies and eighties with a network length of over 40 km and more than 100 trolleybuses in service.

If STCP once considered the possibilities of having a modern tram fleet operating long trunk lines, with the advent and success of buses in 1948, and trolleybuses in 1959, that was now considered absolutely impossible. All the money spent in rail network layout allowed faster tram movement, but showed the ever increasing operational costs of keeping an old fleet of obsolete trams running, that couldn't be afforded for many more years, but would have to be faced in the short term.

The 1970's – When the renovation plan came to its completion, STCP was by far a more modern transport company. New modern electrical sub-stations, a large and modern fleet of buses and trolleybuses, with two available depots and a 3rd one for trams, and still keeping in working condition the foundry, mechanic workshops, carpenters, painters, and permanent way department and the old electrical sub-stations, all the skills could be found. For the first time main frame computers were used to make calculation intensive jobs, like assigning inspectors to check tickets in a pre-established route with an almost complete coverage or for accounting tasks. By 1974 the existing fleet and depot allocation can be seen in the table below:

Lancia double deck trolleybus 139 seen at Carcereira depot when brand new in June 1967. *J. Jordan*

The Lancia double-deckers didn't operate over the top deck of the Dom Luis Bridge for very long. 146 is seen in 1968 when the "via" blinds below the windscreen were still in use on these vehicles. *P. Haseldine*

Yet, not all aims had been achieved: the last Porto long radial line 7 was still tram operated, and lasted till 1977, against all bets, although cut back at Arca d'Água in 1975, due to highway construction, not modernisation, and also the 1967 plans to extend trolley-bus line 10 from Venda Nova to Valongo were abandoned, and trolleybus line 10 was converted to motorbuses in 1977 after 10 years of operation, due to shortage of vehicles. Car traffic was increasing every year, making

1974 - depot allocation		Boavista	Carcereira	Areosa	per type
Trams		100	-	-	100
Trolleybus	single deck	-	18	33	51
	double deck	-	-	50	50
Bus	single deck	-	109	39	148
	double deck	-	78	62	140
Total vehicles per depot		100	205	184	
			389		
Total			489		

BUT 13 leaving Rua de Bolhão terminus on a service to Ermesinde on 11 September 1969. *A. G. Murray*

Lancia 40 at Largo Miguel Bombarda, Vila Nova da Gaia on 7 May 1972. Note the wiring for emergency reverser, later removed. *Michael Russell*

On 17 September 1973 in Rua Alexandre Braga, the city centre terminal point for the trunk routes to the east, three Lancia / Dalfa / CGE double-deck trolleybuses are seen: 149 on line 12 to Gondomar, 145 on an un-numbered working to São Roque, and arriving in the background, 128, which will form a journey to Venda Nova on line 10. *D. Pearson*

buses and trolleybuses average commercial speed slower, and demanding more vehicles to maintain headways. Only on the 13th December 1975 CMP council agreed to a "bus lane", in Avenida da Ponte on the way to Vila Nova de Gaia. City centre streets with two way traffic and badly-parked cars didn't allow easy and smooth running for public transport. Line 7 in its 1977 closure was replaced by buses, as trolleybuses were not available and were no longer considered a priority for replacing trams or new routes. Buses, being cheaper to buy, and easier to operate,

seemed a natural choice. In the 1990's the system was gradually shut down, the last line was closed on 27 December 1997.

Trolleybus network

During the years the following trolleybus lines existed:

9 - Bolhão - Praça Marquês de Pombal - Rua Costa Cabral - Areosa - Águas Santas - Ermesinde (Opened 17 November 1968 as successor of tramline 9. Replaced by diesel buses August 1994)

10 - Bolhão - Bonfim - São Roque - Venda Nova (Opened 10

BUT LETB1 / UTIC / Metrovick No. 5 of 1959 arrives at Ermesinde on 17th September 1973. *D. Pearson*

Double-deck Lancia / Dalfa / CGE 138 leaves Lordelo, the western terminus of line 35, on 17 September 1973. *D. Pearson*

Lancia 117 at Rua Estação / Rua Pinto Bessa on 6 June 1973. *A. G. Murray*

September 1967 as successor of tramline 10. Replaced by diesel buses 9 May 1977)

11 - Bolhão - Bonfim - São Roque - Santa Eulália - São Pedro da Cova (Opened 10 September 1967 as successor of tramline 10/. Replaced by diesel buses 4 March 1995)

12 - Bolhão - Bonfim - São Roque - Santa Eulália - Gondomar (Opened 10 September 1967 as successor of tramline 10//. Replaced by diesel buses 4 March 1995)

14 - Hospital de São João - Alto da Serra – Mondays to Fridays only (Opened 1991, suspended between June and September 1993, extended to São Pedro da Cova in 1996. Replaced by diesel

buses 4 March 1997))

29 - Bolhão - Praça Marquês de Pombal - Rua Costa Cabral - Areosa - Águas Santas - Travagem (Opened 17 November 1968. Replaced by diesel buses August 1994)

31 - Praça Almeida Garrett - Batalha (since 10 December 1972 Avenida D. Afonso Henriques) - Ponte D. Luiz I (upper deck) - Mafamude - Santo Ovídio - Soares dos Reis - Rua Candido dos Reis - Ponte D. Luiz I (lower deck) - Infante - Praça Almeida Garrett (Opened 3 May 1959. Closed 17 April 1978)

32 (1st) - Praça Almeida Garrett - Infante - Ponte D. Luiz I (lower deck) - Rua Candido dos Reis - Soares dos Reis - Santo

There were not many places in the world where trolleybuses ran in parallel, vertically! On the Dom Luis Bridge, there is one Lancia/Dalfa single-decker on the top deck and one on the lower deck seen in this view in 1974. *P. Haseldine*

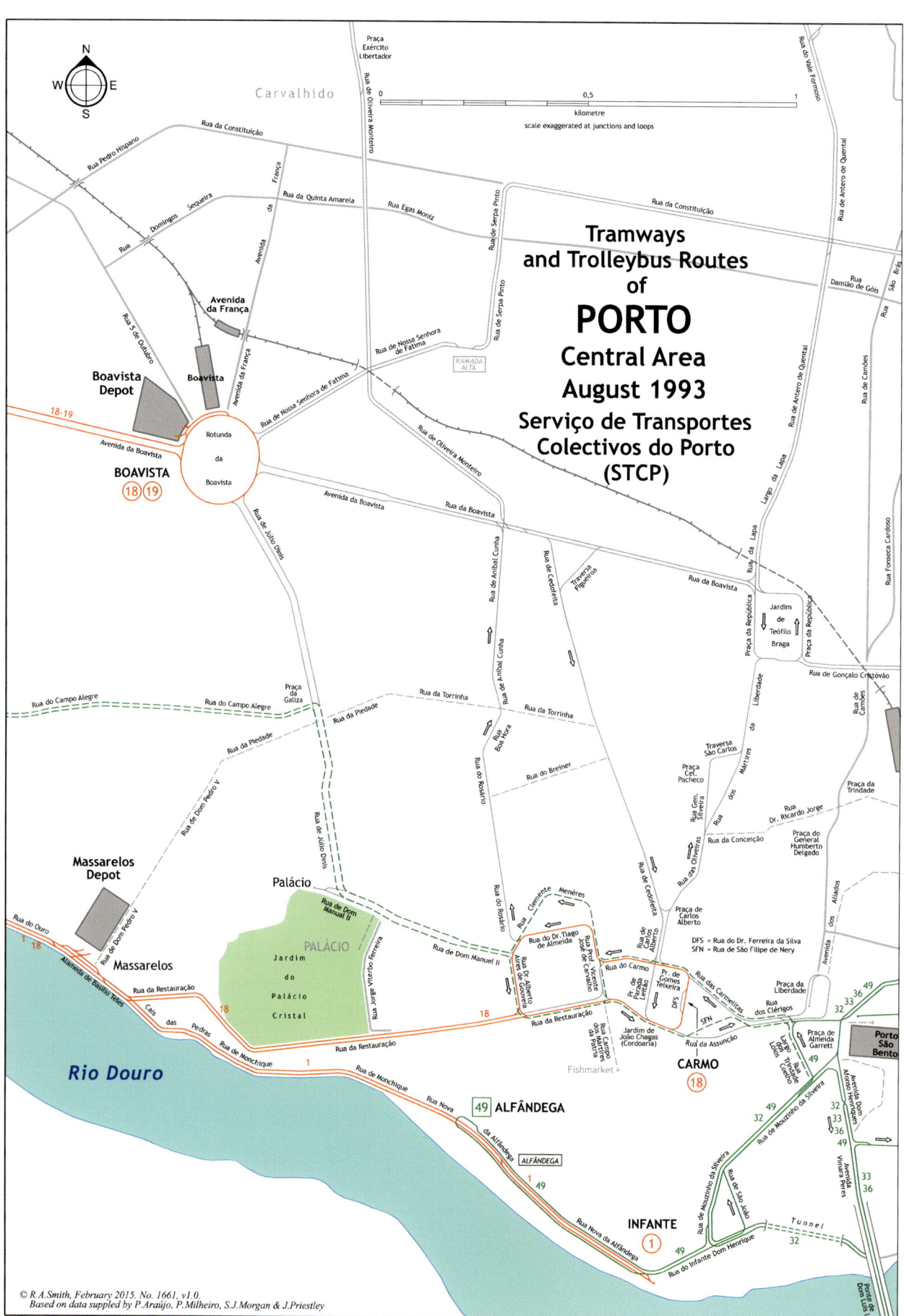

Tramways
and Trolleybus Routes
of
PORTO
Central Area
August 1993
Serviço de Transportes
Colectivos do Porto
(STCP)

Carvalhido

0   0.5   1
kilometre
scale exaggerated at junctions and loops

N
W · E
S

Praça
Exército
Libertador

Rua da Constituição

Rua Pedro Hispano

Domingos

Sequeira

Rua

Rua da Quinta Amarela

França

Rua da

Rua Egas Moniz

Rua de Oliveira Monteiro

Rua de Serpa Pinto

Rua da Constituição

Rua de Antero de Quental

Rua do Vale Formoso

Rua
Damião de Góis

São Brás

Avenida
da França

Avenida da França

5 de Outubro

Rua 5 de Outubro

Avenida da França

Rua de Nossa Senhora de Fátima

Rua de Nossa Senhora de Fátima

RAMADA
ALTA

Rua de Serpa Pinto

Rua de Antero de Quental

Rua de Camões

Boavista
Depot

Boavista

18-19

Avenida da Boavista

Rotunda
da
Boavista

BOAVISTA

(18)(19)

Rua de Júlio Dinis

Avenida da Boavista

Rua da Boavista

Rua de Oliveira Monteiro

Rua de Aníbal Cunha

Rua de Cedofeita

Travessa
Figueiroa

Rua da Boavista

Praça da República

Praça da República

Largo da Lapa

Rua da Lapa

Rua Fonseca Cardoso

Jardim
de
Teófilo
Braga

Rua de Gonçalo Cristóvão

Rua do Campo Alegre

Rua do Campo Alegre

Praça
da
Galiza

Rua da Torrinha

Rua da Torrinha

Liberdade

Rua de Camões

Rua de Aníbal Cunha

Rua da Piedade

Rua da Piedade

Rua da Piedade

Rua de Dom Pedro V

Rua de Júlio Dinis

Rua do Rosário

Boa hora

Rua Boa hora

Rua do Breiner

dos Mártires

da

Rua Gen.
Silveira

Traversa
São Carlos

Praça
Cel.
Pacheco

Rua
Dr. Ricardo Jorge

Praça do
General
Humberto
Delgado

Praça da
Trindade

Massarelos
Depot

Rua de Dom Pedro V

Rua do Ouro

Massarelos

1 18

Alameda de Basílio Teles

Cais
das
Pedras

Rua da Restauração

Rua de Monchique

Rua de Jorge Viterbo Ferreira

Palácio

Rua de Dom
Manuel II

PALÁCIO

Jardim
do
Palácio
Cristal

Rua de Dom Manuel II

Rua do Rosário

Rua Clemente Menéres

Rua Prof. Vicente
José de Carvalho

Rua de Dr.Tiago
de Almeida

Rua Dr. Alberto
Aires de Gouveia

Rua do Carmo

Rua de Cedofeita

Rua de
Carlos
Alberto

Praça de
Carlos
Alberto

DFS = Rua do Dr. Ferreira da Silva
SFN = Rua de São Filipe de Nery

Pr. de
Parada
Leitão

Pr. de
Gomes
Teixeira

SFN

DFS

Rua das Carmelitas

Rua dos Oliveiras

Rua da Conceição

Aliados

Avenida dos Aliados

Praça da
Liberdade

Rua
dos Clérigos

Praça de
Almeida
Garrett

Porto
São
Bento

49

32 33

36

49

49

32 49

32

36

49

Rua de Mouzinho da Silveira

Largo de
Trindade
Coelho

Avenida Dom
Afonso Henriques

Rua de São João

Vímara Peres

33
36

49

18

18

Rua da Restauração

Rua da Restauração

Rua de Monchique

1

18

Rua de Monchique

Rua Nova

Rua Campo
dos Mártires
da Pátria

Jardim de
João Chagas
(Cordoaria)

Fishmarket

CARMO

(18)

Rua da Assunção

Rio Douro

49 ALFÂNDEGA

da Alfândega

ALFÂNDEGA

Rua Nova da Alfândega

32

32

32

INFANTE

(1)

1

49

1

49

Rua do Infante Dom Henrique

Tunnel

Ponte de
Dom Luís I

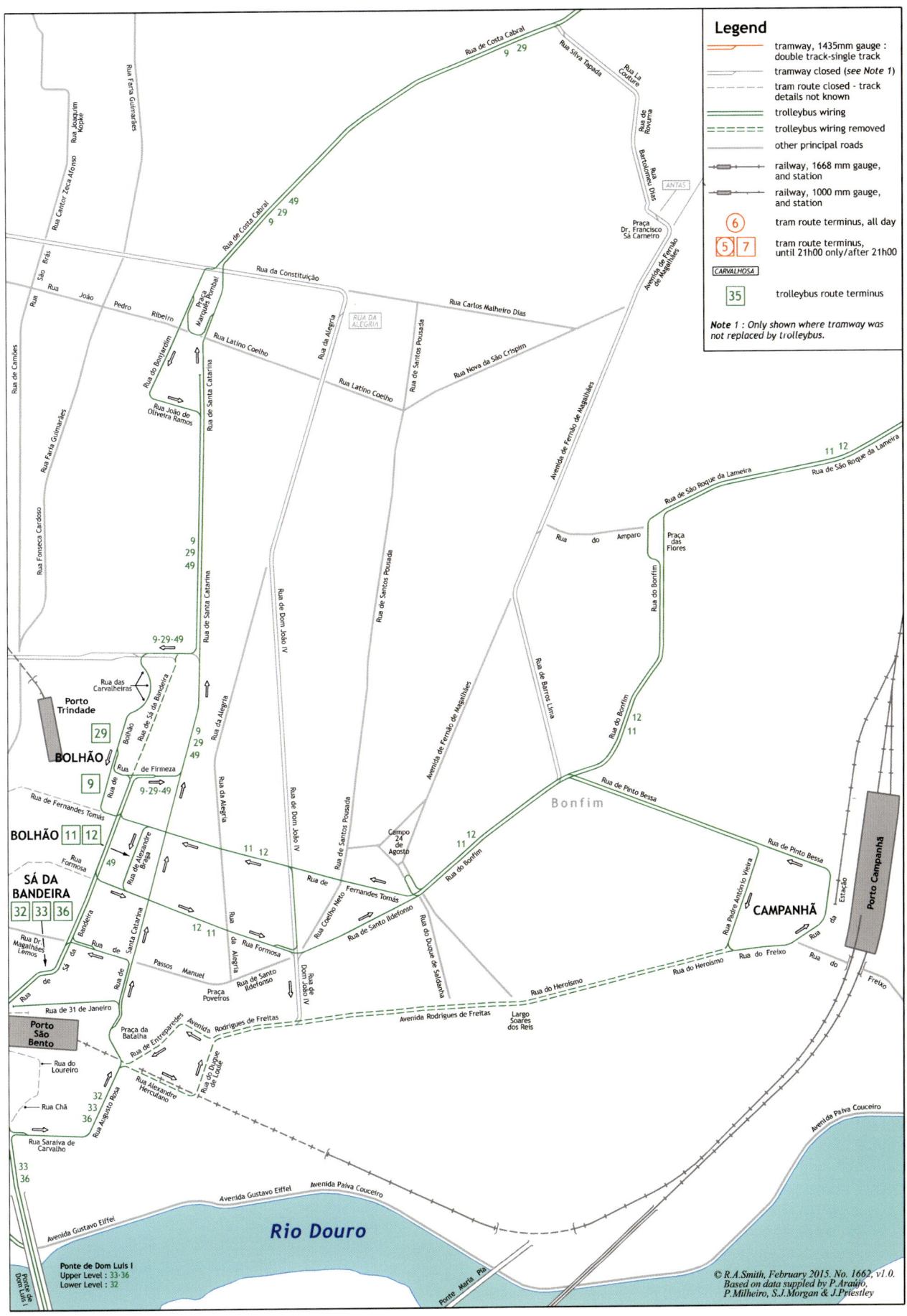

The Tramways
and Trolleybus Routes
of
PORTO
August 1993
Serviço de Transportes
Colectivos do Porto
(STCP)

0   0,5   1   2   3

kilometres

scale exaggerated at junctions and loops

Note A
Original route prior
to harbour extension
in 1938

Leça da
Palmeira

Leixões

MATOSINHOS

① ⑲

GODINHO

**Porto de
Leixões**

*Doca
Mercado*

Senhora
da Hora

Circunvalação

**Monte
dos
Burgos**

Estrada da Circunvalação

Ponte
de Pedra

São Mamede

São Mamede
de Infesta

PRELADA

ARCA D'ÁGUA

19

Praça Cidade
do Salvador

CIRCUNVALAÇÃO

Sub-station
No. 1

Castelo
de Queijo

*CASTELO
DO QUEIJO*

Avenida da Boavista

*ANTÓNIO
AROSO*

*ANTUNES
GUIMARÃES*

Pereiró

Ramalde

Note B
Details of trolleybus
wiring beyond this
point not known.

Rua
Pedro Hispano

Carcereira
Depot

Francos

**See
Central Area
Map**

18-19

Rua Dr.
Joaquim
Costa

Avenida da Boavista

18-19

BESSA

AV. BRASIL

Avenida da Boavista

Avenida Marechal Gomes da Costa

*GOMES
DA COSTA*

Lordelo

See Note B

Rua
Antônio
Cardoso

Avenida
da Boavista

Cadouços

Rua de Diogo
Botelho

Rua   do   Campo

Alegre

Esplanada
do Castelo

Rua da
Senhora
da Luz

Rua   do   Passeio   Alegre   Rua de Sobreiras

Largo
Antônio
Calém

*LARGO
DO CALÉM*

R. do Ouro

Rua do Ouro

SECIL

① 18

Rio Douro

Avenida Dom Carlos I

*PASSEIO
ALEGRE*

Ponte da
Arrábida

**Oceano
Atlântico**

Avenida
Diego
Leite

Rua de Cândido dos Reis

32

(a)

Avenida da República

Vila Nova
da Gaia

33

33
36

33

COIMBRÕES

33

Coimbrões

Rua do Corvo

Barão do Corvo

Conselheiro Veloso
da Cruz

Rua Visconde
das Devesas

32
36

Rua de
Calheiros
Lobo

R. do Barão do Corvo

Rua de
Domingos
de Matos

SANTO OVÍDIO

32 **36**

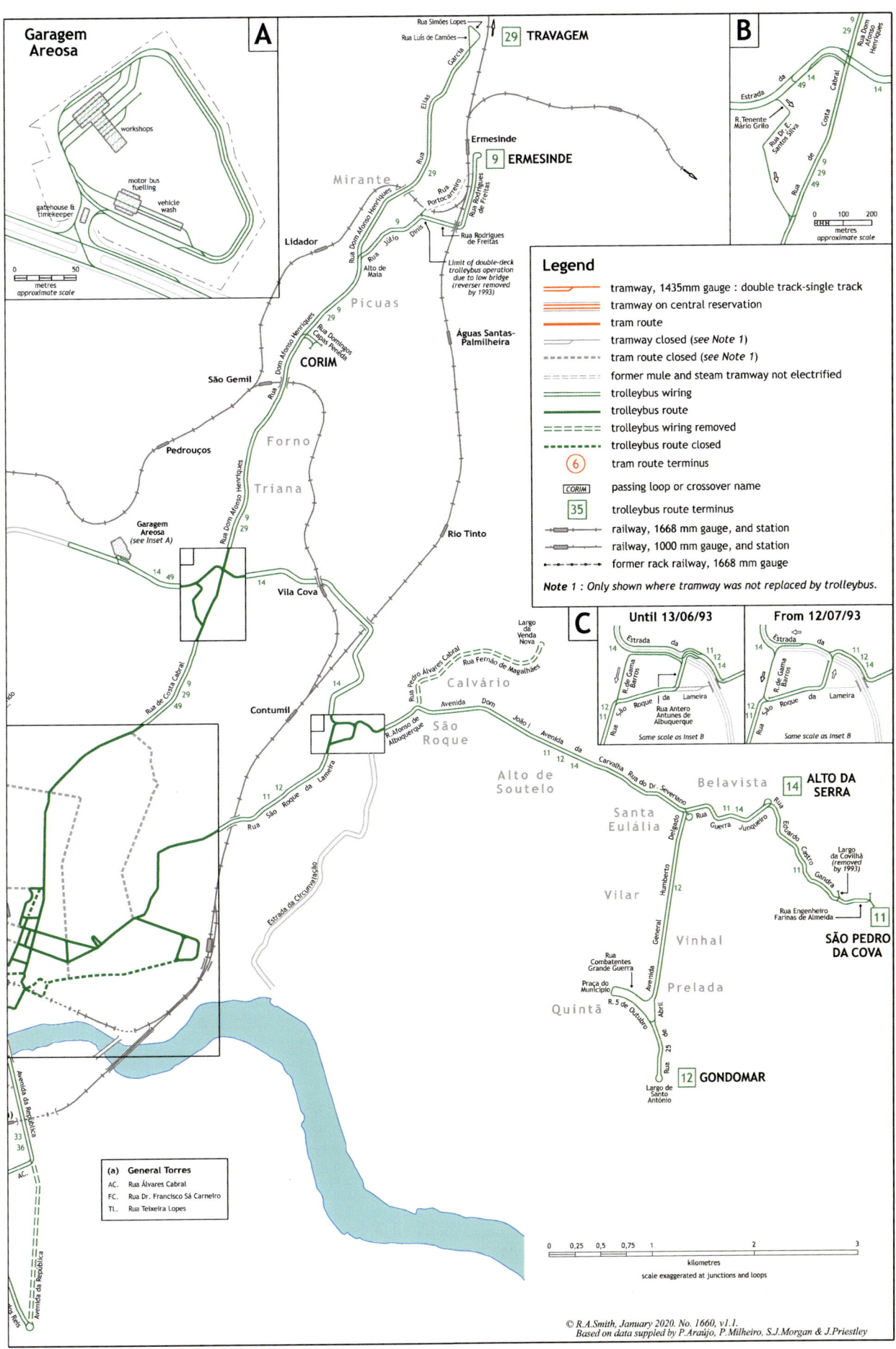

1967-built Lancia/Dalfa/CGE trolleybus 28 is passing Vila Nova de Gaia station and the reverser at Rua Visconde das Devesas on 15 August 1978 while heading for Coimbrões on line 33. *D. Pearson*

Porto 146, one of fifty Lancia / Dalfa / CGE double-deck trolleybuses on the system arrives at the São Pedro da Cova reversing triangle, terminus of line 11 from Bolhao. *D. Pearson*

Double deck Lancia 125 climbing from São Pedro da Cova towards Alto da Serra on its way to the city on 25 May 1980. *Michael Russell*

Lancia 126 is seen high above São Pedro da Cova and is almost at Alto da Serra in 1984. *P. Haseldine*

Ovídio - Mafamude - Ponte D.Luiz I (upper deck) - Avenida D.Afonso Henriques (since 10 December 1972 Batalha) - Praça Almeida Garrett (Opened 3 May 1959. Closed 17 April 1978)

32 (2nd) - Rua Gonçalo Cristovão - Rua Sá da Bandeira - Praça Almeida Garrett (opposite direction via Batalha - Rua Santa Catarina) - Infante - Ponte D.Luiz I (lower deck) - Rua Candido dos Reis - Soares dos Reis - Santo Ovídio (Opened 17 April 1978. On 10 July 1978 the city terminus was changed into Rua Sá da Bandeira (in front of café "A Brasileira") which was now reached via Rua Passos Manuel. On 24 August 1981 and because of road works the line returned temporarily to Rua 31 de Janeiro instead of Passos Manuel but returned to Passos Manuel the year after. Line 33 was closed in July 1993 and on 24 August 1981 into Rua 31 de Janeiro. Closed in July 1993)

33 - Praça Almeida Garrett - Batalha (opposite direction Avenida D.Afonso Henriques, since 10 December 1972 reverse direction) - Ponte D. Luiz I (upper deck) - Devesas - Arco do Prado - Coimbroes (Opened 1 January 1959 as successor of tramline 14. On 17 April 1978 extended from Praça Almeida Garrett to Rua Gonçalo Cristovão via Rua Sta. Catarina, return via Rua Sã da Bandeira. On 10 July 1978 the city terminus was

changed into Rua Sá da Bandeira (in front of café "A Brasileira") which was now reached via Rua Passos Manuel. On 24 August 1981 and because of road works the line returned temporarily to Rua 31 de Janeiro instead of Passos Manuel but returned to Passos Manuel the year after. Line 33 was closed in July 1993)

34 - Bolhão - Bonfim - Campanhã (Opened 17 November 1968. Closed 13 December 1976)

35 - Lordelo - Palácio de Crystal - Carmo - Praça da Liberdade - Batalha - Biblioteca - Campanhã (Opened from Lordelo to Praça da Liberdade 1 January 1959 as successor of tramline 3 and extended to Campanhã 4 March 1960 as successor of tramline 12. Replaced by diesel buses from 17 November 1968 until 17 May 1970 except for peak hour extras. Closed 13 December 1976)

35A - Lordelo - Palácio de Crystal - Carmo - Praça da Liberdade - Batalha - Biblioteca - Campanhã - Bonfim (Opened 11 June 1964 as reinforcement of line 35 and replacement of tramline 11. Closed 17 November 1968)

36 - Praça Almeida Garrett - Batalha (opposite direction Avenida D.Afonso Henriques, since 10 December 1972 inversed directions) - Ponte D.Luiz I (upper deck) - Soares dos Reis -

Lancia / Dalfa / CGE 43 emerges from Avenida de Diogo Leite and prepares to cross the River Douro by the lower level of the Ponte Dom Luís I on a line 32 journey from Santa Ovídio to Rua de Sâo da Bandeira on 17 August 1978. *D. Pearson*

Above: 161, one of ten Caetano / EFACEC articulated vehicles is at Ermesinde railway bridge on 9 July 1986. *Michael Russell*

Above, right: Caetano / EFACEC 73 of 1983/4 seen at Praca Marques de Pombal on 10 July 1986. *Michael Russell*

Right: An atmospheric view of single-deck Lancia 40 coming off the Dom Luis Bridge in 1989. The light shows up the overhead supports well, almost identical supports are now used by the new metro which descends underground at this point. *P. Haseldine*

Santo Ovidio (Opened 3 May 1959 as successor of tramline 13. On 17 April 1978 extended from Praça Almeida Garrett to Rua Gonçalo Cristovão via Rua Sta. Catarina, return via Rua Sá da Bandeira. On 10 July 1978 the city terminus was changed into Rua Sá da Bandeira (in front of café "A Brasileira") which was now reached via Rua Passos Manuel. On 24 August 1981 and because of road works the line returned temporarily to Rua 31 de Janeiro instead of Passos Manuel but returned to Passos Manuel the year after. Line 36 was closed in July 1993). On 10 July 1978 the city terminus was changed into Rua Sá da Bandeira and on 24 August 1981 into Rua 31 de Janeiro. Closed in July 1993)

37 - Praça Almeida Garrett - Mafamude - Santo Ovídio (Opened 3 May 1959. Closed 4 March 1960)

49 - Álfândega - Infante - Praça Almeida Garrett - Bolhão - Praça Marquês de Pombal - Rua Costa Cabral - Areosa - Hospital São João (Opened 1986. The last part, Bolhão - Hospital São João, was the last trolleybus line and replaced by diesel buses 27 December 1997)

Trolleybus fleet
The Porto system had at its maximum a fleet of 126 trolleybuses of six different types. Curiously the later ones were identified by two numbers, a fleet number and a registration number.

BUT (Fleet numbers 1-20 & 21-26)
Trolleybus operation in 1959 started with 20 BUTs (British United Traction) with Park Royal/UTIC bodies with seating for 32. They had a chassis built by Leyland and a 99 kW electric motor from Metropolitan-Vickers.

In 1963 they were followed by another 6 vehicles of the same type, but with UTIC bodies with three doors instead of two and seating reduced to 20. The trolleybuses 1 & 23 are preserved for the museum, the others were scrapped.

Lancia single-deck (Fleet numbers 27-51)
In 1966/7 the single-deck trolleybus fleet was extended by 25 vehicles. The bodies were made by Dalfa. The chassis came from Lancia with a 110 kW GCE electric motor. Seating was for 29. This type was nicknamed "Italianos". Trolleybus 49 is part of the museum collection, the others were scrapped.

BUT 23 is seen in the STCP museum on 19 June 2008. *L Folkard*

Lancia double-deck (Fleet numbers 101-150)
Together with the Italianos, 50 double-deck trolleybuses were acquired from the same suppliers. Seating was for 68 and they had a two-man crew. In 1990 they were converted for one-man operation. In 1995 they were withdrawn. Trolleybus 102 is part of the Porto museum collection and 140 went to the Sandtoft trolleybus museum in England.

Efacec (Fleet numbers 60-74)
15 trolleybuses were acquired in 1983/4 from Salvador Caetano. The 131 kW electric motor came from Éfácéc with equipment from Kiepe. They also had an auxiliary 50 kW Diesel motor from Hatz. No.74 is preserved for the Porto museum but went in 2003 to Coimbra. The others went in October 2000 to Almaty in Kazakhstan.

Efacec articulado (Fleet numbers 160-169)
10 articulated trolleybuses were delivered by the same suppliers in 1984/5. They were technically equal to the smaller type but had electric motors of 209 kW. No.167 is preserved for the Porto museum but went in 2003 to Coimbra. The others went in October 2000 to Almaty in Kazakhstan.

PORTO TROLLEYBUSES FLEET LIST

Fleet number	Chassis	Bodywork	Electrical Equipment	Built New	Acquired	Withdrawn	Notes
1-20	BUT LETB1	UTIC	MV	1958		1984-93	1 preserved
21-26	BUT LETB1	UTIC	AEI	1963		1986-90	23 preserved
27-51	Lancia Esagamma120	Dalfa	CGE	1966-67		1995	49 preserved
101-150	Lancia Esagamma120	Dalfa	CGE	1966-67		199x-95	102/140 preserved
60-74	Caetano 175TR110	Caetano	EFACEC	1983-84		1997	60-73 to Almaty 1170/1165/1166/1159/1160/1167/1172/1171/1163/1162/1164/1161/1168/1169. 74 to Coimbra 71
160-169	Caetano 190TR110	Caetano	EFACEC	1985		1997	160-166/168/169 to Almaty 1154/1157/1155/1152/1153/1150/1158/1151/1156. 167 to Coimbra 70, 2003-2009

Tramways – Ponte Dom Luís I

Car 115 seen with two trolley poles crossing the Dom Luís 1 bridge, thought to be circa 1912, a picture oozing period detail, including containers being carried on ladies' heads. *Photographer unknown*

By Pedro Milheiro and Ernst Kers

The Ponte Luís I, often called the Ponte Dom Luís, connects the central parts of Porto and Vila Nova de Gaia. Although not built for the trams, it had and has a major role in the public transport of Porto and is one of the most important landmarks of the city. This bridge with two decks was built in the 1880's as successor to a suspension bridge. The design was by Téofile Seyrig, who with Gustave Eiffel had founded in 1869 the Eiffel & Cie Company. When still in partnership with Eiffel, Seyrig had designed the Ponte Maria Pia railway bridge a few kilometres upstream across the Douro, which was built in 1876-1877 by Eiffel & Cie. After a dispute Seyrig left Eiffel & Cie in 1879 and joined the Willebroeck company. For building the Ponte Luís I the design of Seyrig (Willebroeck) was preferred above those of Eiffel and four other competitors. The bridge was opened in 1886.

With the construction of the Ponte Luís I, tram tracks were laid on the upper deck of the bridge. These tracks were owned by the Ministerio das Obras Publicas. The idea was that tram companies could use these tracks by paying a fee. Although there was a concession until 1911 given for a mule tram-line to a company named "Caminho de Ferro Americano Praça da Batalha á Estação de Villa Nova de Gaya", this line was never opened. Most probably the gauge of these tracks was 900 mm. Licences were granted to Biel, Morais and Co to construct lines to Santo Ovidio in 1895 and Devesas in 1898. Emilio Biel was an entrepreneur, and the Porto representative of Schuckert.

At the end of the 19th century CCFP came to an agreement with the owners of the concession, which in the meantime was amended for the use of electric trams, to rent the concession with the option to purchase. Already in 1900 the first tracks were laid in Vila Nova de Gaia, the old rail was lifted in 1903.

Probably bureaucratic confusion about a temporary viaduct across the railway near General Torres delayed completing the line over the bridge. Finally on 28 October 1905 the lines to Vila Nova de Gaia were opened. One line went to Sto. Ovídio while the other went to Devesas. CCFP exercised the option to purchase the concession in 1911.

In 1930 the single track on the bridge (situated at the eastern side) was changed to double track. The centres of the track were just 2.43 m apart, which meant that only narrow cars could be used (Carros Antigos, Constructora, Ingles and Brill 23).

On the bridge two trolley poles had to be used to prevent oxidation of the iron by stray electric currents. A number of tramcars were equipped with a second trolley pole for use on the lines to Vila Nova de Gaia. From 1933 the trams could run across the bridge with one trolley-pole.

There was also a line projected with the route Infante - Praça da Ribeira - Ribeira - lower deck of the Ponte Dom Luis I – Vila Nova de Gaia. Although tracks and overhead wires were put in place until the Porto end of the bridge, this line was never opened.

In 1959 the tramlines across the bridge were replaced by trolleybuses. In contrast to the trams the trolleybuses also used the lower deck. In 1993 trolleybus services across the bridge were completely replaced by Diesel buses.

In 2003 following the opening of the adjacent Ponte de Infante all traffic across the upper deck was suspended to enable the reconstruction, which was necessary to build line D of the Metro do Porto. This line was opened in 2005. The upper deck of the bridge is now reserved for the trams of the Metro do Porto and pedestrians. The lower deck is still used by road traffic.

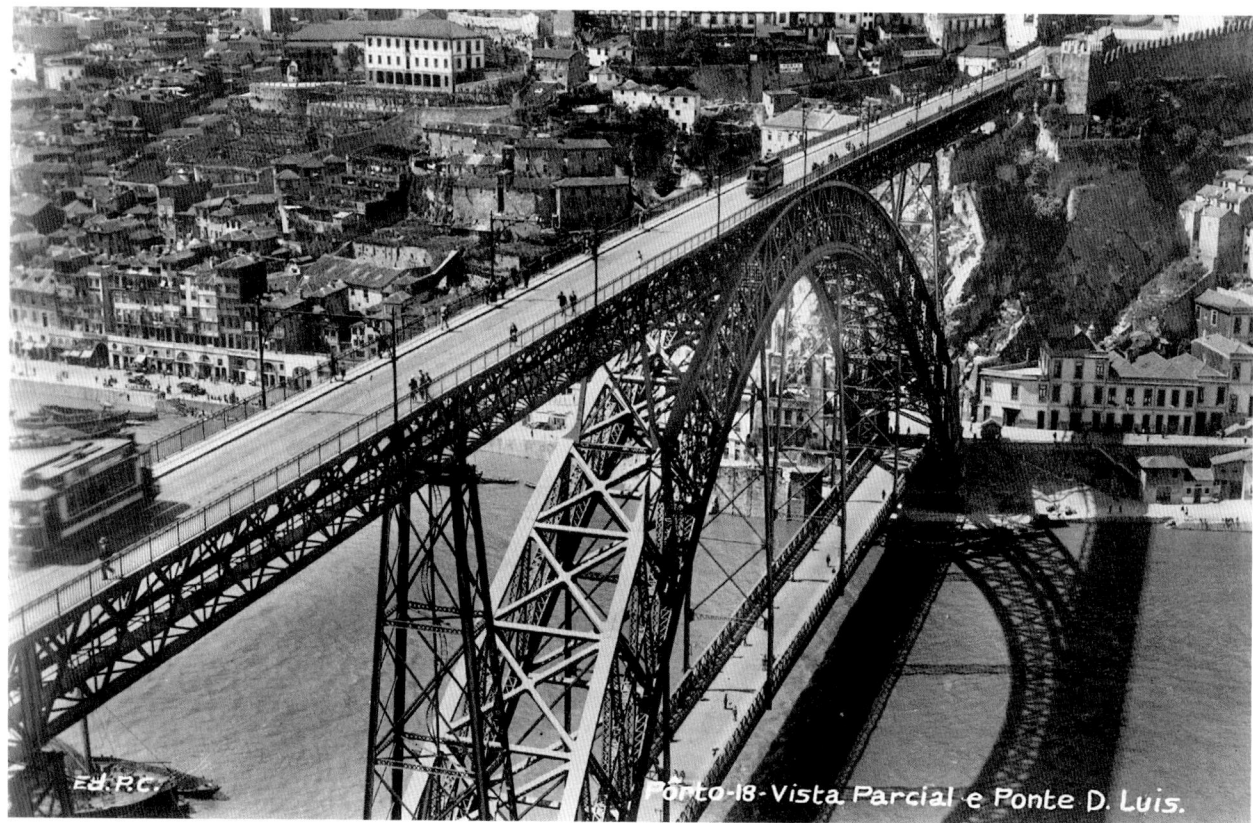

The Dom Luís I bridge with double tracks around 1940 (1933-1947). *Commercial postcard.*

An unidentified coupled set of metro cars seen on the west side of the Dom Luís I bridge on 19 June 2008. The funicular is in view to the right of the picture. *L. Folkard*

Another unidentified coupled set of metro cars also seen on the west side of the Dom Luís I bridge on 19 June 2008. The remains of the earlier suspension bridge may be seen on the far bank just to the left of the arch. *L. Folkard*

On Avenida da República, having left the upper level of the Ponte Dom Luis I on 17th August 1978, BUT LETB1, 11, with Metrovick electrical equipment and bodywork by UTIC, is working line 33 bound for Coimbrões. *D. Pearson*

Porto metro car 041 is seen crossing the Dom Luis I Bridge in 2009 just four years after this line opened. *P. Haseldine*

Metro car 026 has just left
Mercado in 2009 and is entering
Rua Birto Capelo in Mathosinhos.
The Café Electrico on the left with
the red awning has a selection of
vintage tram pictures on its walls.
In the background is the current
lifting bridge. *P. Haseldine*

Metro
do Porto

By Ernst Kers

In the second half of the 1920s the Portuguese government had plans to cover the north of Portugal with a dense network of metre gauge railways. Little of these plans was realised, though it affected strongly the Caminho de Ferro Porto Povoa-Famalicao (PPF). The company was forced in 1927 to merge with another company, Caminho de Ferro de Guimarães (CFG) into the Companhia do Norte. The CFG had opened in 1884 a metre gauge line from Trofa on the broad gauge Minho railway, to Guimarães, which was in 1907 extended to Fafé.

In 1930 the lines and rolling stock of the former PPF were converted to 1000 mm. In 1932 a new line was opened between Senhora da Hora and Trofa making it possible to run direct trains from Porto to Guimarães and Fafé. In 1938 the line was extended in Porto to Trindade. This section had a length of about 2 km including a 500 m long tunnel. The Boavista station became almost redundant for passenger traffic but was retained for freight and as depot and workshops.

In 1938 the broad gauge Linha de Leixões railway was opened, which strongly reduced the importance of the narrow gauge railway for freight transport. Because of the construction of the first dock of the Leixões harbour, the narrow gauge line was cut back from Leça da Palmeira to Matosinhos in 1940.

In 1947 most Portuguese railways, included the Norte, were merged into the CP. and part of the rolling stock was exchanged with that of other metre gauge lines. In the 1950s some diesel railcars came into service but steam locomotives remained responsible several decades more for most of the trains.

Traffic on the Matosinhos line was was suspended in 1965 with official closure the following year. In 1986 the section between Guimarães and Fafé was closed. The same year also freight services on the Porto narrow gauge network were suspended. In the mean time diesel train-sets had taken over all of the passenger services. The line between Póvoa de Varzim and Famalicão was closed in 1995.

In the years that followed the route from Trofa to Guimarães was converted to broad gauge and electrified as part of the project of modernisation of the Minho railway between Porto and Braga. The narrow gauge lines from Porto to Póvoa de Varzim and Trofa were handed over to the Metro do Porto for conversion into standard gauge electric light-rail. Soon after that the lines were closed to accommodate the reconstruction.

The Metro do Porto network

The trunk route
Five out of six lines of the Metro do Porto share a trunk route between Estádio do Dragão and Senhora da Hora. This route runs from east via the centre to the north-west of the city over a distance of almost 10 km and has a total of 14 stations/stops. At Estádio do Dragão is the stadium of FC Porto with the equal name. Between Estádio do Dragão and Campanhã the line is parallel to the railway and partly covered by a concrete structure carrying a highway. Campanhã is the main railway station of Porto. From Campanhã the line is in a 2.3 km long tunnel with three intermediate stations. It comes to surface again at Trindade on the same location where once the terminus was of the narrow gauge railways. Trindade is the hub of the system as here crosses the only line not using the trunk route. From Trindade the route included the tunnel of the former narrow gauge railway. While the narrow gauge railway had four intermediate stops between Trindade and Senhora da Hora, the Metro do

Coupled Metro cars 053 and 001 at Jardim do Morro 19 June 2008. *L. Folkard*

Porto has seven intermediate stations/stops on this part. Near Boavista is the former Avenida da França stop replaced by a subterranean station called Casa da Música. The trunk route is part of the lines A, B, C, E and F, although not all lines use or have always used the whole route. Over the years the usage was changed with time-table changes.

Opening dates:
7 December 2002 Trindade - Senhora da Hora (6.1 km) as part of the opening of line A to Senhor de Matosinhos
6 June 2004 Estádio do Dragão - Trindade (3.8 km)
Line A
About 300 m north of Senhora da Hora line A turns to the west in the direction of Matosinhos. This branch has nine stops. The line does not follow the route of the former narrow gauge railway to Matosinhos. Only between the stops Estádio do Mar and Pedro Hispano the line is (almost) on the same spot. While in the days of the narrow gauge railways the area between Senhora da Hora and Matosinhos was rural, it is now of suburban character. Until 1993 electric trams had been running through Rua Brito Capelo, nine years later this street welcomed the trams of the Metro do Porto.

Opening date: 7 December 2002 Trindade - Senhora da Hora - Senhor de Matosinhos (5.7 km from Senhora da Hora to Senhor de Matosinhos)

Line B
Line B is the successor of the narrow gauge railway to Póvoa de Varzim. Counted from Senhora da Hora the length is about 24 km and there are 22 stops. The narrow gauge railway had nine stations on this part. In Póvoa de Varzim the line terminates on the site of the former narrow gauge railway station. The line has an interurban character. It is the only line with express services calling only at a few stops.

Opening dates:
12 March 2005 Estádio do Dragão - Trindade - Senhora da Hora - Fonte do Cuco - Pedras Rubras (6.7 km from Senhora da Hora to Pedras Rubras)
18 March 2006 Pedras Rubras - Póvoa de Varzim (17.2 km)
July 2017 New stop Modivas Norte opened between Modivas Centro and Mindelo.

Metro car 038 is using the former railway alignment and passing the old station at Senhora da Hora in 2009. *P. Haseldine*

Line C

Line C is the successor of the narrow gauge railway to Trofa, though it only reaches ISMAI. The section between ISMAI and Trofa was part of the project but never realised. Line C diverts from line B in Fonte do Cuco, one stop north of Senhora da Hora. Between the stops Custió and Parque Maia line C diverts from the original route of the narrow gauge railway to join it again at Mandim. This is done to serve better the centre of Maia. Counted from Fonte do Cuco the length is 10 km and there are 10 stops. The old narrow gauge railway had on this part four stations and three more on the part which is realised.

Opening dates:

30 July 2005 Estádio do Dragão - Trindade - Senhora da Hora - Fonte do Cuco - Forúm da Maia (6.0 km from Fonte do Cuco to Forúm da Maia)
31 March 2006 Forúm da Maia - ISMAI (4.5 km)

Nowadays line C has its Porto terminus at Campanhã. When FC Porto is playing line C trams continue to/from Estádio do Dragão.

Line D

Line D is the only line not using the trunk route between Estádio do Dragão and Senhora da Hora. It has a 4 km long tunnel between Polo Universitário and the bridge across the Douro river with 8 stations. For the river crossing the upper deck of the completely renovated Ponte Dom Luís I is used. The upper deck of the bridge and the adjacent Avenida da República in Vila Nova de Gaia had electric trams from 1905 until 1959. The line crosses the trunk route at Trindade. Here is also a 274 m long single track tunnel made between the tunnel of line D and the old tunnel of Trindade, which is now part of the trunk route. This

connecting tunnel is used to get the trams of line D from and to the depot. Line D has a total length of over 8 km and 16 stations/stops.

Opening dates:

17 September 2005 Polo Universitário - Trindade - Câmara Gaia (5.7 km)
10 December 2005 Câmara Gaia - João de Deus (0.4 km)
31 March 2006 Hôspital de São João - Polo Universitário (1.2 km)
28 May 2008 to Dom João II (0.6 km)
15 October 2011 to Santo Ovídio (0.5 km)
Travel time 25'.

Line E

Line E has the same route as line B until Verdes and from there it follows a 1.5 km long branch to the airport with only one intermediate stop.

Opening date: 27 May 2006 Trindade - Senhora da Hora - Fonte do Cuco - Verdes - Aeroporto (1.5 km from Verdes to Aeroporto) with opening of the station Verdes on line B (Red) between Crestins and Pedras Rubras. Later line E was extended via the trunk route to Estádio do Dragão.

Line F

This line leads from Estádio do Dragão in eastern direction to Fânzeres and serves the northern part of Gondomar. The line has a 950 m long tunnel between the stops Nau Vitória and Levada to cross the Minho-Douro and Leixões railways. The official western terminus of line F is Senhora da Hora, but in practice part of the trams of both the lines A and E continue to and from Fânzeres.

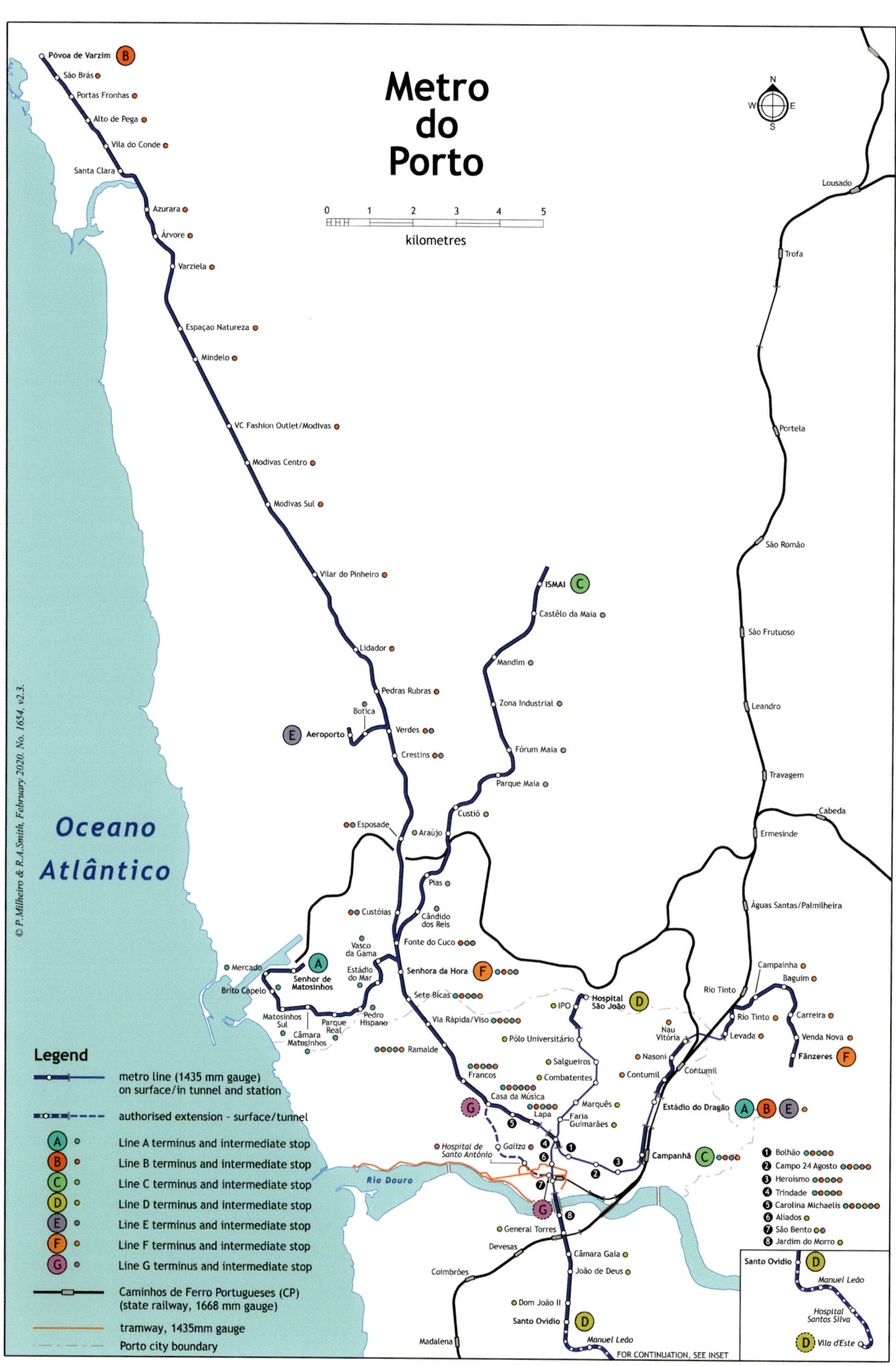

Metro do Porto

Póvoa de Varzim B
São Brás
Portas Fronhas
Alto de Pega
Vila do Conde
Santa Clara
Azurara
Árvore
Varziela
Espaçao Natureza
Mindelo
VC Fashion Outlet/Modivas
Modivas Centro
Modivas Sul
Vilar do Pinheiro
Lidador
Pedras Rubras
Botica
Aeroporto E
Verdes
Crestins

ISMAI C
Castêlo da Maia
Mandim
Zona Industrial
Fórum Maia
Parque Maia
Custió
Araújo
Esposade
Pias
Cândido dos Reis
Custóias
Fonte do Cuco
Vasco da Gama
Mercado
Estádio do Mar
Senhor de Matosinhos A
Senhora da Hora F
Sete-Bicas
Brito Capelo
Matosinhos Sul
Parque Real
Pedro Hispano
Câmara Matosinhos
Via Rápida/Viso
Pólo Universitário
Ramalde
Francos
Salgueiros
Combatentes
Casa da Música
G
Marquês
Lapa
Faria Guimarães
Hospital de Santo António
Galiza
Campanhã
G
General Torres
Devesas
Coimbrões
Dom João II
Santo Ovídio D
Manuel Leão
Madalena

IPO
Hospital São João
D
Nau Vitória
Nasoni
Contumil
Contumil
Rio Tinto
Rio Tinto
Levada
Campainha
Baguim
Carreira
Venda Nova
Fânzeres F

Estádio do Dragão A B E
Campanhã C

1 Bolhão
2 Campo 24 Agosto
3 Heroismo
4 Trindade
5 Carolina Michaelis
6 Aliados
7 São Bento
8 Jardim do Morro

Rio Douro

Lousado
Trofa
Portela
São Romão
São Frutuoso
Leandro
Travagem
Cabeda
Ermesinde
Águas Santas/Palmilheira

Oceano Atlântico

© P.Milheiro & R.A.Smith, February 2020. No. 1654, v2.3.

Legend

metro line (1435 mm gauge) on surface/in tunnel and station

authorised extension – surface/tunnel

A Line A terminus and intermediate stop
B Line B terminus and intermediate stop
C Line C terminus and intermediate stop
D Line D terminus and intermediate stop
E Line E terminus and intermediate stop
F Line F terminus and intermediate stop
G Line G terminus and intermediate stop

Caminhos de Ferro Portugueses (CP) (state railway, 1668 mm gauge)

tramway, 1435mm gauge

Porto city boundary

Santo Ovídio D
Manuel Leão
Hospital Santos Silva
Vila d'Este D
FOR CONTINUATION, SEE INSET

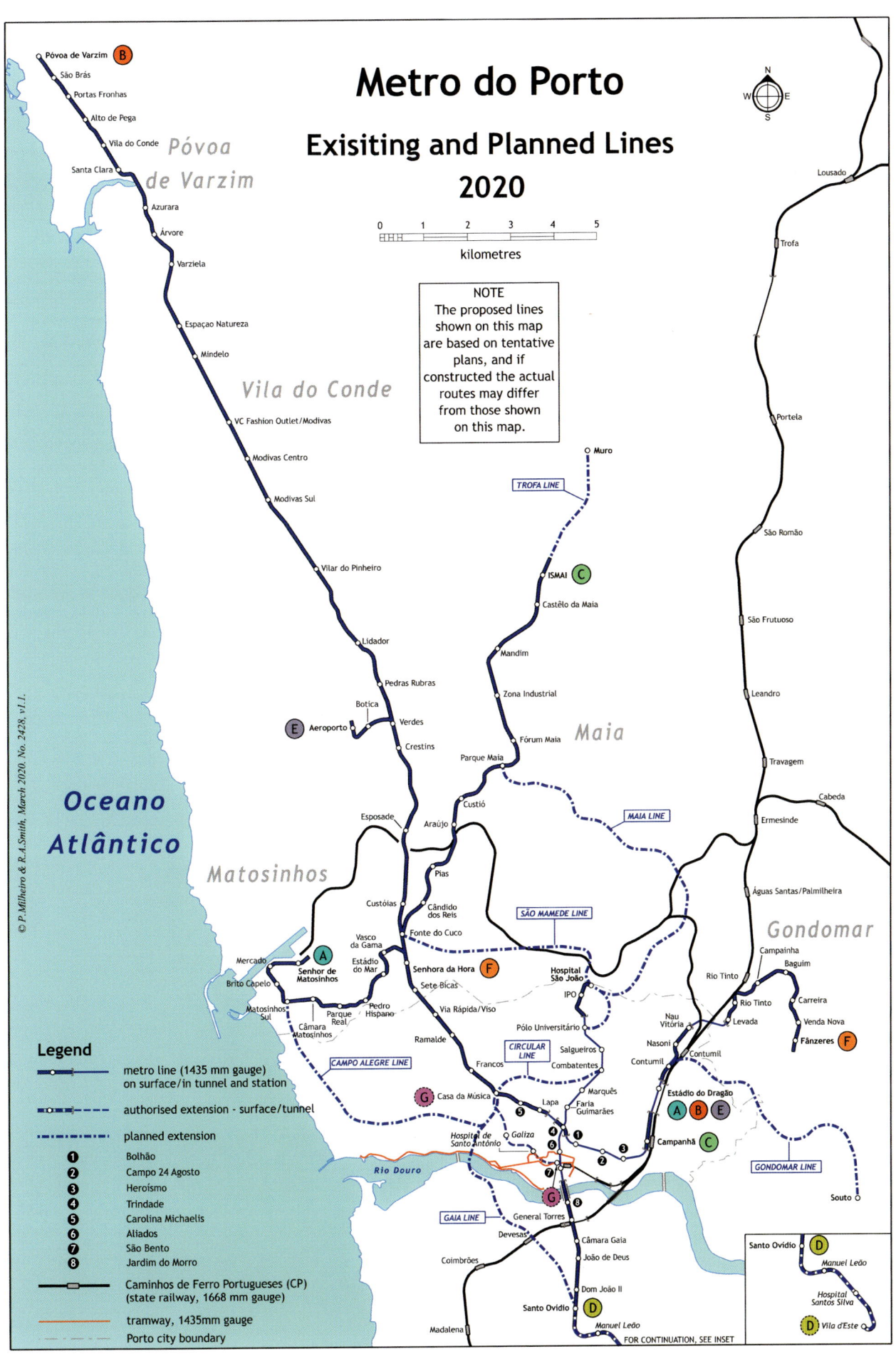

Metro do Porto

Exisiting and Planned Lines
2020

Póvoa de Varzim

Vila do Conde

Oceano Atlântico

Matosinhos

Maia

Gondomar

NOTE
The proposed lines shown on this map are based on tentative plans, and if constructed the actual routes may differ from those shown on this map.

0 1 2 3 4 5
kilometres

© P.Milheiro & R.A.Smith, March 2020. No. 2428. v1.1.

Legend

metro line (1435 mm gauge) on surface/in tunnel and station

authorised extension - surface/tunnel

planned extension

1 Bolhão
2 Campo 24 Agosto
3 Heroísmo
4 Trindade
5 Carolina Michaelis
6 Aliados
7 São Bento
8 Jardim do Morro

Caminhos de Ferro Portugueses (CP) (state railway, 1668 mm gauge)

tramway, 1435mm gauge

Porto city boundary

TROFA LINE
MAIA LINE
SÃO MAMEDE LINE
CIRCULAR LINE
CAMPO ALEGRE LINE
GONDOMAR LINE
GAIA LINE

Rio Douro

Stations (selected): Póvoa de Varzim, São Brás, Portas Fronhas, Alto de Pega, Vila do Conde, Santa Clara, Azurara, Árvore, Varziela, Espaço Natureza, Mindelo, VC Fashion Outlet/Modivas, Modivas Centro, Modivas Sul, Vilar do Pinheiro, Lidador, Pedras Rubras, Botica, Verdes, Crestins, Aeroporto, Esposade, Araújo, Pias, Custóias, Cândido dos Reis, Fonte do Cuco, Vasco da Gama, Estádio do Mar, Senhor de Matosinhos, Mercado, Brito Capelo, Matosinhos Sul, Câmara Matosinhos, Parque Real, Pedro Hispano, Senhora da Hora, Sete Bicas, Via Rápida/Viso, Ramalde, Francos, Casa da Música, Lapa, Marquês, Faria Guimarães, Galiza, Hospital de Santo António, General Torres, Devesas, Câmara Gaia, João de Deus, Dom João II, Santo Ovídio, Manuel Leão, Coimbrões, Madalena, Muro, ISMAI, Castêlo da Maia, Mandim, Zona Industrial, Fórum Maia, Parque Maia, Custió, Hospital São João, IPO, Pólo Universitário, Salgueiros, Combatentes, Nasoni, Contumil, Campanhã, Estádio do Dragão, Nau Vitória, Rio Tinto, Levada, Campainha, Baguim, Carreira, Venda Nova, Fânzeres, Souto, Águas Santas/Palmilheira, Ermesinde, Cabeda, Travagem, Leandro, São Frutuoso, São Romão, Portela, Trofa, Lousado

Inset: Santo Ovídio, Manuel Leão, Hospital Santos Silva, Vila d'Este

FOR CONTINUATION, SEE INSET

Opening date: 2 January 2011 Fânzeres - Estádio do Dragão - Trindade - Senhora da Hora (7 km from Fânzeres to Estádio do Dragão)

Characteristics

The network is extended over a number of different types of areas. This means that the character of the system varies over the network. In the central part of the city tunnels are used. The stations in this area have the character of real metro stations. Passengers are not allowed to cross the tracks. This full metro character exists for all stations on the trunk route from Estádio do Dragão up to and including Trindade, for all stations on the tunnel part of line D and for the Casa da Música (trunk route), Sto.Ovídio (line D) and Nau Vitória (line F) stations. On all other stations and stops passengers are allowed and in most cases have to cross the tracks at level.

The remaining part of the trunk route and most of the lines B and C have no level crossing with other traffic than pedestrians. On the lines B and C are a few railway type automatic half barrier crossings.

In the suburban areas of Matosinhos (line A), Vila Nova de Gaia (line D) and Gondomar (line F), but also in the built-up areas of Vila do Conde, Póvoa de Varzim (both line B), Maia (line C) and near Hôspital de São João (line D) and Botica (line E) the tracks are on reserved space, often in the central or side of roads, with traffic lights protecting road crossings.

In the centres of Matosinhos (line A) and Maia (line C) are streets where the trams share the space with pedestrians. This is also the case on the top deck of the Ponte Luís I and the adjacent area of Jardim do Morro.

Future

The success of the Metro do Porto resulted in ambitious plans to extend the system, though difficult economic and financial problems put realisation of these plans on hold until funds became available in 2016. Proposed extensions are:

• Extension of line D in southern direction from Sto.Ovídio to Manuel Leão - Hospital Santos Silva - Vila d'Este. This extension will be 3.1 km long of which about 600 m will be on a viaduct, 1100 m in tunnel and 1400 m at surface. The Manuel Leão station will be underground, both the others at surface. Between Hospital Santos Silva and Vila d'Este a depot will be constructed for about twenty trams with equipment for cleaning and maintenance as well as accommodation for staff.

• New line G in the central area of the city: Casa da Música - Galiza - Hôspital Sto.António - Liberdade (São Bento). This line will be about 2.7 km long and fully underground. At Casa da Música a connection will be made with the trunk route for depot trips. Between Liberdade and the existing São Bento station of line D a pedestrian tunnel will be made for easy interchange.

Both extensions should be inaugurated by 2022.

Rolling stock

The Metro do Porto has two types of trams, both made by Bombardier. Operations started with Eurotrams, a 100% low floor type that Bombardier inherited from ABB/Adtranz and now branded as one of the Flexity Outlook models. There are 72 units, which are used on the lines A, D, E and F. During peak hours they also do service on line C. Each tram is 35 m long and 2.65 m wide. The maximum speed is 80 km/h and they can go through curves with a radius of 18 m. They can be coupled to a maximum of two units, which is normally done during daytime on weekdays on line D and during peak hours on the lines A, C and F. Capacity is for 80 passengers seating and a maximum of 215 standing.

For the longer routes the Metro do Porto acquired 30 trams of the Flexity Swift type. These trams are 37 m long and 2.65 m wide. The maximum speed is 100 km/h and the minimum radius 25 m. Their low floor range is 70%. Seating is for 100 passengers with capacity for 148 standing. They are used on the lines B and C and can be coupled to a maximum of two units, which is done during peak hours.

In 2020 eighteen metro cars were ordered from CRRC Tang Shan in China for € 49.6m which includes five years maintenance.

Metro car 103 at Fonte do Cuco on 1 May 2014. *C. F. Isgar*

PORTO | Elevador de Guindais

By Ernst Kers

The first funicular of 1891

On June 4 1891 a funicular was opened at Guindais by the Parceria dos Elevadores do Porto. The lower end was at Ribeira near the lower deck of the Ponte Luiz I, the upper end in Rua de Augusto Rosa, near Praça da Batalha. The line was divided in two parts. One car made journeys over the whole length of the line, which was 412 m. On the lower and steep part of the line was next to the main track a second track with a length of 114 m. When the main car had to travel the lower steep part, a second car on the second track counter-balanced. A steam engine

A general view taken from the Dom Luis I bridge of the Elevador dos Guindais, seen to the left of the picture on 19 June 2008. *L. Folkard*

housed in a building at the top of the second track propelled the cars by cables. The main car had three cabins on a joined truck. The cabins were kept horizontal over the whole trip. The second car had a single step-type cabin.

Two years after opening (June 5 1893) a serious accident occurred resulting in a runaway of the second car. Although repairs were done, the line was not opened again.

The second funicular of 2004

On the same location a new funicular was built, which was opened on February 19 2004. This funicular is of the common design: single track with a crossing-point in the middle and two cars counter-balancing each other.

The cars have a capacity of 25 persons each. This funicular has a length of 282 m, the elevation is 71 m. The lower half is in the open air and has an inclination of 55%, the upper half is in a tunnel and almost level. The lower station is at the same place as the first funicular. The upper station is in Rua de Augusto Rosa near Rua de Saraiva de Carvalho. The cabins of both cars remain horizontal during the whole trip. The funicular is operated by the Metro do Porto.

Car on the lower section of the Elevador dos Guindais on 1 May 2014. *C. F. Isgar*

A perfect composition of one of the Elevador dos Guindais cars with a coupled set of metro cars on the Dom Luis 1 bridge in 2009. *P. Haseldine*

The Sintra Tramway

By Bob Lennox

Car 7 in yellow livery is seen on 1 September 2001 passing Adega de Viscondes Salreu Banzao wine cellar.

Michael Russell

Car 4 with an unidentified trailer is approaching the station in the early 1950s, note the streamlined livery. *Courtesy of the National Tramway Museum, photographer N N Forbes.*

The town of Sintra, named by the Romans as Cintia, is situated in what is known today as a "high amenity area". Probably its greatest asset is its location, sitting on the shoulder of a mountain range which rises in the Atlantic Ocean, surfaces at Cabo da Roca (the most westerly point of Continental Europe), and forms a crescent, trapping the sea breezes and creating a pleasant microclimate. This moderates the summer heat and ensures mild winters.

It is strategically placed, being within commuting distance of the Lisboa metropolitan area, but also has easy access to bathing beaches, the lush evergreen forests of the mountain, and the rich agricultural hinterland of the coastal plain, which includes the demarcated wine-growing region of Colares.

During the latter part of their occupation of the Iberian Peninsula, the Moors built a fortress above the town on one of the summits and garrisoned it as the most westerly outpost of their European empire. The Portuguese Royal Family claimed another of the highest peaks, commissioning the exquisite Pena Palace, which dominates the surrounding countryside high above the town. Its former hunting grounds are now an exotic and well-planted public park. For generations, European Royalty, aristocracy, poets and artists have made Sintra and its environs their home and playground, establishing estates and architectural follies which are tourist attractions today.

The first rails to reach the town from Lisboa were of the Larmanjat system in 1873, invented by the French engineer, Jean

Car 5 and tower wagon at Ponte Redonda. *Courtesy of the National Tramway Museum, photographer W.G. Thomas*

Larmanjat. Originally sponsored in Portugal by the Duke of Saldanha, the Government granted an operating concession to the UK based Lisbon Steam Tramways Company Ltd.

This unlikely pioneering marvel employed a single steel central rail (18kg/metre) which supported and guided a steam locomotive and its carriages. The central rail bore the main weight, but the ensemble balanced by means of additional side

The only available picture of Azenhas do Mar terminus with car 1.
B. Lennox collection

Car 3 in blue livery taken between 1957 – 1960. *G. Krambles – B. Lennox collection*

wheels running on two 20 cm wide timbers set 104 cm apart, all held in place by wooden sleepers.

The locomotives' driving wheels acted on these side timbers providing the propulsion, and the weight bearing on them could be adjusted mechanically according to the gradient and the traction required. This "track" was set in, and flush with, the public road surface, avoiding the construction and expense of a private right-of-way. The passengers sat back to back in knifeboard seating along the centre of the carriages, to concentrate the centre of gravity on the rail. Sixteen locomotives were ordered to provide motive power from the Scottish firm of Sharp Stewart in Glasgow, although there is some doubt as to whether they were all delivered. Coaches were supplied by Brown & Marshalls of Birmingham, UK.

Two lines were constructed from a common terminus in Lisboa (Portas de Rego), the 26 km line to Sintra and another of 54 km to Torres Vedras, further to the north. The Sintra route passed through Sete Rios, Benfica, Amadora, Queluz, Cacém, Rio de Mouro, and Ranholas, encountering many steep gradients along the way.

The Sintra line was inaugurated at 09 00 on 2 July 1873 by a special train consisting of locomotive "Lumiar" and four coaches conveying special guests. Public service started three days later. Other named locomotives are known to be "Lisboa" and "Cintra" (the former spelling of the town's name). Journeys were scheduled to take 1 hour 55 minutes, with 1st and 3rd class single tickets costing 550 and 400 Reis respectively.

From the start, passengers complained of the rough ride, and the locomotives suffered from a lack of adhesion, especially in wet weather, considerably extending the journey times. It is even said that on such occasions, passengers were obliged to get out and push. Breakdowns, derailments and even overturned carriages took place, some passengers were injured, and patronage understandably fell away. Despite reducing ticket prices and increasing the service to three round trips per day, both lines closed in 1875 and the Company went bankrupt in 1877.

Six of the carriages were subsequently purchased by the Porto tramways company, (CCFP) and rebuilt in their workshops to conventional standard gauge (1435mm), and saw service as trailers between Porto and Matosinhos on the steam tramway there, surviving into the era of electric traction and lasting until at least 1912. Quite a remarkable career.

Sintra had to wait until April 1887 before it was reached by a conventional railway, first from Alcântara, (some distance from Lisboa city centre) and then in June 1891 from Rossio, when a tunnel was driven to a new station, built above the city square of that name. This line was built to the Iberian standard gauge of 1,668 mm, initially using single track with loops at stations, then doubled from Campolide to Cacém in 1895 and eventually to Sintra in 1949. The new railway terminated a kilometre short of the historic Sintra town centre because of the steep and difficult terrain surrounding it, and the main station was established at this point with facilities for both passenger and freight.

Proposals to extend the railway to the important market town of Colares, and eventually the coast, were discussed as early as 1886, but stalled possibly because of the steep drop (170 metres in 3 km) behind the town, which presented somewhat of an obstacle to conventional railway construction. The terrain here dictated that narrow gauge should be considered as the only option, as it could follow the steep and tortuous course necessary to navigate the hill contours and reach the coast.

In 1895, numerous plans were drawn up, with variations of the main route, including aspirations to reach the towns of Magoito and even Ericeira further up the coast. In addition to the main line, there were also proposals for branches to São Pedro and to the village of Almoçageme, high above Colares on the coastal road to Cascais. Later on, plans emerged for an extension to Cascais, Boca do Inferno and Estoril but in the event, these were never built.

In November 1898, Sintra Council granted local businessmen Nunes de Carvalho and Emidio Pinheiro Borges a 99 year concession to build and operate the proposed line, the new company being called "Companhia do Caminho de Ferro de Cintra à Praia das Maçãs", which was later incorporated into the simpler working title "Cintra ao Oceano" (Sintra to the Ocean). However a crucial decision was made in view of the recently electrified tramways in Porto to the north and Lisboa to the east. Instead of steam traction as originally proposed, electricity was chosen to be the motive power.

The Sintra mountains are formed by a mass of magma that arose from deep in the Earth's crust as a result of volcanic activity relatively recently in geological times. The area is still slowly rising. Until about 300 years ago, the floor of the valley though which the tramway was to pass was below sea level and this finger of the sea reached as far inland as Varzea de Sintra, below the town. A beach existed at Galamares, whose name is derived from "Lagamares" (lake of the sea).

The roads in the area clearly follow what was once the coastline of this indentation of water. Long before the building of the tramway, the main route from Sintra to the coast, today known as the "Queen's" or "Royal" road, followed a course which clings to the hillside for most of its length and descends more gently to sea level through Colares.

In 1901 construction started, beginning with an exchange siding at the railway station, a short line climbing to the historic town centre, and the sinuous drop to a location on the confluence of two small rivers known as Ribeira. The main depot and a thermal power station were constructed here, using the river water to feed the boilers. The water supply was enshrined in a contract that is still in force today - the feed to the artificial reservoir at the back of the depot is still maintained.

Three steam reciprocating engines, each of 450 horsepower, were supplied by the German company Franz Scheiffer, with boilers by Belleville of France and three electric generators provided by Westinghouse of the USA, who also installed the power distribution system. An overhead crane built by Gustav Eiffel in France ran the length of the power station. The

A group of cars including trailers 14 and 11 with motor car 3 at the Sintra railway station terminus at some time in the early 1960s. *Brian G Dutton/Online Transport Archive*

electricity generated at Ribeira was also used for street lighting in Sintra, São Pedro and Colares, this being a pioneering scheme in Portugal. The route then followed the existing road (now Estrada Nacional EN247) along the flanks of the river valley, a distinctive border being created by the planting of a line of plane trees, most of which survive today.

The prospect of electric railed vehicles running through the village of Colares had alarmed its more conservative residents, who imagined their children perishing under the wheels of this

Motor car 6 with trailer 13 passing the depot on 2 September 1962. *W C Janssen/Online Transport Archive*

Car 7 on private hire in 2008 at the Adega de Viscondes Salreu Banzao wine cellar. *B. Lennox collection*

newly imported phenomenon. They banished the course of the line to the other side of the river valley from their village. This decision had a detrimental effect on the village's development as its commerce, banks and supermarkets were developed on the tramway side. Colares village today remains mainly residential.

As a result of this intervention, the line diverged and climbed away from the main road, taking a short cut through what was then open countryside and establishing a halt for Colares at a location some distance away from the actual settlement. The route continued through some fields to emerge on the main road again (Estrada Nacional EN375) at Banzão, continuing to follow the roadside to the terminus on the Atlantic Ocean at Praia das Maçãs (Beach of the Apples).

This was named after the fruit annually washed downriver from orchards and deposited on the sands by the tide. In former times, when Praia was at the mouth of the sea inlet, pirates were a menace on the high seas. During the period when Portugal was briefly reintegrated politically with Spain (1581 - 1640), King Phillip II of Spain built a castle in Praia on the rocky outcrop on the south side of the inlet in an attempt to control these pirates, which were threatening his trade with the New World. The foundations of this construction can still be seen.

The line was constructed using lightweight Vignoles rail spiked

directly to wooden sleepers laid in ballast on its own private right-of-way, mainly by the side of the road. Exceptions were the short branch from the railway station to Sintra Vila, as the historic centre was known, a short section through the main street of the town, and a few metres through Ribeira, which were laid in grooved rail along the public highway.

Single track and loops prevailed, with the loops corresponding to passing places dictated by the timetable, and not necessarily corresponding to centres of habitation, a phenomenon common to conventional railway construction, but anachronistic when viewed today. The loop at the half-way point in the line at Galamares is today some way from the shops, cafes and restaurants there.

In addition to the trams, which provided a passenger service, the line was intended to handle freight, for which six goods wagons were acquired. These were employed transporting coal from the rail exchange sidings in Sintra to the company's own thermal power station at Ribeira, and returning the considerable quantities of ash produced from the boilers back up to Sintra.

A small timber shed was built adjacent to the tramway at Banzão which housed a weighbridge and a siding into the basement of the adjacent wine cellar. This included a loading bay for wine barrels. Facilities were also provided here to handle

General view of Banzao depot. The white building further up the streets is the Adega de Viscondes Salreu Banzao wine cellar. The track in the depot continues across a weighbridge through to the cellar. *B. Lennox collection*

parcels and other small goods items.

On 13 November 1903, the first trials using the newly delivered trams were carried out on the steep section between Ribeira and Sintra. Another test followed on the 22 November as far as Colares, carrying a group of 39 people, including the Company Directors, Contractors and assorted local VIPs. By contemporary accounts, the cars acquitted themselves well. Although initial hopes were to begin a service at Christmas, it was not until 31 March 1904 that the line to Colares was inaugurated, followed by the rest of the line to Praia das Maçãs on the 10 July 1904.

The cars were built by J.G.Brill of Philadelphia, USA, who were supplying vehicles to most tramway systems on the Iberian Peninsula at that time. Details follow in the Fleet description, but they consisted of four open crossbench motor cars and four open crossbench trailers for summer use, together with three closed motor cars and two closed trailers for winter. Curiously, the trams were of smaller dimensions than those purchased for the other Portuguese city systems.

Despite their size, the requirement to haul trailers and wagons over the steep grades of the line obliged the fitting of two 50 horsepower motors to each car, making them amongst the most powerful trams in the country at that time. Prior to this, life had proceeded in the area at a most leisurely pace, nothing exceeding the speed of a horse (or more likely a mule). The impact that these new vehicles must have made on the local population can only be imagined.

While the cars were of the high quality construction expected of the prestigious Brill Company, it has been written that the line was built frugally using lightweight materials, the overhead being particularly basic. Poles were of timber, sparsely positioned, and at curves the wire was suspended with the minimum number of hangers, resulting in frequent dewirements. It is noted that a specialist from another tramway operator had to be called in, the overhead system being redesigned and restrung, especially on curves. As there are few straight sections along the course of the route, this must have been a major revision and a setback for the new company.

Apocryphal tales describe the working of the wagons on the

steep, winding section between Ribeira and Sintra, with their loads of coal and ash. Motor trams would each haul two wagons, one being propelled and the other being towed. Collisions with passenger cars on the many blind curves have been described in previous histories of the line.

However a deft system was developed to facilitate turning round at Praia. The tram and trailer would slow down approaching the terminal loop, which was conveniently laid out on a slight slope. A member of the crew (usually the points boy) would pull out the coupling pin which secured the trailer to the motor car, while they were still both in motion. The brakesman would slow the trailer down, allowing the motor car to separate and pull ahead. The points boy would then jump off, run ahead and once the motor car had passed, changed the points on the terminal loop, directing the trailer into it. The motor car would then brake, allowing the trailer to roll past, overtaking it in the loop, and arriving at the terminus first. The motor car then arrived, the two cars now being in the right order to depart. This slick manoeuvre remained one of the operational showcases of the line, much to the admiration of its regular passengers.

Initially, the exchange siding in Sintra was accessed by a level crossing on the main railway line close to where it emerges from a tunnel. This somewhat hazardous crossing was removed and replaced by a new connection laid in at the far end of the station, curving round the railway buffer stops. Whilst this arrangement avoided conflicts with trains on the busy main line, it involved a reversal at the top of a steep incline on the Vila branch, with the towing car and one of the wagons actually over the summit of the grade. Securing this load on the steep hill with only handbrakes and subsequently starting off into the siding was apparently a work of art for the crew, albeit a challenging one.

Surprisingly, despite providing an important rail link to the coast, linking otherwise remote communities, and providing transport for both passengers and goods, the company did not do well, and went into liquidation late in 1913. In 1914 a new company was formed, taking over as the Companhia Cintra-Atlântico, S.A.R.L.

It is recorded that, in November 1921, the line was used to transport members of the Sintra fire brigade and their firefighting equipment in an emergency, when the restaurant in the Hotel Real Belle-Vue in Praia caught fire. It is noted that the journey from Sintra to the beach took 25 minutes. If it is understood that the usual running time is 45 minutes, then the flight of the firefighters through the normally tranquil countryside must have been quite a prospect to behold.

The tramway continued to stagger from crisis to crisis, however in 1926 the owners promoted the dynamic Camilo Farinhas to head the company, and he took the operation to new heights.

The village of Azenhas do Mar (watermills of the sea - named after the number of water-driven flour mills found there) perches beautifully on a clifftop promontory like some elegant decoration on a wedding cake, just over a kilometre along the coast northwards from Praia das Maçãs. The two villages had always been rivals; however plans to link them by extending the tramway had been in people's minds for some time. It is said that a local dignitary was most disparaging of this prospect and declared that the public well at Azenhas would run with red wine before the trams would ever reach there.

On 31 January 1930, however, the tramway was indeed extended through the main street of Praia and then on roadside

Driver Carlos Farinha inches car I cautiously across the hazardous blind Mucifal crossing. The locals habitually ignore the traffic lights here. *B. Lennox collection*

reservation to Azenhas. In order to support the myth, and in good humour, it was arranged that a substantial supply of local red wine was piped through the well fountain on the opening day. Unfortunately no photographs of this event have come to light; in fact photos of trams running on the extension to Azenhas have proved to be very rare indeed.

Shortly afterwards, the timber shed at Banzão was replaced by a substantial and elegant station building and depot in the art deco style, comprising a large passenger waiting room, public toilets, refreshment kiosk, ticket office, and a small two-bedroom house for the staff. Parcels and small goods traffic was catered for by a new office with loading bays on both the street and inside the depot.

An inspection pit for maintenance was included in the one-road depot, the track continuing into the basement of the adjacent winery as before. A splendid roof terrace was included in the construction, with magnificent views of the Sintra mountains, where many staff social events were held on balmy summer evenings. Under these improvements, it was also planned to double the track between Banzão and Praia, some land being acquired which can still be seen today, but this interesting project was never pursued.

The depot and workshops at Ribeira upper level were totally enclosed (they had been open sided until now), and a row of houses built for staff, including grand quarters for the manager's

own use. A telephone system was installed, and an inspector's booth and open passenger shelter built at Galamares, as well as a closed passenger shelter at Colares. Finally, a staff cabin was constructed at the beach terminus (including a sink with running water) and seating provided for passengers as part of a new retaining wall there.

During the Second World War, like most other countries, Portugal suffered shortages of materials and fuel. The tramway came under increased pressure as road traffic dried up. Instead of holidaymakers, the hotels of Praia were filled with refugees from the war, including many from Holland and Luxembourg. At this time, as Portugal was a neutral country, it was naturally a centre for spying and espionage between the Allies and the Axis powers. Lisboa was famous for being a hotbed of secret negotiations and intrigue. It has come to light that the Sintra tramway played an intriguing but important part in this scenario, as the trams had a virtual monopoly of all travel between the town and the beach. The crews were in a special position to observe and gather information about who was moving about, and also overhear conversations.

A strange story has come to light. The Allies' SOE (Special Operations Executive) had a secret base in Lisboa, and recruited a refugee from Luxembourg housed in a hotel in Praia to travel back and forth on the trams, feeding back any information gained from the crews or passengers. It is recorded that, on a night

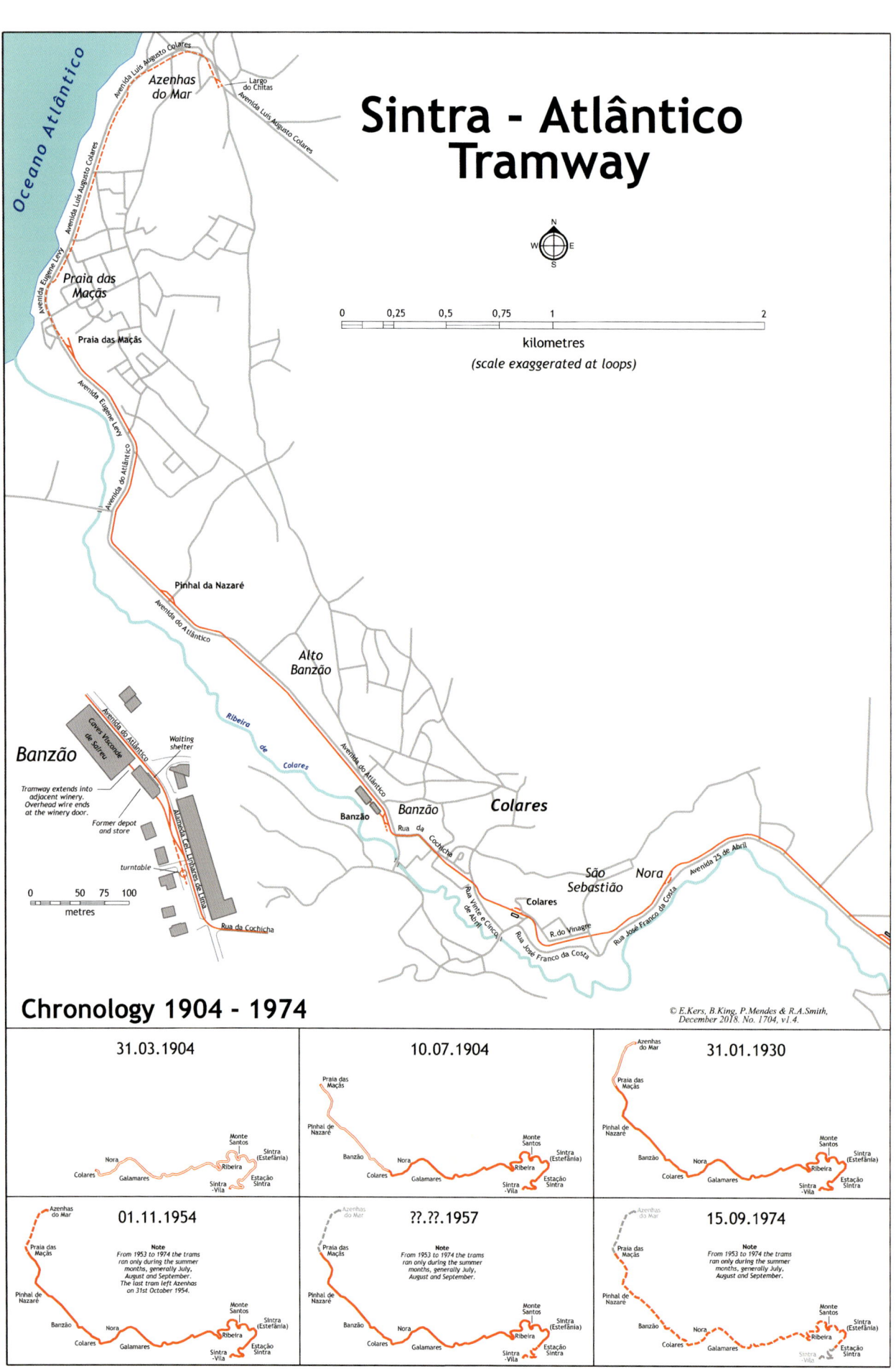

Sintra - Atlântico Tramway

Oceano Atlântico

0   0,25   0,5   0,75   1   2
kilometres
(scale exaggerated at loops)

Azenhas do Mar

Avenida Luís Augusto Colares

Largo do Chitas

Avenida Luís Augusto Colares

Praia das Maçãs

Praia das Maçãs

Avenida Eugene Levy

Avenida do Atlântico

Pinhal da Nazaré

Alto Banzão

Ribeira de Colares

Avenida do Atlântico

Banzão

Avenida do Atlântico
Conde Visconde de Soíreu

Waiting shelter

Tramway extends into adjacent winery. Overhead wire ends at the winery door.

Former depot and store

turntable

Alameda Cel Linares de Lima

0   50   75   100
metres

Rua da Cochicha

Banzão

Banzão

Rua da Cochicha

Colares

Colares

São Sebastião   Nora

Avenida 25 de Abril

Rua Vinte e Cinco de Abril

Rua José Franco da Costa

R. do Vinagre

Rua José Franco da Costa

© E.Kers, B.King, P.Mendes & R.A.Smith, December 2018. No. 1704, v1.4.

Chronology 1904 - 1974

31.03.1904

Monte Santos · Sintra (Estefânia) · Nora · Ribeira · Colares · Galamares · Estação Sintra · Sintra-Vila

10.07.1904

Praia das Maçãs · Pinhal de Nazaré · Banzão · Nora · Colares · Galamares · Monte Santos · Sintra (Estefânia) · Ribeira · Sintra-Vila · Estação Sintra

31.01.1930

Azenhas do Mar · Praia das Maçãs · Pinhal de Nazaré · Banzão · Nora · Colares · Galamares · Monte Santos · Sintra (Estefânia) · Sintra-Vila · Estação Sintra

01.11.1954

Azenhas do Mar · Praia das Maçãs · Pinhal de Nazaré · Banzão · Nora · Colares · Galamares · Monte Santos · Sintra (Estefânia) · Ribeira · Sintra-Vila · Estação Sintra

Note
From 1953 to 1974 the trams ran only during the summer months, generally July, August and September. The last tram left Azenhas on 31st October 1954.

??.??.1957

Azenhas do Mar · Praia das Maçãs · Pinhal de Nazaré · Banzão · Nora · Colares · Galamares · Monte Santos · Sintra (Estefânia) · Ribeira · Sintra-Vila · Estação Sintra

Note
From 1953 to 1974 the trams ran only during the summer months, generally July, August and September.

15.09.1974

Azenhas do Mar · Praia das Maçãs · Pinhal de Nazaré · Banzão · Nora · Colares · Galamares · Monte Santos · Sintra (Estefânia) · Ribeira · Sintra-Vila · Estação Sintra

Note
From 1953 to 1974 the trams ran only during the summer months, generally July, August and September.

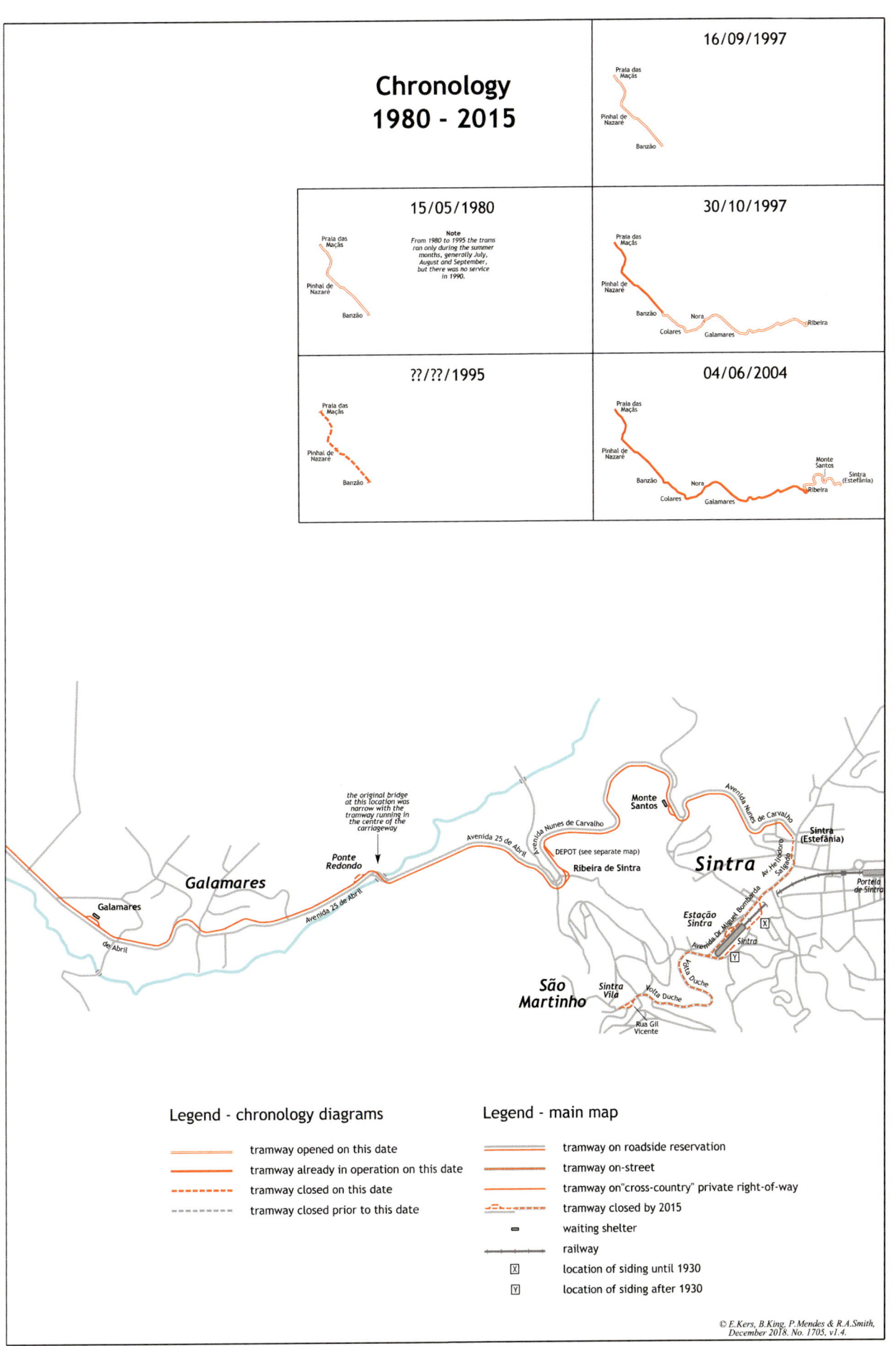

Chronology
1980 - 2015

16/09/1997

Praia das Maçãs
Pinhal de Nazaré
Banzão

15/05/1980

Note
From 1980 to 1995 the trams ran only during the summer months, generally July, August and September, but there was no service in 1990.

Praia das Maçãs
Pinhal de Nazaré
Banzão

30/10/1997

Praia das Maçãs
Pinhal de Nazaré
Banzão
Nora
Colares
Galamares
Ribeira

??/??/1995

Praia das Maçãs
Pinhal de Nazaré
Banzão

04/06/2004

Praia das Maçãs
Pinhal de Nazaré
Banzão
Nora
Colares
Galamares
Ribeira
Monte Santos
Sintra (Estefânia)

Galamares
Galamares
de Abril
Avenida 25 de Abril
the original bridge at this location was narrow with the tramway running in the centre of the carriageway
Ponte Redondo
Avenida 25 de Abril
Avenida Nunes de Carvalho
DEPOT (see separate map)
Ribeira de Sintra
Monte Santos
Avenida Nunes de Carvalho
Sintra (Estefânia)
Sintra
Portela de Sintra
Av Heliodoro Salgado
Estação Sintra
Avenida Dr. Miguel Bombarda
Sintra
Vista Duche
São Martinho
Sintra Vila
Volta Duche
Rua Gil Vicente

Legend - chronology diagrams

— tramway opened on this date
— tramway already in operation on this date
-- tramway closed on this date
-- tramway closed prior to this date

Legend - main map

═ tramway on roadside reservation
═ tramway on-street
— tramway on "cross-country" private right-of-way
-- tramway closed by 2015
▭ waiting shelter
┼ railway
X location of siding until 1930
Y location of siding after 1930

© E.Kers, B.King, P.Mendes & R.A.Smith, December 2018. No. 1705, v1.4.

walk along the clifftop path, he heard great activity and observed substantial movement and lights out in the waters of Praia Grande (the next major beach along from Praia das Macãs). He heard a number of heavy trucks on the beach, engaged in loading (or unloading). The next day, he returned to find tracks on the sand, confirming the extent of the operation. This was duly reported back to the Allies.

The impact of the war also had a major, but positive, effect on the tramway, underlining its importance as a strategic link in the area. This was manifest after the war's end by the sanctioning and construction of three new car bodies, substantially larger than the existing closed saloons, each with an increased seating capacity of 24 (from 17 on the closed original cars), and based on the 200 and 700 series trams already operating in Lisboa. It is now evident that there had always been a close contact between the Lisboa and Sintra tramway management.

Although Ribeira depot had its own carpenter's shop and craftsmen, a project of this scale was beyond their means. Staff from Santo Amaro in Lisboa were brought to Ribeira during the months of July and August, when they would normally be taking their holidays, along with material, to construct the new cars. The summers of 1945, 1946 and 1947 saw the appearance of a new car body each year.

Despite the addition of the larger cars and the improvements carried out to the infrastructure, the tramway never surpassed its peak carrying capacity achieved during the war, and its last year in which it turned out a profit was 1947. By 1950 it was making a substantial loss, its lack of revenue being blamed on the increasing general prosperity and private car ownership.

In 1951 a new administration took over, and purchased the company's first five diesel buses. They proposed to replace some tram services by bus, with a longer term aspiration to replace the trams completely, and began by operating a new bus service between Lisboa and Azenhas do Mar. In 1953, buses took over from the remaining tramway operation during the winter, with the trams running only in the peak bathing season from mid-July until the end of September each year. The attractive section of tramway along the clifftops from Praia to Azenhas, which had suffered from bus competition from its own company since 1951, was finally closed in 1954.

The Sintra town branch from the station to the Vila ceased operating on Sundays and public holidays as from 1956, owing to complaints from motorists, and was withdrawn completely in 1957 following road widening works along its principal thoroughfare, the Volta do Duche.

However, despite the closures, it was around this time that a political decision was made to recognise and retain the remaining tramway as an item of significant historical importance, and financial assistance was made to the Company to secure the operation. As a result of this intervention, new and heavier rail was installed on the difficult Sintra Estefânia - Ribeira section, and traction poles were replaced by cast cement lattice masts throughout most of the line. A new rotary converter, capable of powering the entire tramway (if necessary), was purchased and installed in the former parcels office in Banzão depot to supplement the supply, together with a new feed from the local power company.

A diesel locomotive, built by the German company of Deutz, was acquired from the Leixões docks near Porto, enabling track and overhead maintenance to continue during the winter when the power was switched off.

The former power station building, a substantial masonry construction in the lower level of Ribeira, was converted into a bus garage, offices and workshops. The steam driven electrical generation plant had been previously disposed of, three rotary motor-generator sets replacing it, taking their power direct from the local electricity company. These and their associated switchgear were relocated upstairs in the upper level behind the tram depot, providing more space for work on the buses.

At this time, the casino in Sintra was closed, and amalgamated with the casino in Estoril. It is said that as a condition of its closure, a share of the profits would be used to subsidise tourist projects in the Sintra area. This has occasionally included the tramway, and this source of funds has played an unseen part in the tramway's survival over the years.

As the 1960s dawned, the tramway entered a "Golden Age". With Praia das Macãs, Praia Pequena (small beach) and Praia Grande (large beach) being the nearest Atlantic bathing beaches to the Lisboa metropolis, they became increasingly popular as leisure time and prosperity expanded. Demand was such that line capacity was increased by installing two new passing loops at Ponte Redonda and Casal da Nora, allowing a 15 minute frequency, connecting with trains at Sintra station.

Open motor and trailer sets were packed to capacity, with huge standing loads being regularly carried. A figure of 198 passengers (including the crew of four; motorman, conductor, trailer brakesman and points boy) has been regularly quoted as a record load for an open motor and trailer set. Even the diminutive Brill saloon motor car No 3 was used in regular service, inevitably pulling a trailer.

However, with the opening of the new Salazar (now known as the 25th April) suspension bridge across the River Tagus in 1966, this brought easier access to the longest sandy beaches in Western Europe, beginning at Costa da Caparica south of the river on the Atlantic coast, and served by the Transpraia narrow gauge railway. Illegally-built summer houses, campsites and caravan sites along the coast further boosted that area's popularity. The Praia resort and its neighbours began a decline, from which they have never really recovered. To add to the gloom, the Sintra-Atlântico company was taken over by a road transport firm, Eduardo Jorge (of Lisboa "Carros de Chora") in August 1967. By all accounts, the new management was unsympathetic to the tramway and virtually ran it into the ground.

The trams were now looked on as a necessary liability, and its infrastructure and vehicles seemed increasingly anachronistic at a time when the country as a whole was beginning to develop and modernise. Its fate was sealed shortly after the "Carnation Revolution" of 25 April 1974, when the Salazar Regime (then headed by a new dictator, Marcelo Caetano) was overthrown in a bloodless coup by the young captains of the army. The revolution was so named as during the uprising, the people put carnations in the barrels of the soldiers' guns.

This was as a result of wars attempting to retain Portugal's African colonies against a rising tide of violent nationalism there. The flower of Portuguese youth was being routinely sacrificed for what was generally perceived by both the army and the population at large as a lost cause. The new political order changed everything. Eduardo Jorge and the Sintra tramway was nationalised and became part of the Rodoviária Nacional (national bus network).

1974 was to be the last summer season when the Sintra

Car 6 having broken an axle rests after a hair-raising rescue journey back to the depot on two of its own wheels and a permanent way bogie. *B. Lennox collection*

tramway operated in its traditional form. In September, after the last cars entered Ribeira depot for their winter rest, a meeting was convened by the staff. In those heady days after the Revolution, the former management of virtually all industry had been replaced by worker's committees who effectively ran the operations.

They recognised that the tramway was actually in a perilous state, with track and electrical plant largely on its last legs. It has to be said at this point that the Sintra trams had always been regarded with great affection by its own staff and passengers alike. A family atmosphere embraced everyone involved. By this time each car had developed its own individual characteristics - almost a "personality". Many people in the Sintra area had either professional or family links with the tramway, and the cars were held dear by all, including several generations of Lisboa holidaymakers, whom they had transported during happy times.

At the meeting it was suggested that the tramway be "mothballed", or suspended, until such times as an investor of some kind could be found, allowing the entire system to be revamped, making it safe and sound for the future. Continuing in its current condition was to invite the possibility of a disastrous accident, with the tramway being forced to close by an outside agency, in a way that it could perhaps never be reopened. The motion was discussed and reluctantly approved, but in all probability good sense prevailed.

There followed a careful inventory and gathering of spares that would be needed in the future, and these were dutifully stored under lock and key in the substation at the back of Ribeira depot. Grass and small pine trees began to grow amongst the ballast,

the wires started to sag, and an air of complete abandonment pervaded the tramway. The encroaching vegetation was not the only enemy at the doors. Some properties backing on to the line were surreptitiously extended, small land grabs were made, sleepers and copper wire began to disappear, and certain sections of the tramway looked like it had completely vanished, the former alignment being difficult to identify.

It was during this time that, according to the company archives, many requests to purchase trams and equipment were received, principally from museums and individuals from the USA. The letters still exist today, confirming that large sums of money were offered. It is a credit to the integrity and honesty of the Portuguese in charge of the Sintra tramway during this period that all these requests were politely declined, not a nut or a bolt leaving the premises.

The Rodoviária company did make some serious attempts to revive the tramway during this period, spending some serious time and cash studying the possibility. The line was surveyed, and Carris engineers from Lisboa produced a plan to reinstall the overhead to current standards. Some photos exist in the files of some metre gauge ex-Belgian Vicinal cars then operating second-hand in Northern Spain, together with correspondence relating to their operating requirements, purchase price and transport to Sintra. However the political climate changed again, and enthusiasm for the scheme dried up.

It was not until 1980 that a move was actually made to resurrect at least part of the system. Banzão depot with its small inspection pit, together with the 3 kilometre section to the beach, was a perfectly self-contained section of line that

Trailer operation is limited to private hires owing to a lack of run-round facilities at Estefania, Sintra terminus. *B. Lennox collection*

could be reopened without significant investment. It still had its rotary converter, the track was still largely intact, and there were enough sections of overhead existing that could be salvaged and restrung to equip the line.

Accordingly, open motor car 1, closed saloon car 3, open trailer 9 and closed trailer 10 were repainted and tested at Ribeira, the last time that the substation there was used. They were then transported by road to Banzão, together with the diesel, overhead tower wagon, a small closed tool wagon and a permanent way hand trolley. The tools for the job were all to hand.

On the 10 June 1980, the UK-based Tramway Sponsorship Organisation, with around 35 members, visited both Ribeira and Banzão, hiring the two motor cars together with their trailers for several memorable trips along the newly re-opened line. Although Rodoviária had held an official launch prior to this visit which appeared in the Portuguese media, this visit, which included the original author of this book, John Price, ensured that the enthusiast world knew of the tramway's revival.

From then on, every summer, in July and August, a half-hour public service was operated between Banzão and Praia, usually by open car no 1, six days a week, with the crew resting each Monday. It was truly a delightful little operation, which retained the "small town rural tramway" atmosphere of the original line, albeit truncated. Occasionally in inclement weather, the diminutive Brill saloon car No 3 would deputise, always a pleasant surprise if one were to venture into the area on speculation. This arrangement continued and worked quite reliably until 1993 by which time the track and pointwork had deteriorated to the extent that several derailments occurred.

Expertise and materials to improve the track came from the Lisboa Metro, and the tramway was able to continue.

In 1995, a major initiative from Sintra Council, under the leadership of Mayoress Edite Estrela, envisaged refurbishing the tramway between Ribeira and Praia das Maçãs. In anticipation of this project, open crossbench motor 6 and closed saloon 4 were extracted from Ribeira depot and transported to the Carris workshops in Santo Amaro, Lisboa with a view to their assessment and restoration. In a fateful coincidence following on from practice abroad, the Rodoviária bus company was privatised, the franchise for the Sintra, Cascais and Oeiras areas being won by the Scottish-based Stagecoach bus operator, which of course included the Sintra tramway.

In 1996, work on the refurbishing of the line began, with Council contractors clearing the route of obstacles, assorted outhouses, and undoing the work of Mother Nature, which by this time had taken over large swathes of terrain. However it came as something of a shock to the Sintra Council to find that the land on which they were working now belonged to Stagecoach as a consequence of the recent privatisation. Work stopped while discussions followed, and after some internal political difficulties were resolved, an agreement was reached whereby Stagecoach would collaborate with the project, and were granted a concession from the Council to operate the tramway.

Work recommenced rebuilding the right of way, excavating a trench two metres wide, one metre deep and 9.2 kilometres long (the length of the tramway to be reopened). This was filled with a crushed rock base which was graded and consolidated to form the foundation of the line. New drainage, road crossings, retaining walls, kerbing and over 200 neatly paved entrances to private houses across the tramway were then meticulously installed, all being carried out by the Council contractor Sanestradas.

On 25 April 1997, the first loads of rails, sleepers and ballast began to arrive in Sintra, being obtained second-hand from Portuguese Railways. The rail was lifted from the Beira Baixa line, close to the Spanish border, which was being upgraded. Points were salvaged from the former Sabor metre gauge line in the Douro Valley, which had closed some years previously.

New sleepers were also purchased, being laid in the ratio of one new timber to every three second-hand timbers, the railway contractor Ferrovias undertaking the tracklaying. On 30th April, the first spike on the new line was driven in without ceremony at Praia. The Council also funded and constructed two new electrical substations, one at Banzão and the other at Ribeira, as well as new steel masts, bracket arms, feeders and overhead line, using the Portuguese firm of EFACEC as contractors. Six road crossings were equipped with traffic lights and warning bells, activated automatically by the approach of a tram.

Simultaneously, Stagecoach began renovating the depot and office facilities at Banzão, and prepared the two motor trams and two trailers for further service. These had lain gathering dust for two years, but had also suffered from the flooding of an adjacent river, and required substantial attention as well as a repaint. During the summer of 1997, no power was available to test the trams so a generator was purchased from Bolton Trams Ltd. in the UK.

This enabled the cars that had lain stored in Ribeira since 1974 to be tested and assessed. As a result, the truck from closed car 2, which was found to be in the best condition, was removed, transported to Banzão and put under open car 1, which had water-damaged motors and badly worn wheels, having run regularly without any serious maintenance for 15 years.

The Council had an incentive to complete the project prior to a local election, and work on the tracklaying, substations, overhead and tram refurbishment continued rapidly throughout the summer of 1997. The project was sufficiently advanced that by August 31st, the new substation at Banzão was switched on and the trams were tested along the new line to the Beach. Some weeks afterwards, the excitement continued with the arrival of car 4, beautifully restored by Carris in the Lisboa workshops of Santo Amaro.

The Carris team brought their own electric jacks to unload the car from the lorry, but at that time Banzão didn't have sufficient power to energise them, creating a panic. Fortunately the local Fire Brigade, who had a station around the corner, were contacted, and were able to provide their portable generator, solving the problem. Car 4 was the first wide saloon car to run in Sintra since 1974.

Fortunately, Stagecoach had inherited an employee, José Mindouro, who had driven and maintained the trams since 1962. He was joined by the former works foreman, Fernando Jorge, who came out of retirement, both of them driving and carrying out the routine maintenance. A third tram driver was hired from Carris, and former conducting staff were drafted in from the Stagecoach bus operation, so that on 16 September a two-car public service could be operated between Banzão and Praia. The entire line to Ribeira was inaugurated on 30 October, with Edite Estrela, Mayoress of the Council, Brian Souter, Chairman of Stagecoach, and the Duchess of Bragança and her family as special guests.

This was a significant achievement, considering the moribund

At Ponte Redonda Sintra Council Mayoress Dr Edite Estrela looks down on car 3, which is sporting inaugural flags featuring the emblems of Sintra Council and Stagecoach. *B. Lennox collection*

state of the tramway just seven months previously, and took many people by surprise, the virtual resurrection of a dead tramway being an extremely rare event. The project embraced the construction of 9.2 kilometres of track and overhead, two new substations, a refurbished depot, office and staff quarters, 3 restored trams (plus 2 trailers), the designing and printing of new tickets and timetables, advertising, vehicle insurance, staff training, new rules and regulations, and the setting up of an administration and accounting system, all in a relatively short time. From being a dusty store, Banzão depot suddenly brimmed and hummed with life.

However, Banzão had limited facilities for maintenance, being very cramped, and cars could only be lifted off their trucks outside in the yard due to its low ceiling. The following winter, the disused track into the former bus garage in Ribeira low level was reconnected and extended, allowing use of a small pit and the overhead crane there. This considerably improved the maintenance facilities, and also permitted the relative luxury of lifting car bodies indoors. The service continued into the winter months, the first time the Sintra trams provided a year-round service since 1952.

This was terminated abruptly a few months later during January 1998, when the same river that had damaged cars in Banzão the previous winter flooded again during heavy rains, this time damaging a bridge carrying the line, and washing away part of the embankment, including an overhead mast. Tramway operation was restricted once again between Banzão and Praia until the bridge was rebuilt, the full service to Ribeira resuming in May. The rebuilt bridge incorporated a wider aperture for the river, eliminating the danger of flooding at this point. The trams stored in Banzão would never be damaged by water again.

In Santo Amaro, Lisboa, car 6 received a full overhaul and repaint, and prepared for service. It had been chosen for restoration as it was thought to be in the worst condition of the fleet, however on dismantling, was found to be in better condition than expected, and managed to retain 80% of its original timber, a fitting tribute to its builders, J.G. Brill. The previous year, car 4, a mere youngster from 1947, hardly required any new timber at all, and was found to be in extremely sound condition.

Beautifully restored, car 6 was unloaded at Ribeira on 7 April, and was duly tested by the Carris staff, before being handed over to Stagecoach. This then provided the tramway with four motor cars, two open and two closed. Four more drivers and a conductress (all in their early twenties, giving a youthful aspect to the tramway for the first time) were recruited locally, trained and tested to enable a midweek service of two cars and a weekend service of three to be operated during summer 1998. The three-car service proved to be over-optimistic and ran for only a few weekends, two cars being sufficient for the rest of the summer.

Unfortunately passenger numbers were disappointing, not least because the landward terminus at Ribeira was two kilometres away from Sintra town. The fares, set at 500 escudos for a single ticket (Ribeira-Praia) and 300 escudos (Banzão-Praia), were substantially higher than the parallel bus, and perceived as expensive by many locals. To promote the flagging service, two ex-London Routemaster open top buses were purchased, and ran on a special route between Cascais, Sintra and Ribeira to feed the tramway. While the bus service did relatively well, it did not result in many extra passengers for the tramway, despite combined tickets and an advertising campaign.

Fortunately, many tourist coach operators ply their trade in the area, as Sintra, Cabo da Roca (the most westerly point in Continental Europe) and the Adega do Colares (Colares winery in Banzão) attract many visitors. Through a series of travel agencies, regular private hires were tied in to these coach trips, both in summer and winter, providing the tramway with a steady and important stream of income, at that time better than the "farebox".

In addition, the Miramonte hotel, on the tramway close to the beach, specialised in off-season walking holidays, and hired a tram on a weekly basis throughout the winter. This was supplemented by evening hires for "wine and fado" parties, taking the Miramonte revellers by tram to a night-spot in the village of Galamares, famous for its fado (a melodic and emotional brand of Portuguese folk-singing). This often resulting in exhilarating rides through the deserted and darkened countryside in the early hours of the morning, something that probably never happened during the Sintra-Atlântico days.

In retrospect, the restoration of the tramway had been a joint project between two entities with completely different outlooks and expectations - Sintra Council for cultural and political reasons and Stagecoach as a business proposition - an unlikely partnership. Although some serious and imaginative efforts were made to render the tramway profitable, the "core business performance" fell short of expectations, mainly because of its physical isolation from its main source of passengers.

In 2001, Stagecoach decided to sell the tramway, and it was bought by Sintra Council, who then operated it directly themselves. It was recognised that, in its current form, reliant on subsidy, its future would always hang in the balance, but its significance as a National treasure was recognised. The crucial decision was taken to reinstate the important Sintra - Ribeira section, which was without doubt the most expensive and difficult part to re-engineer, with its sharp curves, continuous steep grade and tight clearances. This was to prove to be an "umbilical cord", providing the sustenance that the tramway needed to survive as a healthy entity once again.

In April 2004, the tramway's hundredth year, the important section between Ribeira and Estefânia (the beginning of the private right of way some five minutes' walk from Sintra station) was reopened. Relatively new rail and points (with hardly any wear) from the former carriage sidings at Sete Rios had been gifted by the Lisbon Metro, for the token price of one euro. This section included many tight curves and retaining walls, as well as substantial drainage works, but brought the terminus within a short walk from the rail station in Sintra. This, combined with a reduction of the single ticket price to one euro, (recently raised to three euros but still less than the parallel bus route), ensured that the fortunes of the tramway were utterly transformed, 60,000 passengers being carried to the beach during the first summer. Large queues can be counted on to form at the terminus on summer days, many waiting for over an hour for the opportunity to ride by tram to the beach.

Since then a new service has operated using two single cars giving an approximately hourly frequency from each terminus crossing in the Galamares loop, once again finding itself in the operational centre of the line. In winter one car provides a two hourly service. Journey time is 45 minutes, just like in the old days. Private hires are generally undertaken during the week. Because there are no run-round facilities at Estefânia, trailers are rarely seen in regular service, but can be used on busy hires by towing to the terminus by one car, then transferring to a motor car that has followed it up the hill.

In 2006, the trams stored in Ribeira depot were transported to a Council facility some kilometres away to allow for the old complex to be completely transformed. Cracks in some of the structures of the upper level betrayed slippage in the ledge on which they stood. The depot buildings and former staff housing were demolished, and the entire hillside was removed, excavating to bedrock found below the road level. Thousands of tons of rock and spoil were shifted in this massive and costly operation.

In the void thus created, ten solid pillars were built up using circular concrete tunnel segments, reinforced in steel and filled with cement. New foundations were built up around them and the ground reinstated as if nothing had happened. On the newly prepared site a tram shed of modern appearance was constructed, containing a new track fan with four roads, a new pit and workshop area as well as new offices and space to accommodate the substation. Initially constructed without doors like the depot it replaced, a series of petty thefts and attacks with graffiti resulted in doors being fitted retrospectively.

The exception was the original house built and lived in by

A view of Ribeira depot being rebuilt showing the extent of the works.
B. Lennox collection

Cars 3, 6, 1 and 4 in a depot line-up in 2009. *P. Haseldine*

the former Director, Camilo Farinhas, which was retained and restored, and tastefully included in the new construction, with its terrace and superb views of the tramway below and the mountains above. Maintenance of the trams, formerly carried out in Banzão, was then transferred to the new facilities at Ribeira, using a team from Carris who travel from Lisboa twice a week.

The passing loop at Ribeira still retained its grooved rail and points, laid in setts from the time when it formed the main highway before the road was widened. It dated from the construction of the line in 1903 and these rails were now considerably worn. They were replaced by rail from the Lisboa Metro and the setts were restored tastefully. The former power station building on the lower level was also completely refurbished, becoming the Museum of Science, and the connection to the tramway (both physical and operational) was removed. The new substation, installed in this building in 1997, was resited in the upper level.

The trackwork on the Ribeira - Praia section had been built in a hurry in 1997 with second-hand material and required constant maintenance. Electric trams built in 1903, with their simple suspension and heavy motors, soon distort the geometry of conventional sleeper tracks. The joints between straight rails work loose, the adjacent sleepers sink into the ballast, and cause cars to bounce over them alarmingly. Joints on curves are pushed sideways out of alignment, exaggerating the angle between the rails, and create a "lateral lurch" as the wheels pass over them.

During the winters of 2008 and 2009, the tramway was closed to permit the track to be rebuilt and improved. A large quantity of sleepers and many of the rails were replaced, with rail joints on curves being welded. The right of way was further enhanced with the building of more retaining walls and better drainage. This cured a perennial problem where the runoff from heavy rainfall on adjacent properties would bring sand, soil, gravel and loose rocks down from the hillsides, blocking the tramway at various locations. The upgraded track formation was then aligned, levelled and the rail joint foundations packed, giving a vastly improved ride.

A derelict house, Vivenda Alda, at the Sintra Estefânia terminus, was rebuilt as a waiting room, ticket office and exhibition centre for the tramway. During this period, closed trailer 10, Brill motor car 3 and closed saloon 2 received substantial restoration work, car 2 returning to service having lain idle since 1974.

Since 2001, the tramway has been operated within Sintra Council by the Division of "Patrimonio Imovel e Móvel" (Buildings Heritage), and cherished like family jewels. The small number of operating staff are both dedicated and flexible. In addition to driving and taking fares, they clean the cars' interiors and window glasses, sweep the tracks clear of falling leaves and generally carry out more ancillary tasks than is normally expected of regular tram crews, including painting of cars and other general repairs. This work continues during the quieter winter season, and the proud traditions of Sintra Atlântico are thus maintained.

In an area of arboreal beauty (the line runs through several kilometres of pinewoods), winter storms can bring trees down and landslips bought on by heavy rains can block the line. Fortunately the tramway can rely on the wider Sintra Council for services and manpower when problems like these occur.

In 2009 the management of the tramway was transferred to the Division of Tourism within the Council, with plans to develop the tramway still further. The return of the tramway to the station, and even the historic centre in the Vila has been often suggested, although there are now heavy traffic flows through the narrow streets of the town, making this prospect somewhat of a challenge. The Portuguese have always been resourceful, and have surprised many by their bold decisions. Never say never.

The story of the Sintra trams has had as many twists, turns, ups and downs as the line itself. Stoically enduring hard times, it has survived and prospered and seems to have had somewhat of a charmed existence. During the inaugural journey in 1904, it is said that the car ran over and killed a black cat, implying that a degree of bad luck would follow. However the tramway has also had more than its fair share of guardian angels, and is now a remarkable and unique transport icon. The joys of riding a Brill crossbench car through the pinewoods to the beach really cannot be enjoyed to the same extent anywhere else.

The Tramway Today

"Heritage" tramways in Portugal tend to blossom or stagnate at the whim of local and national political strategy at the time, as in many other countries. Although subject to major investment in its infrastructure (re-extended from Ribeira to the edge of Sintra in 2003, and the complete rebuilding of Ribeira depot in 2008), the line lost the lower part of the depot to a "Museum of Science" during the rebuilding, and also lost the Banzão depot to a local micro-brewery and bar complex in 2018.

On the positive side, motor car 2 has been restored and returned to service after being stored for over 40 years, and two of the open motor cars now feature a quasi-historical livery, car 6 in blue and car 7 returned to yellow. The summer service approximately every hour starting at 10:20, although this is subject to alteration at short notice. A less frequent schedule is maintained in winter, using a single car.

Complaints are frequent that advertising is scarce, and some visitors have stated that they had difficulty in finding the terminus. Despite this, the tramway is popular, although the capacity is limited, and many tourists are either turned away or face a long wait. For those fortunate enough to find a place aboard, the journey winding through the green valley in the shadow of the Sintra hills on a centenarian "elétrico" is a memorable experience.

Fleet Description

The most famous and iconic vehicles associated with the Sintra tramway are the open crossbench motor cars. It is most likely that these were originally numbered 1, 3, 5 and 7, and similar trailer cars 9, 11, 12, 13. They were built by J.G. Brill of Philadelphia, USA in 1903, and are identical to vehicles which ran in Lisboa some years previously as horsecars and which survived into the electric era as trailers there. They each have 8 benches providing seats for 32, four of whom ride with the motorman and four on the rear platform bench. The central four benches have reversible backrests.

They can be distinguished from other contemporary Brill crossbench electric cars by having end bulkheads with two sash windows instead of the usual three, overall lighter construction, and bullseye glasses set into the clerestory end panels. Brackets survive on the bullseye glasses that held removable coloured lenses, but these lenses have long since been removed. The sash

Car 7 seen on test in Galamares after its restoration in 2001.
B. Lennox collection.

Interior of car 7 after restoration in 2001. *B. Lennox collection*

windows drop into pockets in the bulkheads. Patterned canvas roller blinds can drop to protect passengers from sudden rain showers or strong sunlight.

These cars also have wooden safety straps that drop to elbow height along the length of the car on both sides that act as restraining bars and arm rests, together with chains across the front and rear platform steps. These were rarely used except when carrying groups of young children, but recent Health and Safety regulations see them used now in daily service. Full-length hinged footboards are provided on both sides, and provided extra standing space which was well-used during busy times; however current regulations restrict standing passengers to four on the rear platform.

Like all Brill open cars in the USA and exported worldwide, the side seat valances have a distinctive shape, employing three-dimensional curves which look elegant. They are in fact the main reason why these cars have survived so long, ensuring a robust attachment of the main body pillars to the substantial timbers which form the underframe. The secret of their longevity is that they are cast in a bronze alloy, solid but malleable enough to allow a certain flexing of the upright pillars, while securing the integrity of what would otherwise be a relatively delicate body format.

The closed motor cars, also likely to have been originally numbered 2, 4, and 6, together with similar trailers 8 and 10, were built by J.G. Brill in 1903 as 14 foot 8 inch saloon cars with open platforms and five side sash windows. When built, these windows could drop into pockets in the car body sides, followed by the top sections, creating a semi-open car on warm days. Sprung roller blinds in the window bays can drop and act as sunshades.

These cars are of smaller and lighter construction than contemporary electric cars being supplied by Brill at this time. As they also have bullseye lights in the centre of the clerestory end panels (similar to the open cars), it has been surmised that these are also a horsecar design adapted for electric traction. Initially built with open platforms, vestibule windows were added sometime during the 1920s.

The original 7 motor cars (4 open and 3 closed) were each equipped with Brill 21E trucks of 6 foot 6 inch wheelbase and 35 inch diameter wheels. Motors, controllers and circuit breakers were supplied by Westinghouse and came from their factory in France. The motors were of 50 horsepower, and are today as originally supplied except that their plain armature bearings were replaced by roller bearings in 1955. Surprisingly the motors still retain their white metal nose bearings, with which they hang on the axles.

Service braking was by conventional gooseneck handbrakes with columns mounted outside the dash plates. When the closed cars were vestibuled, the new windscreens obstructed the sweep of the handbrake handles, and they were replaced by vertical wheels (except on car 6). The hand brakes were augmented by Westinghouse-Newell electro-mechanical track brakes. When these are applied, the polarity of the motors is reversed, generating current which slows the forward motion of the car and energising magnets suspended above the track. These magnets drop and are allowed to drag a short distance along the rail, also retarding the speed of the car. This movement forces open mechanical callipers between the truck frames which push auxiliary brake shoes against the wheels by the force of the drag. When finely tuned, this gives a much smoother and more effective stop than conventional electro-magnetic track brakes.

The tyres (the replaceable rims of the wheels) are of similar size to the four-motor bogie cars which operating in Lisboa for many years, and there is evidence that new tyres were jointly sourced by the two companies, making a considerable saving in what was possibly the single biggest replacement cost of tram spare parts. The original Westinghouse camshaft controllers were replaced in 1955 by British Thomson Houston type B18s, which were obtained second-hand by Lisboa from Dundee in the UK, but never used there and subsequently sold to Sintra. The

original Westinghouse circuit breakers were also replaced by BTH examples in 1955.

The original trolley bases and heads are of the Ohio Brass Company's double-acting spring type, designed originally to flip over at termini like a bow, without needing turned by a conductor, who walked the securing rope round to the rear end, his rope fixed to the trolley head by a ring on a sleeve that turned on the shaft as he walked round. This is confirmed by photographs showing the Praia terminus with a high wire, and the trolley pole almost vertical.

These trolley bases are also different from later designs in that they are not secured to the spigot on which they turn, other than by gravity. This gives rise to an unfortunate tendency for the entire trolley assembly to lift off the roof if it gets caught in the wire, and fall to the ground in the wake of the car. Most of these bases have now been replaced by conventional trolley bases obtained from Lisboa.

Gongs are of the "Dedenda" type and mechanical sanders are of the "Dumpit" pattern. Couplers are of the Van Dorn design. Additional safety chains and hooks were provided on the car collision fenders but rarely used. Current is supplied at line voltage (550 volts DC) to the lighting in the motor cars and to the trailers via cables suspended from the canopies and connected using a simple one pin plug and socket.

As a result of careful sanding down of paintwork, discoveries while dismantling for restoration, old tinted postcards and staff memories, it appears that the early livery was cadmium yellow (also known as traction orange in the USA) relieved by primrose yellow. This seems to have been reversed at some stage, resulting in the lighter primrose yellow taking prominence. Roofs were aluminium/silver, with white destination boxes, and lining out was in orange or black depending on the panel colour.

This was superseded in 1956 by unlined matt sky blue with large red numerals, shaded in black. In 1962, the livery was again altered to a cherry red and ivory, lined out in white. Subsequently, coloured photos have come to light which confirms some of the former colour schemes and details.

There is evidence that some trailer cars were originally motored and some motors were originally trailers. Differences include trolley planks, the position of the handbrake column, and the ceiling, which is ribbed on the original motor cars, and panelled on the trailers. Trailer 3 became motor 6, the original trailer 8 became motor 3, motor 4 became the second trailer 8, motor 5 became trailer 14, and motor 6 became trailer 15. Trailer 12 was originally an unidentified motor car.

Three surplus cars were obtained from Lisboa in the 1990s. The former 615 was regauged and repainted into Sintra livery and used on works duties and occasional private hire. Its performance is very similar to the Sintra motor cars but the riding is much better. Lisboa 709 was also regauged and repainted, and numbered 100, but not used, and car 703 is stored "as bought".

Works Fleet

The most significant member of the works fleet is a venerable overhead inspection and maintenance trailer dating from the building of the line, numbered S1. This has survived in service, and indeed has proved invaluable in the restoration and subsequent maintenance of the tramway overhead. Consisting of an extendable steel-reinforced timber tower, revolving timber working platform with folding safety rails, and contained in a

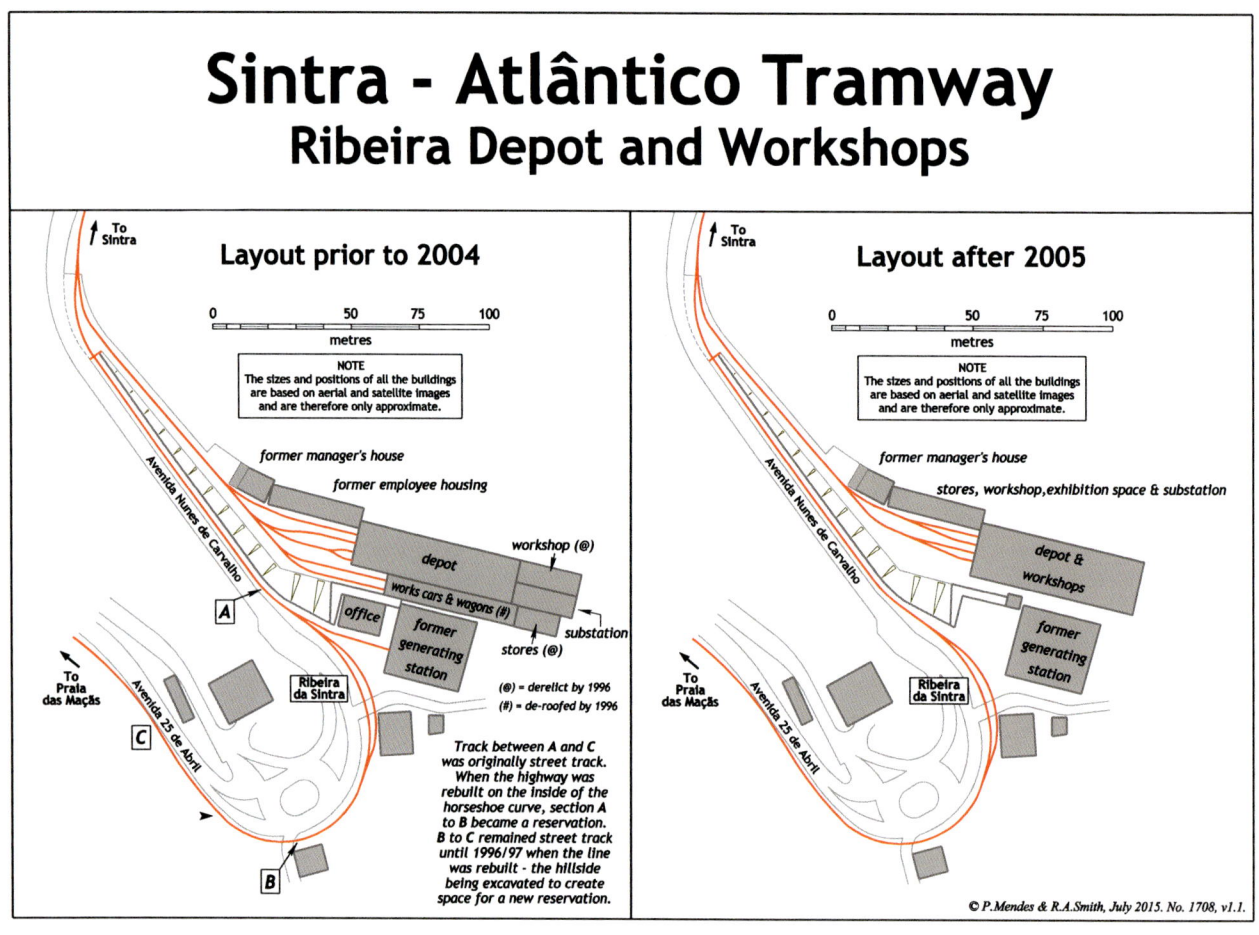

Sintra - Atlântico Tramway
Ribeira Depot and Workshops

Layout prior to 2004

To Sintra

0   50   75   100
metres

NOTE
The sizes and positions of all the buildings
are based on aerial and satellite images
and are therefore only approximate.

former manager's house

former employee housing

depot

works cars & wagons (#)

workshop (@)

office

A

former generating station

stores (@)

substation

To Praia das Maçãs

C

Ribeira da Sintra

Avenida Nunes de Carvalho

Avenida 25 de Abril

(@) = derelict by 1996
(#) = de-roofed by 1996

*Track between A and C
was originally street track.
When the highway was
rebuilt on the inside of the
horseshoe curve, section A
to B became a reservation.
B to C remained street track
until 1996/97 when the line
was rebuilt - the hillside
being excavated to create
space for a new reservation.*

B

Layout after 2005

To Sintra

0   50   75   100
metres

NOTE
The sizes and positions of all the buildings
are based on aerial and satellite images
and are therefore only approximate.

former manager's house

stores, workshop, exhibition space & substation

depot & workshops

former generating station

To Praia das Maçãs

Ribeira da Sintra

Avenida Nunes de Carvalho

Avenida 25 de Abril

© P.Mendes & R.A.Smith, July 2015. No. 1708, v1.1.

four-wheel truck with iron grid platforms protected by quaint tin canopies, this artefact must qualify as one of the strangest and possibly oldest service vehicles in regular use on a commercial tramway anywhere.

It has compact storage for tools and equipment in both the tower base and in wooden trunks on each platform. It further boasts a mechanical handbrake, a vice, and is usually adorned by a ladder and a roll of wire hanging on special hooks, a veritable workshop on wheels. The tower and platform are raised by cranking a handle which turns a large drum, round which a steel rope is wound, the assembly being raised by a series of pulleys. A simple ratchet secures the platform at the required height. The top platform is swung round by means of a rope from the ground, although an adept operator can also position the platform from above by applying some force on the tower beneath him.

The line had always been envisaged to handle freight as well as passenger traffic. For this purpose, six wagons were supplied by F. H. Bagge & Sons, Barcelona, Spain in 1903. Their main purpose was to carry coal from the exchange siding in Sintra station to the thermal power station at Ribeira, and the return loads of ash which was transferred at the station for onward transport by the broad gauge. In addition to coal and ash, other agricultural commodities found their way on to the Sintra tramway, including quantities of manure (much to the chagrin of the passengers).

Out of the original six freight wagons, two still survive in operable condition.

A small "J" type four-wheel general purpose tool van was also restored in 1997. Originally used for parcels traffic, it had sliding doors on both sides, and substantial internal shelves, which were useful to store heavy permanent way items and tools. Both the wagons and the van had mechanical hand brakes.

Last but not least is a diesel-mechanical locomotive of 0-4-0 wheel arrangement, built by the German company of Deutz in 1932 (works number 23074.) It was part of a batch of 176 similar standard-gauge light shunters designated OMZ 122 R. These two-axle locomotives were originally fitted with 40 hp two-cylinder upright two-stroke diesel engine, weighed 16 tonnes in working order and were capable of maximum speed up to 13 km/h. At some stage in its career, it was converted to 5' 6" (1,668 mm) Iberian gauge and sold to the Port of Lisboa Docks Authority. It was then re-gauged and sold to the metre gauge docks system in Leixões, to the north of Porto. In 1955, it was sold to Sintra Atlântico, when its four buffers were removed, and couplings altered to suit the existing Van Dorn units on the Sintra trams. Its original Deutz engine was replaced with a Leyland unit. It is one of the last of its class to be still active and has always been referred to by the French term "Dresine". Re-conditioned and repainted in 1997, it received a substantial overhaul in 2011.

With its unsophisticated but reliable works fleet and manual equipment, combined with experienced and dedicated staff, the Sintra tramway has proved relatively self-sufficient in an age of increasingly complicated technology. Its works fleet is as anachronistic as its passenger trams, but acts as a perfect complement and support.

COIMBRA

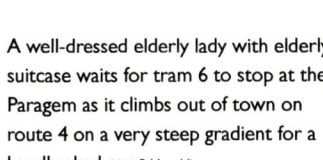

Trams

A well-dressed elderly lady with elderly suitcase waits for tram 6 to stop at the Paragem as it climbs out of town on route 4 on a very steep gradient for a handbraked car. *P. Haseldine*

By Pedro Costa, translation by Andy Steel, contributions by Ernst Kers

Coimbra is Portugal's historic university town; education is its main industry, and the students traditionally wear a curious black cloak swung over the shoulder, with a strip of coloured ribbon to denote their faculty. Baedeker, writing in 1901, said that the situation of Coimbra had long been a theme for the praise of poet and traveller, combining as it did the charm of Northern Europe with that of a sub-tropical climate: "The sea-pine and the poplar are neighboured by the date-palm; the slopes are covered with vines. agaves, eucalyptus, pines and orange trees, and the curious costume of the guitar-strumming students combines with numerous historical associations to invest Coimbra with a unique charm". With the exception of the students' costumes, none of this has changed. The students now only wear the traditional costumes for special occasions.

The town of Coimbra is situated about 200 km north of Lisboa and 100 km south of Porto. It occupies an irregularly-shaped hill overlooking the Rio Mondego. The Baixa (lower town), built along the river, contains the principal shops and other commercial activities, the university city occupies the nearer hilltop, and the higher ground beyond is mainly residential. The main shopping street (Rua Ferreira Borges) is barely 6 to 7 m wide in parts, and must be one of the narrowest in the world to have had a tram service. The path from the Baixa to the Alta (upper town) to the university is sufficiently steep that a part has earned the local name of 'Rua Quebra Costas' (Rib-Cracker Street). This side of the hill is far too steep to make a tramline possible and the trams had to go around the hill to the other side where the slope is less steep, but still challenging.

There are two railway stations. In 1864 the railway between Lisboa and Porto was opened. This gave Coimbra good connections with the two largest cities in the country. The main line station, Coimbra-B (Bifurcação = Junction), commonly known as Estação Velha, is located about 2 km from the old city, so a branch was later added to serve the town station commonly known as Coimbra-A or Estação Nova. This branch was extended on street track through the town centre to Serpins.

Rail Road Conimbricense (RRC)

On 8 February 1873 the Baron of Trovisqueira (also involved in the original line in Porto) asked for a concession for a Caminho de Ferro Americano (mule tram) line between the railway station and the city. On 27 February the City Council agreed and set 10 conditions for the linha Americana. The concession had been sublet to a company named Matosinhos e Companhia and the RRC was constituted to manage the line.

The construction work on the line began in April 1874. The official inauguration took place on September 15, with music, fireworks, a parade of three trams filled with everybody important and, according with Portuguese tradition, a banquet.

The provisional timetable showed 4 journeys daily connecting with the trains of the Companhia Real dos Caminhos de Ferro, at the railway station.

On September 17, 1874 the first section was opened between the Rua Ferreira Borges and the railway station. The line was extended according to the original plans by May 8, 1875 from Rua Ferreira Borges via Largo da Portagem to Praça do Comércio, which was between Estação Nova and Praça 8 de Maio. The offices of Rail Road Conimbricense were on the Praça do Comércio while the depot of the cars and the stables of the mules were in the Rua da Sofia, opposite to the Quartel (barracks) da Graça. The RRC transported also freight for which

there was a siding to the goods yard at the railway station.

In 1877, the RRC proposed the construction of a new line between the Praça 8 de Maio and the Alta. However this idea was not followed up. Instead a Mr. Joaquim Augusto dos Santos started on 14 October 1882 a service with Ripert cars between the Baixa and Alta.

For more than ten years the tramline flourished well, but on 18 October 1885 the railway between Coimbra-B (Estação Velha, Old Station) and Coimbra-A (Estação Nova, New Station) was opened. This competition between the mule trams and the steam trains must have caused a severe drop in the number of passengers and amount of freight for the tram company. On 30 September the tram company reduced the ticket prices. The central station of the RR Conimbricense was moved from Praça do Comércio to the Praça 8 de Maio, allowing the latter to function as dispatch goods station. On 27 October the RRC split its line operationally. Trams now ran between Coimbra-A and Praça 8 de Maio and between Coimbra-B and Praça 8 de Maio. Apparently the short section between Coimbra-B and Praça do Comércio through the very narrow Rua das Solas was not used anymore.

In March 1886 the tram had a temporary revival for a few weeks when ground under the new railway was washed away by the notorious floods of the Mondego river.

But without the line to the Alta and having a route parallel to the railway branch, the small tram company inevitably lost the battle with the large railway company. The end of the RR Conimbricense came at the beginning of April 1887.

Network of the RRC

Station Coimbra-B - Rua Figueira da Foz - Rua da Sofia - Praça 8 de Maio - Rua Ferreira Borges - Largo da Portagem - Avenida Emídio Navarro - Rua Adelino Veiga - Praça do Comércio.

Rolling stock of the RRC

Little is known about the earliest trams in Coimbra. They were manufactured by Starbuck, United Kingdom, and were finally assembled by Mr. António Gonçalves Gabriel, owner of a locksmith in the Rua Quebra-Costas. It is likely that the assembly of the cars was done in the workshops of the tram company as it would have been challenging to transport an assembled tramcar from the Rua Quebra-Costas to anywhere. According to the press at the time they were similar to those that existed in Porto. However it's unclear if this referred to the tramcars of the CCAFPM or those of CCFP. The date suggests the latter, but the Baron of Trovisqueira was involved with the first company. At least one of the cars was a double decker with seats on the roof. In 1884 the RRC had 5 passenger cars, 4 freight wagons and 12 mules for pulling the vehicles. What happened with the tramcars after closure is uncertain, but according to a contemporary newspaper they went to the Caminho de Ferro Americano de Torres Novas à Alcanena. This line existed only from 1887 (year of closure of the Coimbra tram) to 1893. The subsequent history of the cars is unknown, but four of them might have gone to the PPF (Porto - Póvoa de Varzim - Famalicão) narrow gauge railway. If so, then they were probably only scrapped around 1947.

Companhia Carris de Ferro de Coimbra (CCFC)

After 1887, the city of Coimbra no longer had a system railed urban transport at a time when around the world the development of electric trams began. When large cities electrified their tram networks, the availability of second-hand horse trams and rails was high. Small cities and towns took the opportunity to get very cheap horse trams. In Coimbra in October 1902, a retired colonel Sr. Augusto Freire de Andrade applied for a concession for an Americano urban rail system in various parts of the town: Coimbra-B station, the Alta, main streets of the Baixa, and other locations. The City Council of Coimbra, remembering the failure of the former company, was not enthusiastic. Nevertheless in January 1903 the City Council gave the provisional concession for a period of 30 years for the construction of a new mule tram network. Augusto Freire de Andrade immediately created a new company with the name CCFC. The new company started work and adopted the same gauge as Lisboa (900 mm). It will not have been coincidence that the CCFC came to existence shortly after the CCFL had ended its horse and mule tram operations and had mule trams and rails for sale.

On January 1, 1904 the first line was opened between Largo da Portagem and Coimbra-B station. The second line opened about a month later on 4 February connecting the Largo Portagem with Rua Infante D. Augusto in the Alta, in front of the University.

Soon it was realised that mule trams were not the ideal type of urban transport on the steep gradients to the Alta. The mule trams departed from Largo Portagem with two mules, upon arriving near the bottom of the Avenida Sá da Bandeira two more mules were attached. At the Arcos do Jardim a further two mules were added, thus climbing the Calçada Martim de Freitas.

In December 1904 the concessionaire proposed replacement of animal traction by electric traction. The Coimbra City Council, aware of the advantages of electric trams, immediately accepted the idea and agreed to award a grant for implementation of the project. The City Council promptly sent a proposal to the government for grant approval and the first studies were made concerning importing the necessary materials for electric traction.

Once it had been decided to adopt electric traction, in August 1906 a new company was established, the Companhia Carris de Ferro de Coimbra, SARL, that would enable the construction and exploitation of the electric network and which could fulfil all the requirements of the concessions held. A Company Prospectus was published, trying to demonstrate a promising future for electric traction reserved to those who signed up for shares. However in the stock market demand was weak, which meant that because of lack of capital the progress of the project was slow.

In November 1906, there was a great setback, the unexpected death of the big promoter and president of CCFC, Lt. Col. Eduardo Augusto Freire de Andrade.

The CCFC continued with importing materials and presenting projects to realise the new electric network, but the financial problems became evident and in March 1908 the Company informed the authorities that "the installation work on the electric traction was interrupted due to financial difficulties." The company was still trying to get support from the forty largest apartment building owners and industrial taxpayers. But in May 1908, the City Council of Coimbra was informed of the decision to suspend the operations of the mule trams, which happened on August 31, 1908. In May 1909 the CCFC surrendered its rights for operation of electric traction in favour of the Câmara Municipal de Coimbra. The CCFC still formally existed as a company until February 1916.

Network of the Carris de Ferro de Coimbra

The network was completely single track with a few passing loops, although it is not known exactly how many and where. The depot was located across the Rua da Figueira da Foz, near the Estação Velha. The "Carris de Ferro de Coimbra" had three lines:
• Estação Nova - Praça 8 de Maio - Rua da Sofia - Rua Figueira da Foz - Estação Velha (Coimbra-B).
• Estação Nova - Praça 8 de Maio - Praça da República - Rua Oliveira Mattos - Rua Castro Matoso - Arcos do Jardim - Calçada Martim de Freitas - Rua Infante D. Augusto (The current Rua Larga is not exactly, but very close to the same location, but at that time the area was still one with narrow medieval streets) - Universidade.
• Universidade - Rua Infante D. Augusto – Calçada Martim de Freitas - Arcos do Jardim - Rua Castro Matoso - Rua Oliveira Mattos - Praça da República - Rua da Sofia - Rua Figueira da Foz - Estação Velha (Coimbra-B).

In 1907 one of the schedules announced 25 daily trips between Coimbra-A station and Universidade, 9 trips between Coimbra-A station and Coimbra-B station and 9 trips between Universidade and Coimbra-B station. The manager who signed some of the notices announcing the schedules was Sr. Joaquim A. dos Santos, the same person who in 1882 launched the "Ripert" bus routes between the Baixa and the Alta.

Rolling stock of the Carris de Ferro de Coimbra

Few details are known about the trams of the CCFC, not even how many there were. Taking into account the size of the network there were probably no more than about half of a dozen. No doubt all were second hand bought from the CCFL in Lisboa with both open and closed types. There must have been also a few freight cars as the company transported mail bags between the Central Post Office and the Coimbra-B station and coal to the Gas factory at Rua da Sofia which provided the lighting to the city.

This early and slightly imperfect view is of a mule car pulled by six mules heading for the university. It is believed to have been taken between 1904-1908. The car is thought to be an American import, later rebuilt by the SMC as a trailer and withdrawn in 1954. It was subsequently retained and later put aside for eventual restoration. It is understood that the car was once part of the Lisboa fleet.

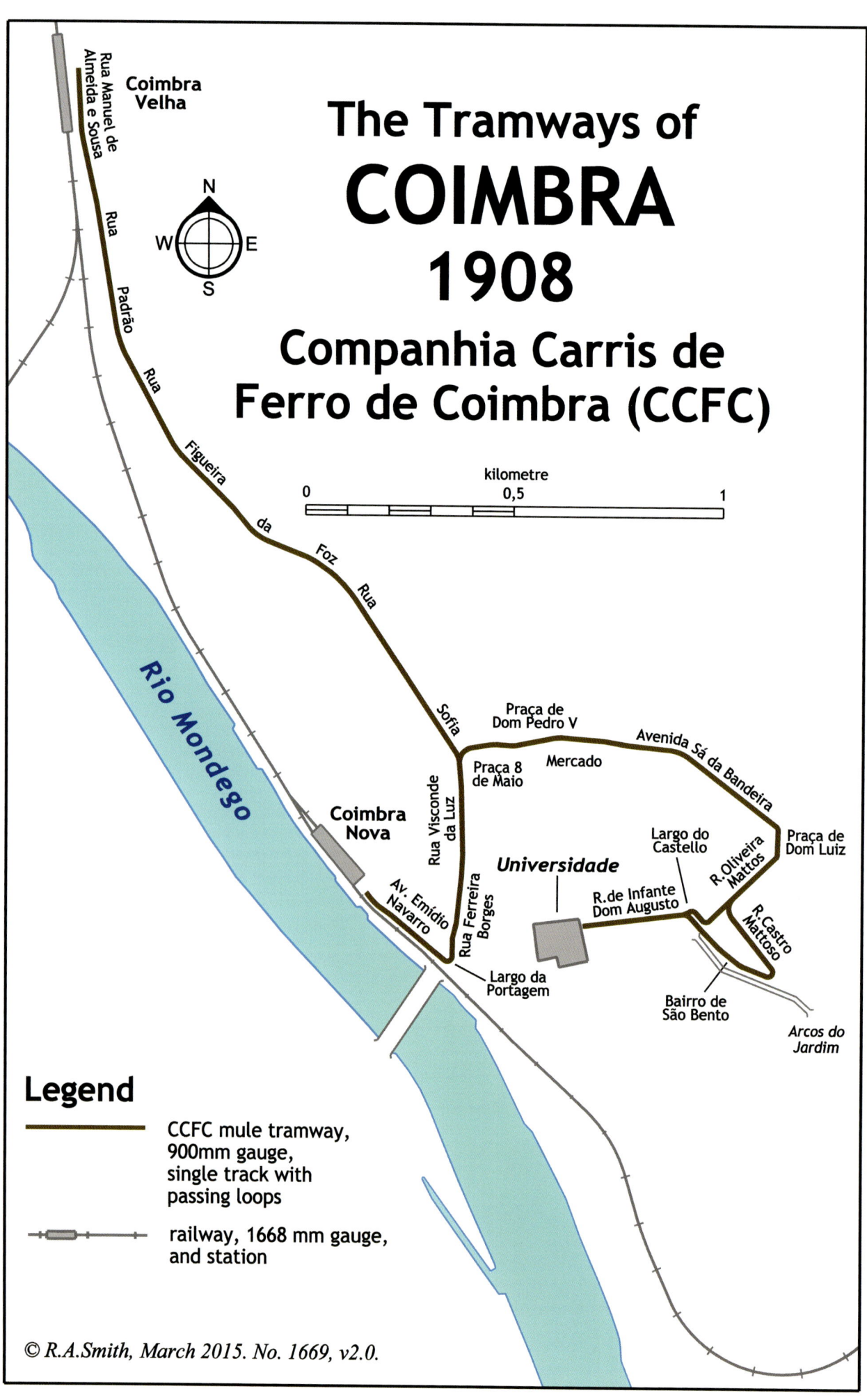

The Tramways of
COIMBRA
1908
Companhia Carris de Ferro de Coimbra (CCFC)

Coimbra Velha

Rua Manuel de Almeida e Sousa

Rua Padrão

Rua Figueira da Foz

Rio Mondego

Coimbra Nova

Rua Sofia

Praça de Dom Pedro V

Praça 8 de Maio

Mercado

Avenida Sá da Bandeira

Rua Visconde da Luz

Universidade

Largo do Castello

Praça de Dom Luiz

R. Oliveira Mattos

Av. Emídio Navarro

Rua Ferreira Borges

R. de Infante Dom Augusto

R. Castro Mattoso

Largo da Portagem

Bairro de São Bento

Arcos do Jardim

kilometre

0   0,5   1

Legend

CCFC mule tramway, 900mm gauge, single track with passing loops

railway, 1668 mm gauge, and station

© R.A.Smith, March 2015. No. 1669, v2.0.

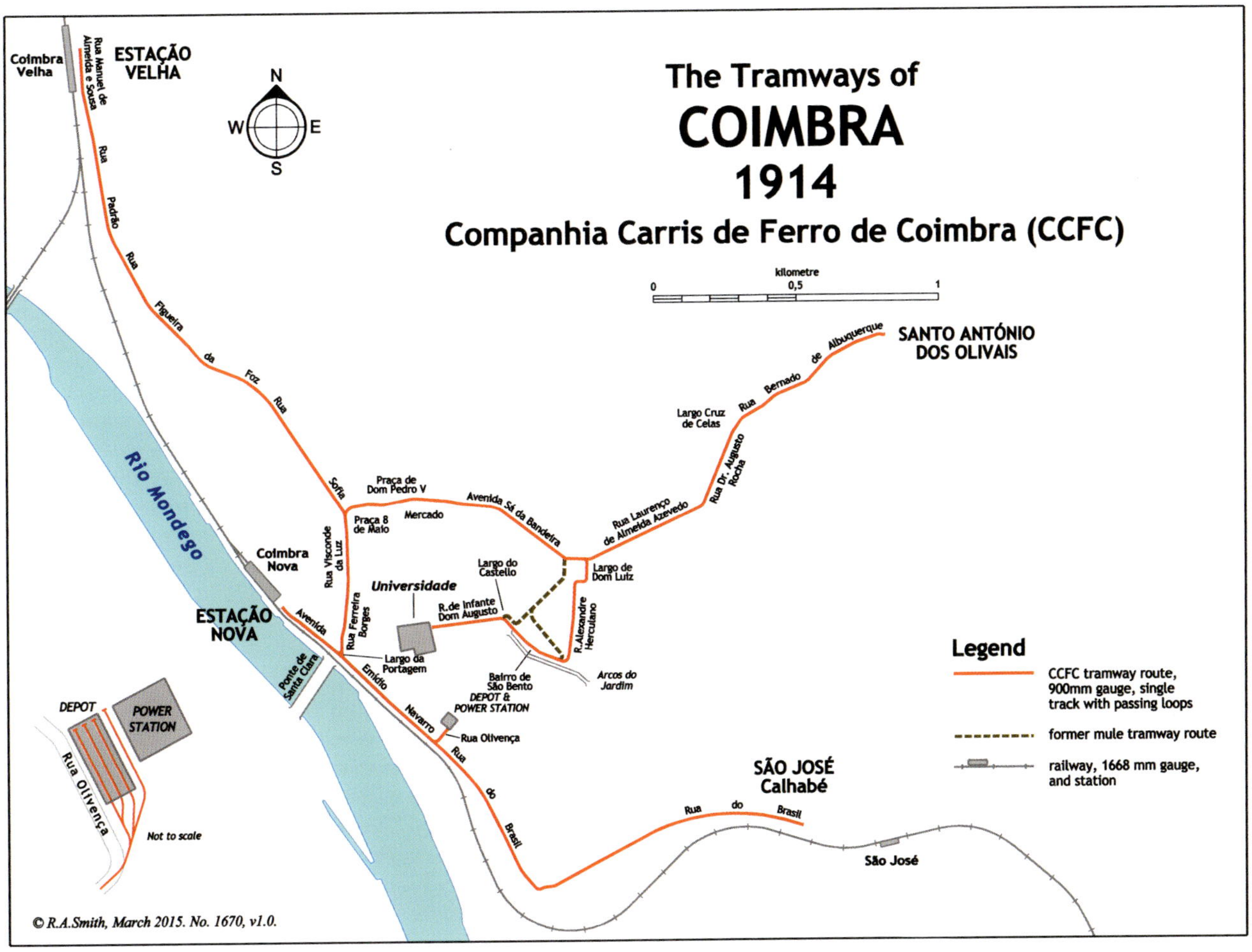

The Tramways of
COIMBRA
1914
Companhia Carris de Ferro de Coimbra (CCFC)

Coimbra Velha
ESTAÇÃO VELHA
Rua Manuel de Almeida e Sousa

kilometre
0   0,5   1

SANTO ANTÓNIO DOS OLIVAIS

Rua Padrão
Rua Figueira da Foz
Rua Bernado de Albuquerque
Largo Cruz de Celas
Rua Dr. Augusto Rocha

Sofia
Praça de Dom Pedro V
Mercado
Praça 8 de Maio
Avenida Sá da Bandeira
Rua Laurenço de Almeida Azevedo

Coimbra Nova
Rua Visconde da Luz
Largo do Castello
Largo de Dom Lutz

ESTAÇÃO NOVA
Avenida
Rua Ferreira Borges
Universidade
R.de Infante Dom Augusto
R.Alexandre Herculano

Emídio
Largo da Portagem
Bairro de São Bento
DEPOT & POWER STATION
Arcos do Jardim

Ponte de Santa Clara
Navarro

DEPOT
POWER STATION
Rua Olivença
Rua Olivença
Not to scale

Rio Mondego

do
Brasil

SÃO JOSÉ Calhabé

Rua do Brasil
São José

Legend
— CCFC tramway route, 900mm gauge, single track with passing loops
- - - former mule tramway route
━ railway, 1668 mm gauge, and station

© R.A.Smith, March 2015. No. 1670, v1.0.

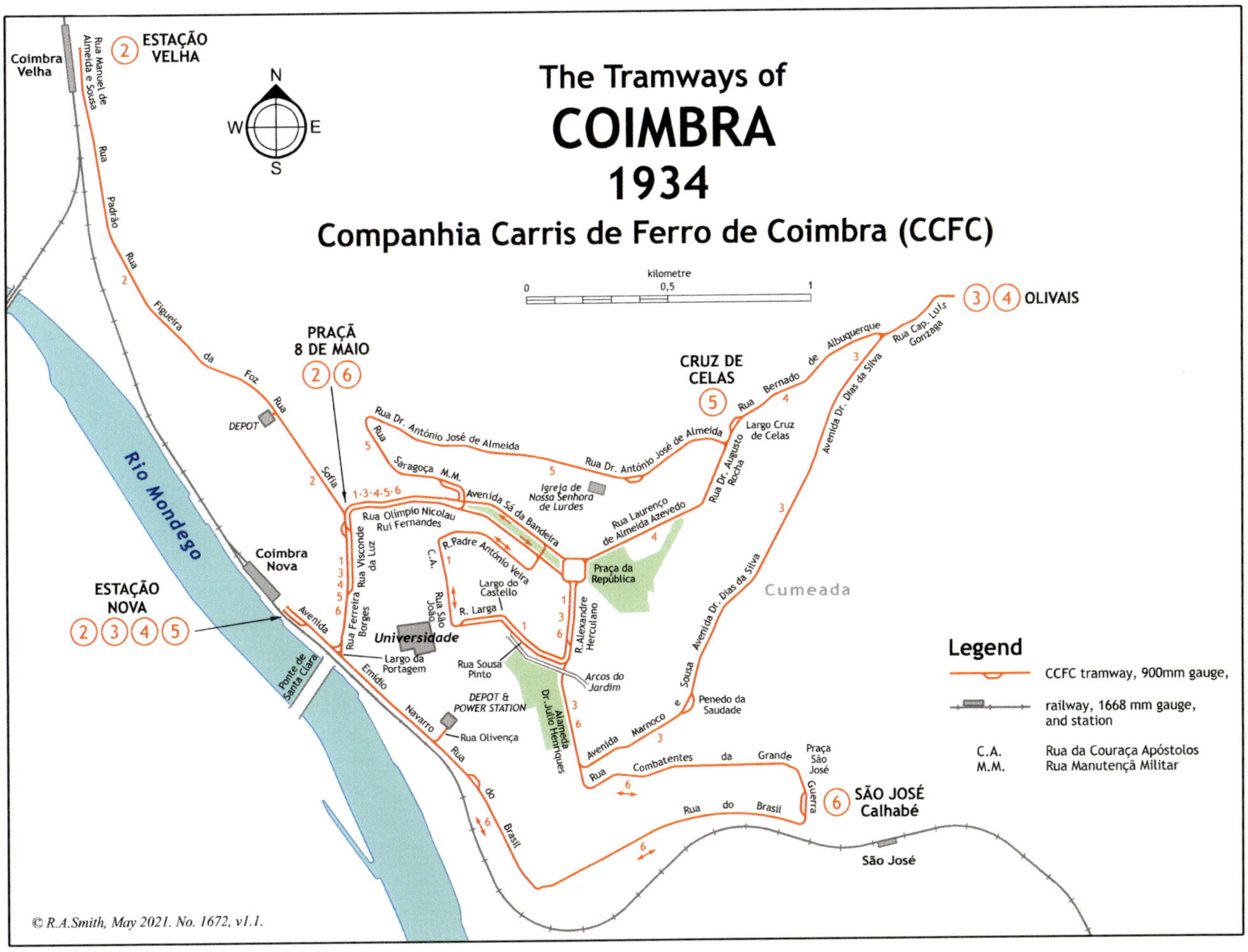

The Tramways of
COIMBRA
1934
Companhia Carris de Ferro de Coimbra (CCFC)

Coimbra Velha
② ESTAÇÃO VELHA
Rua Manuel de Almeida e Sousa

kilometre
0   0,5   1

Rua Padrão
Rua Figueira da Foz

③ ④ OLIVAIS
Rua Cap. Luís Gonzaga

PRAÇA 8 DE MAIO
② ⑥
DEPOT

CRUZ DE CELAS
⑤

Rua Bernado de Albuquerque
Avenida Dr. Dias da Silva

Rua Dr. António José de Almeida
Saragoça M.M.
1·3·4·5·6
Avenida Sá da Bandeira
Igreja de Nossa Senhora de Lurdes
Rua Dr. António José de Almeida
Largo Cruz de Celas
Rua Dr. Augusto Rocha

Rua Olímpio Nicolau Rui Fernandes
R. Padre António Veira
Praça da República
Cumeada

Coimbra Nova
Rua Visconde da Luz
C.A.
Largo do Castello
R. Larga
Rua Laurenço de Almeida Azevedo

ESTAÇÃO NOVA
② ③ ④ ⑤
Rua Ferreira Borges
Universidade
Largo da Portagem
Rua Sousa Pinto
R.Alexandre Herculano
Arcos do Jardim
Alameda Dr. Julio Henriques
Avenida Dr. Dias da Silva
Sousa
Penedo da Saudade

Ponte de Santa Clara
Emídio
DEPOT & POWER STATION
Rua Olivença
Navarro

Rio Mondego

do Brasil

Avenida Marnoco
Rua Combatentes da Grande Guerra
Praça São José
SÃO JOSÉ Calhabé
⑥

Rua do Brasil
São José

Legend
— CCFC tramway, 900mm gauge,
━ railway, 1668 mm gauge, and station
C.A.   Rua da Couraça Apóstolos
M.M. Rua Manutença Militar

© R.A.Smith, May 2021. No. 1672, v1.1.

The extension to Tovim under construction in 1953-4. *Brian G. Dutton/Online Transport Archive*

Câmara Municipal de Coimbra (CMC) The Necessity for Electric Trams

With the success of the electric trams in Porto and Lisboa and the failure of the CCFC to electrify the Coimbra mule tram network, the city council decided to take matters into their own hands. Work on a new urban transport system on rails with electric trams was started, which would prove to be a constant in the day to day life of the city of Coimbra, for the next 69 years.

The tram system of Coimbra, had the distinction of being the first built by a local authority and not by a private company as in the cities of Lisboa, Porto and Sintra.

The Bidding for Electric Traction

The tender for the electric traction installation project was sent in 1909 to five competing suppliers. The CMC asked for a technical opinion from Eng. William Clark (then director of CCFL). In August 1909 the contract was awarded to Dick Kerr & Co., however they declined to sign the contract on the terms which the Coimbra City Council required. Faced with this situation, the contract was re-awarded to Thomson-Houston-Iberica in September 1909. They had to deliver:
• Five electric trams.
• Installation of rails and overhead lines (following the example and materials used by the CCFL on the routes: Estação Velha (Coimbra-B) - Alegria; Estação Nova (Coimbra-A) – Universidade; and Estação Nova - Sto. António dos Olivais (via Celas).

• A new depot at Rua de Alegria with enough space for eleven trams, a repair workshop and a paint shop.
• A steam-powered generating plant alongside the new depot for electric trams.

Early in 1910, the installation work for electric traction began. In June 1910, the first electric trams arrived. On 24 October the first notice of the recruitment of ten motormen and ten conductors was published. The training of the drivers was carried out by staff of Porto and Lisboa tramways.

The Republican revolution of October 1910 and the embargo placed by the Companhia dos Caminhos de Ferro Portugueses on the electric traction works near the Coimbra-B station caused some delay to the final installation of electric traction, but on November 25, 1910, the first experimental runs were carried out with excellent results.

Inauguration of the Electric Trams

On 1 January 1 1911 the tram services were inaugurated.

In September 1911, the company AEG Thomson Houston Iberica, was awarded the contract to acquire two more electric trams similar to the first. These were delivered in 1912.

On 28 October 1912, the works for a new line between Alegria and Calhabé were started. This opened on May 24, 1913. Between 1913 and 1925 there were no changes in the tram network. The reasons were obvious: the effects of World War I and its consequences on the national and international economies. Shortly after the war serious social problems emerged; strikes, sabotage to the electric tram service, etc.

Some mule tramcars were re-gauged to 1000 mm. In February 1916 the Coimbra City Council allowed a night service with a mule tram to transport mail bags from the Central Post Office Station to the Coimbra-B Railway Station, which also carried passengers.

After experiments in March 1919 with an old mule tram as a trailer for the electric trams, in order to increase passenger capacity, trailer operations commenced

Serviços Municipilizados de Coimbra (SMC)

On 27 April 1920 the Serviços Municipalizados de Coimbra was created. This new organisation had autonomy from the City Council, providing the businesses of Electric Trams, Water and Gas. With the management of the Municipal Services the electric tram system in Coimbra reached the golden period.

Projects for the Future

In its 1921 Annual Report, the SMC drew attention to the need to develop the network of electric trams, with the acquisition of new rolling stock, with the construction of new lines, with installing double track, if space allowed, on the busiest parts of the network, with extending the fleet and improving the power plant. Most of these plans were realised in the second half of the 1920's and most of the rest by 1934. The major exception was the line to Santa Clara, never built because the Companhia dos Caminhos de Ferro Portugueses would not grant the necessary license to cross the Lousã railway in Largo da Portagem. In 1947 it would become the route of the first trolleybus line.

Decline

At the end of the nineteen-thirties, the first anti-tram sounds emerged. In the annual report of the SMC for 1938 was written:

Car 8 at Tovim on 10 June 1964. *Courtesy of the National Tramway Museum, photographer J.H. Price*

Car 1 at Largo da Portagem on 4 September 1962. *W C Janssen/Online Transport Archive*

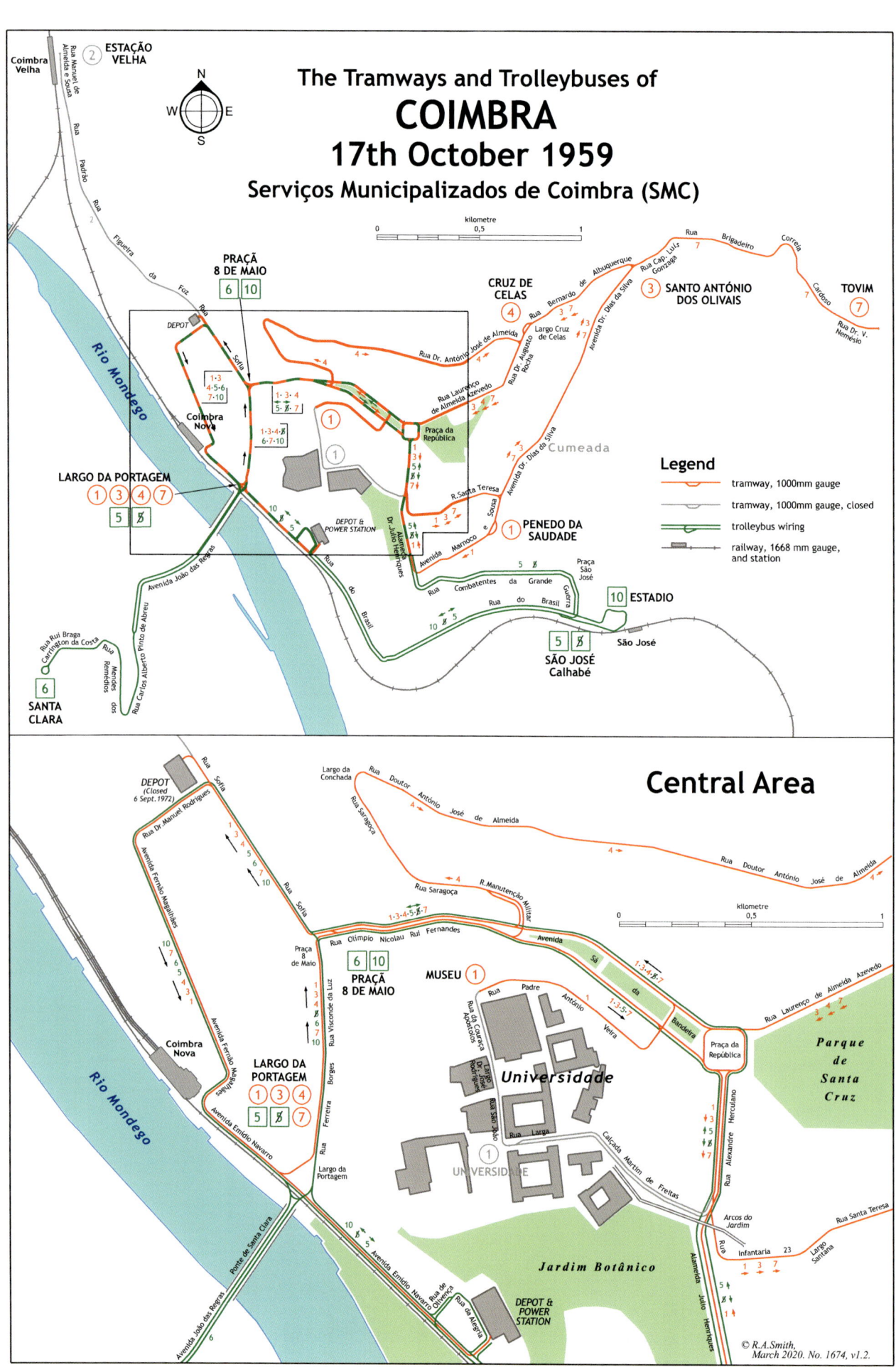

The Tramways and Trolleybuses of
COIMBRA
17th October 1959
Serviços Municipalizados de Coimbra (SMC)

Legend

tramway, 1000mm gauge

tramway, 1000mm gauge, closed

trolleybus wiring

railway, 1668 mm gauge, and station

Central Area

© R.A.Smith,
March 2020. No. 1674, v1.2.

'... while a growth of the transport network is desired, it is also less advisable to realise this with trams.' In the same year the SMC ordered three buses. These were used in 1940 to commence a bus-line to Taveiro. Thus began the process that led to the abandonment of electric trams in Coimbra.

From 1944 the demolition of a large part of the medieval Alta gradually pushed the trams out of this area. This part of the Alta was demolished in the 1940's to give place to the new buildings of the university in the style preferred by totalitarian regimes, wiping out a large part of the heritage of the old medieval city. Later trolleybuses returned here, between the new buildings of the University. The first trolleybus line had appeared in 1947. It went from Portagem to Santa Clara at the other side of the river. From the success of the trolleybus line arose the idea to replace the trams by trolleybuses and buses, but soon the conclusion was made that the operation was too expensive if done swiftly. In fact it would take more than 30 years. Two tramlines were replaced (line 5 by trolley buses in 1951 and line 2 by buses in 1954), but also a few, small extensions of the tram network were realised. More replacements of trams by trolleybuses followed in the 1960's.

The maintenance of the rolling stock and permanent way was continued at a high level. Even in the early 1970's the tracks of line 4 on the Montes Claros route were renewed. During the 1950's and early 1960's seven older trams received a new body and even around 1967 another four older trams also received new built bodies. After withdrawal of the other older cars not required anymore because of reduction of the tram services, this meant that during the last years the remaining fleet consisted of fifteen similar looking trams.

Car 12 at Santo Antonio on 10 September 1962.

W C Janssen/Online Transport Archive

The final years

In 1978, a new connection was installed on Line Number. 4. This allowed direct access (without the need to reverse) from the inward Avenida Sá da Bandeira into the Rua da Manutenção Militar. But the end was inevitable, and the occasion was the works to install a new sewer system in the Baixa. New and modern Volvo buses were already in service from 1976. The SMC wanted to acquire new trolleybuses.

On 2 May 1979 the sewer works entered into the Baixa. That day the tram routes were curtailed to Manutenção. Line

Car 16 at Estação Nova on 4 September 1962. *W C Janssen/Online Transport Archive*

Car 4 at Largo da Portagem on 10 September 1962. This car retains its original clerestory body. *W C Janssen/Online Transport Archive*

Car 12 in Avenida Sá da Bandeira on 4 September 1962. *W C Janssen/Online Transport Archive*

Car 14 turning from Rua Santa Teresa into Avenida Dr. Dias da Silva on 10 September 1962 displaying in the windscreen a plate "Carro Directo" which indicated either a limited stop or a direct service. *W C Janssen/Online Transport Archive*

3 - Sto. António dos Olivais, reversed using the cross-over at Manutenção that in 1963 had replaced the cross-over that existed at the west end of Rua Olímpia Nicolau Rui Fernandes. Line 4 - Manutenção-Cruz de Celas became a one-way circular line using the connection added in 1978 giving direct access from Av. Sá da Bandeira to Rua da Manutenção Militar. Returning to the Alegria depot could now only be done with police escort. All cars that operated during the day (three on line 3, two on line 4 and sometimes a sixth that had been out on trial) formed a

parade of trams with a police vehicle in front using floodlights for the journey against the traffic flow to Alegria. The last time that this occurred was in the night of 8 to 9 January 1980 close to 1 am.

The final years of the Coimbra tram can be summarised with the words of the late Gerard Stoer, published in the Dutch book "Trams 1981":

"As predicted the rest of the network of trams in the city of Coimbra was abolished on 9 January 1980. Any tram enthusiast with the luck to have experienced it, will say that one of the

Under a stormy sky, car 15 is seen on the Tovim route on 7 October 1968. *W C Janssen/Online Transport Archive*

most atmospheric small urban tram systems in the world disappeared. The network consisted of a small number of complicated single line routes through narrow streets, which were worked in one-way. The curious effect of this tram system was that those travelling in a tram hardly ever saw another tram but an enthusiast sitting on a strategically located terrace, could see passing a procession of trams without end ..."

Network of the CMC and SMC

From 1911 to 1920, the tramways were run by the Municipal Council. The CMC operated the following lines , which were identified by colours.
• Green: Estação Nova - Portagem - Praça 8 de Maio - Praça da República - Arcos do Jardim - Ladeira do Castelo - Rua Infante D.Augusto (now Rua Larga, but at that time the area was still one with narrow medieval streets) – Universidade and back by the same route.
• Red: Alegria (depot) - Av. Navarro - Portagem - Praça 8 de Maio - Rua da Sofia - Rua Figueira da Foz - Estação Velha and back by the same route.
• White: Estação Nova - Portagem - Praça 8 de Maio - Praça da Republica - Largo Cruz de Celas - Olivais and back by the same route.
• Pink: Portagem - Av. Navarro - Rua do Brasil to Calhabé and back by the same route.

The first three lines were opened on 1 January 1911, the pink line followed on 24 May 1913. All routes were single track with a few passing loops. As there were only seven trams, there would have been at most two trams at the same time working on a route.

On 15 April 1928, the work of installing double track began between the Rua Ferreira Borges and the Arcos do Jardim to c ope with the increasing number of trams. The double track came into use on 16 December 1928.

On 1 November 1928 a new tram line from Estação Nova to Montes Claros was opened. The line was soon extended to the Largo da Cruz de Celas.

On 24 February 1929, the connection of Calhabé line to the Arcos do Jardim (Via the Rua dos Combatentes) opened, creating the circular line to Calhabé.

On 19 May 1929 the new route between Arcos do Jardim and Cumeada-Olivais opened, passing along the Penedo da Saudade, then following Av. Dias da Silva reaching its terminus at Olivais.

With the extensions of the network also route numbers were introduced, although the use of colours was retained too. Since 1929 there were six tramlines:
• 1 (green): Estação Nova - Portagem - Praça 8 de Maio - Praça da República - Arcos do Jardim - Ladeira do Castelo - Rua Infante D.Augusto (now Rua Larga, but at that time the area was still one with narrow medieval streets) - Universidade and back by the same route.
• 2 (red): Alegria (depot) - Portagem - Praça 8 de Maio - Rua da Sofia - Rua Figueira da Foz - Estação Velha and back by the same route.
• 3 (white): Estação Nova - Portagem - Praça 8 de Maio - Praça da República - Arcos do Jardim - Penedo da Saudade - Olivais and back by the same route.
• 4 (yellow): Estação Nova - Portagem - Praça 8 de Maio - Praça da República - Cruz de Celas - Olivais and back by the same route.

• 5 (blue): Estação Nova - Portagem - Praça 8 de Maio - Rua Dr. José d´Almeida until Montes Claros and back by the same route. By 1 January 1930 this line had been extended to Cruz de Celas.
• 6 (pink): Praça 8 de Maio - Praça da República - Arcos do Jardim - Rua dos Combatentes - Calhabé - Rua do Brasil - Av. Navarro - Portagem - Praça 8 de Maio. Circular line operated in both directions.

Some more sections of lines were constructed in 1932:
• The connecting of Olivais (now the very end of Avenida da Silva Dias) to the Church of St António dos Olivais, an extension of the lines 3 and 4.
• The construction of the line on Rua Abílio Roque (today Rua Padre António Vieira); this line left the Avenida Sá da Bandeira, and up across the street mentioned above (Rua Abílio Roque) becoming the Museum line. First numbered 8, it was in 1933 indicated as line 1-Museu.
• Extending the double track from Arcos do Jardim to the Universidade, which began operating in early September.

On July 1, 1934, the line connecting the Museu to the Universidade via Ladeira de São João was opened, combining the two lines 1 (Universidade and Museu) into one with a loop through the Alta. The gradient of that street was steeper than the gradient of the Calçada de S.André on line 12 in Lisboa. Some people bet that the electric car would not be able to climb up the hill. But the first tramcar, fleet number 8, did climb the slope.

On January 19, 1941, the numbering of several lines was altered. The services of the lines 3 and 4 were combined getting the number 3 for the counter clockwise direction and the

number 3/ for the clockwise direction. The lines 5 and 6 became 4 and 5 respectively. Until 1944 the network was:
• 1 (green): Estação Nova - Portagem - Praça 8 de Maio - Praça da República - Arcos do Jardim - Ladeira do Castelo - Rua Infante D.Augusto (now Rua Larga, but at that time the area was still one with narrow medieval streets) - Universidade - Ladeira de São João - Museu - Rua Abílio Roque - Av.Sá da Bandeira - Praça 8 de Maio - Portagem - Estação Nova. The loop was operated in both directions.
• 2 (red): Praça 8 de Maio - Estação Velha in both directions.
• 3 (white): Estação Nova - Portagem - Praça 8 de Maio - Praça da República - Arcos do Jardim - Penedo da Saudade - Sto. António dos Olivais - Cruz de Celas - Praça da República - Praça 8 de Maio - Portagem - Estação Nova.
• 3/ (red): Opposite direction of line 3.
• 4 (blue): Estação Nova - Portagem - Praça 8 de Maio - Rua Dr. José d´Almeida - Cruz de Celas in both directions.
• 5 (pink): Praça 8 de Maio - Praça da República - Arcos do Jardim - Rua dos Combatentes - Calhabé - Rua do Brasil - Av. Navarro - Portagem - Praça 8 de Maio. Circular line operated in both directions.

On 26 July 1944 line 1 was divided again into two sections because of the demolition of the medieval Alta to construct the new buildings of the University. In the later years of both branches (Universidade and Museu) the termini were changed several times depending on the progress of the redevelopment.

On 27 April 1947 an experiment was started by abandoning line 3/, turning the Montes Claros route of line 4 into a one-way clockwise service going inward via Rua Lourenço Almeida de

Car 5 with original body is seen at Largo do Portagem on 7 May 1964 with all its windows fully open indicating a particularly hot day. *L. Folkard*

Car 12, originally an open car, rebodied in 1957, and Belgian car 18 are seen at the aqueduct at Arcos do Jardim on 8 May 1964. *L Folkard*

Azevedo and changing line 5 to a one-way circular counter-clockwise line. With this long stretches of single track were changed from two-way to one-way working, but with the disadvantage that many people had to make much longer trips for half of the time. The reactions from the public were very negative and soon the changes were reversed.

On 10 March 1951 trolleybuses took over tramline 5 and on 3 September 1954 the trams on line 2 were replaced by buses. These were the only two lines on which trailers were used. But 1954 brought more changes. On 28 May the line from Sto.António dos Olivais to Tovim was opened and on 30 October the Baixa route Estação Nova - Portagem - Praça 8 de Maio, which was operated in both directions by the lines 1, 3, 3/, 4 and 7, was replaced by a one-way loop through the Baixa: Manutenção - Rua da Sofia - Rua Dr.Manuel Rodrigues - Av.Fernão de Magelhães - Estação Nova - Portagem - Praça 8 de Maio - Manutenção. The official Baixa terminus of the lines 1, 3, 3/ and 4 became Portagem. Until 1959 the network was:
• 1 (Universidade): Portagem (Baixa loop) - Praça da República - Arcos do Jardim - (Old) Hospital da Universidade and back by the same route.
• 1 (Museu): Portagem (Baixa loop) - Manutenção - Av.Sá da Bandeira - Rua Padre António Vieira (terminus at the top of this street)
• 3: Portagem (Baixa loop) - Praça da República - Penedo da Saudade - Olivais - Cruz de Celas - Praça da República - Portagem (Baixa loop).
• 3/: Opposite direction of line 3.
• 4: Portagem (Baixa loop) - Rua Dr. José d'Almeida - Cruz de

Celas and back by the same route.
• 7: Portagem (Baixa loop) - Praça da República - Cruz de Celas - Olivais - Sto.António dos Olivais - Tovim and back by the same route.

On 18 October 1959 the operations were changed to one-way working loops for several lines going through streets with single track. Also line 1 (Universidade) was changed to line 1 (Penedo de Saudade). A variation of trolleybus line 5 took over the University branch. To accomplish these changes a single track was laid in Rua Santa Teresa. For the lines 3 and 4 this meant a repeat of the 1947 experiment, but now made definitive. But in 1960 a new line 8 was introduced giving two-way working again on the Cruz de Celas and Cunheda routes. The result was:
• 1 (Penedo de Saudade): Portagem (Baixa loop) - Praça da República - Arcos do Jardim - Rua Santa Teresa - Penedo de Saudade - Arcos do Jardim - Praça da República - Portagem (Baixa loop) The Penedo de Saudade loop was worked in one-way only.
• 1 (Museu): Portagem (Baixa loop) - Rua Padre António Vieira (terminus at the top of this street) and back by the same route.
• 3: Portagem (Baixa loop) - Praça da República - Rua Santa Teresa - Olivais - Cruz de Celas - Praça da República - Portagem (Baixa loop).
• 4: Portagem (Baixa loop) - Rua Dr. José d'Almeida - Cruz de Celas - Praça da República - Portagem (Baixa loop). Short workings until Largo da Conchada kept using Rua da Manutenção for the return trip.
• 7: Portagem (Baixa loop) - Praça da República - Rua Santa Teresa - Cumeada - Olivais - Sto.António dos Olivais - Tovim and

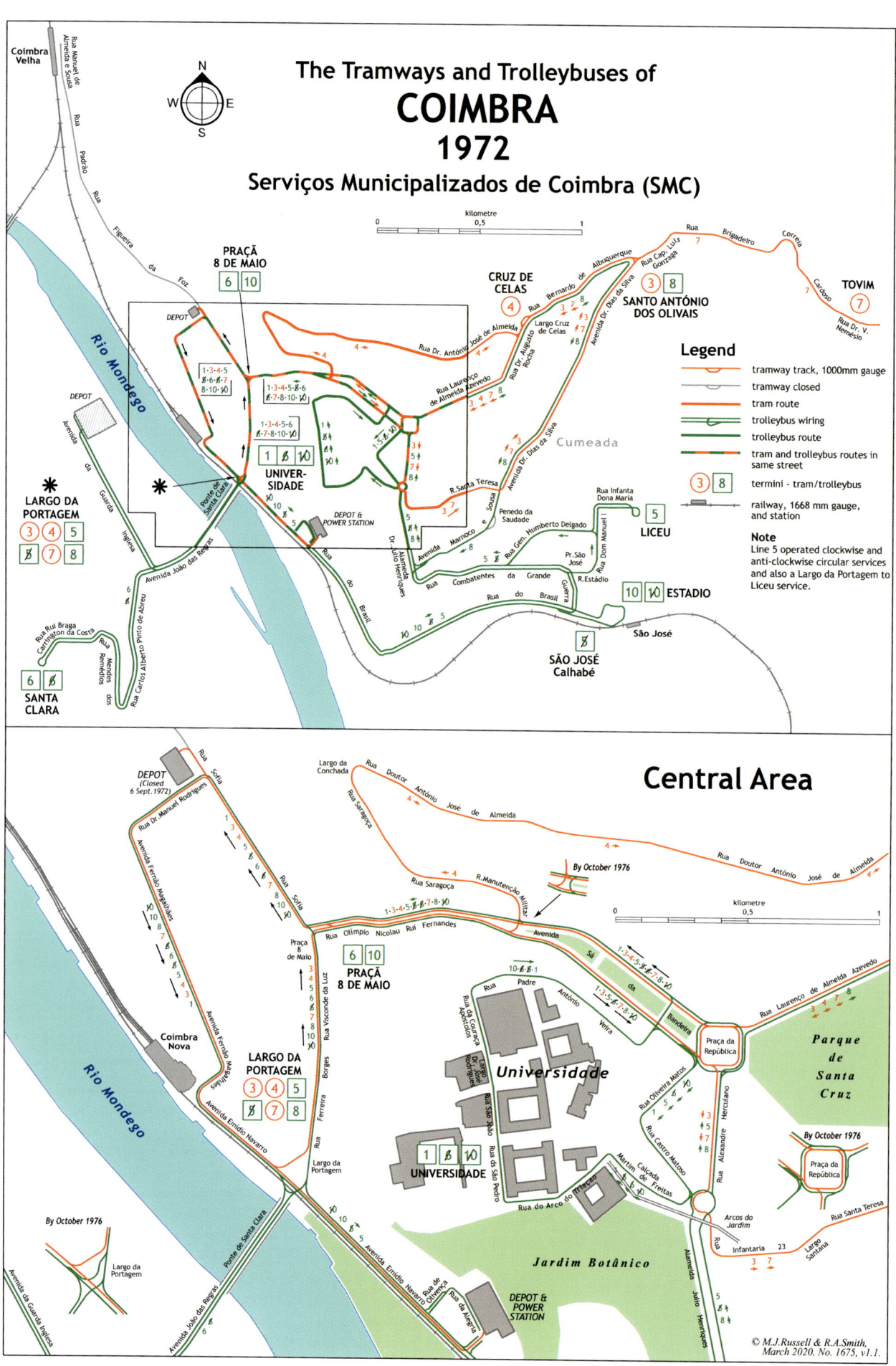

The Tramways and Trolleybuses of
COIMBRA
1972
Serviços Municipalizados de Coimbra (SMC)

Central Area

Legend

tramway track, 1000mm gauge
tramway closed
tram route
trolleybus wiring
trolleybus route
tram and trolleybus routes in same street
termini - tram/trolleybus
railway, 1668 mm gauge, and station

Note
Line 5 operated clockwise and anti-clockwise circular services and also a Largo da Portagem to Liceu service.

© M.J.Russell & R.A.Smith,
March 2020. No. 1675, v1.1.

Car 13, with rebuilt crossbench body, arrives at Tovim on 10 June 1964. *Courtesy of the National Tramway Museum, photographer J.H. Price*

Car 12 ascends Rua Saragoca in 1967 with in the background a section of unused track on the steep gradient left over from the previous two-way running on this section. *J. Jordan*

back by the same route. but in the opposite direction via Cruz de Celas instead of Cumeada.
• 8: Praça da Republica - Cruz de Celas - Olivais - Penedo de Saudade - Arcos do Jardim - Praça da Republica.

The only later changes were closures:
• 10 May 1961 line 1 (Museu), replaced by another variation of trolleybus line 5.
• 18 June 1967 line 1 (Penedo de Saudade)
• 21 August 1967 line 8. Replaced by a new trolleybus line with the same route. With this change two-way working on single track routes was reduced to Olivais - Tovim on line 7. On the Olivais loop now trams worked counter-clockwise and trolleybuses clockwise.
• 19 February 1977 line 7 closed.
• 2 May 1979 the Baixa loop because of sewer works.
• 9 January 1980 the remaining sections of the 3 and 4. With this the tram services in Coimbra came to an end. Trolleybuses took over the lines 3 and 4.

Depots
At the opening of the system the depot was in Rua da Alegria. There was capacity for eleven cars, along with the repair shop, paint-shop and material store. This capacity sufficed until the mid-1920s. On June 28, 1928 a new depot was opened. This was located on the site of the old Gas factory at the end of the Rua da Sofia and had space for 24 trams, apparently to be prepared for an extension of the fleet and network to a size that never was realised. The Sofia depot had a traverser which served six parallel lines. The workshops were retained in Alegria and this location was also used as depot for the buses and trolley-buses. With the dismantling of the power station the capacity of Alegria as a depot could be increased. In 1972 the buses and trolley-buses moved to Guarda Inglesa, a new depot at the other side of the Mondego river. This made it possible to move the trams back to Alegria and close the Sofia depot in that same year.

Power supply
Like all other Portuguese electric tram systems, in Coimbra the production of the electric energy was initially an internal activity. The Coimbra power plant was on the same site as the Alegria depot. It consisted of two large water-tubed boilers from Babcock & Wilcox with 300 square metres of heating surface; two high speed vertical cylinder steam engines from Belliss & Morcom of 390 HP rating, coupled to two dynamos of 180 Kilowatts by General Electric. Additionally there were other items including a surface condenser with cooling tank, a switchboard to distribute the energy and a powerful accumulator battery with its booster designed to compensate for substantial fluctuations of the load to be expected on a service in a city as hilly as Coimbra.

In July 1929 direct electrical power was taken from the power station of União Eléctrica Portuguesa (UEP), bypassing the power plant at Alegria, which then produced energy for electric trams only in case of failure or interruptions to the power supply from the UEP. From 1933 to 1941 the power station at Alegria worked only for conservation purposes. However, in February 1941 when a cyclone hit the country, the Alegria power station produced energy for the trams and public lighting for a few weeks while the power lines of the UEP were repaired. In April 1943, the SMC began to receive power from a second supplier, the Companhia

Eléctrica das Beiras (CEB). The power station of Alegria was dismantled and the boilers were sold in November 1943.

Rolling stock of the CMC/SMC
The maximum fleet that existed in Coimbra, was 20 electric trams, 2 trailers and 2 freight trailers. The cars were originally red, but this colour was soon changed to yellow. Most cars had only hand and rheostatic brakes. In later years all cars had line switch attachments to the controllers, an unusual feature on such a traditional tramway.

Electric trams
Series 1-8 Brill semi-convertible
These eight trams were built by Brill and employed Brill's patented semi-convertible design. They were of the small type with at each side six windows which could slide up into pockets in the roof. They had seats for twenty with twelve on two-and-one Brill "Winner" seats and the others on longitudinal benches for two in each corner. All seats were covered in twill rattan woven cane. All cars had drop-windscreens and folding gates. The Brill 21E trucks had a wheelbase of 1.98 m. General Electric supplied the controllers and the 37 hp motors of the GE58 type.

The order (no.16980) for the first five trams was received by Brill on 16 October 1909 and the cars despatched from Philadelphia on 19 January 1910. They arrived in Coimbra in June 1910. Brill usually despatched overseas orders in a "knocked-down" state to save shipping space. It is likely that this happened with the Coimbra trams too and they would have had to be assembled after arrival. Test running started in November 1910 and the services were started with these five trams on 1 January 1911. They had hand operated slipper brakes.

Brill received the order (no.18033) for two more trams on 27 October 1911 and despatched them from Philadelphia on 22 December 1911. These trams, numbered 6-7, entered service in May 1912.

Tram no.8 was only ordered in 1926. Brill received the order (no.22372) on 3 March and despatched the car from Philadelphia on 15 June 1926. The window gear was simpler than of 1-7.

Six of these trams received new bodies of the "Standard Coimbra" style in the 1950's and 1960's. The other two, 2 and 8, were scrapped in 1969 and the 1970's respectively.

Series 9-13 Carros Pneumónicos - Fumistas - Carros Grandes
These were originally 10-bench open cross-bench cars on 21E trucks. Brill received the order (no.22571) on 3 April 1927. The first two entered service the same year, the other three in 1928. They were highly unpopular, this in contrast to Lisboa where open cars were very popular. The open Coimbra trams gained the nickname Carros Pneumónicos (Pneumonia cars). The SMC altered them in the 1930s to convertible cars with very deep windows reaching down to seat level, which could be used as open cars during the summer season and closed cars in the Winter. When operating as open cars it was allowed to smoke inside, which resulted in the new nickname Fumistas. This twice yearly conversion of removing and installing windows however had to be done by the workshops. In 1941 they were modified again, to make it possible for passengers to open and close the windows, now of normal depth, themselves. As the cars had eight bays and 26 seats they were significantly larger than the old semi-convertible cars, they were now called Carros Grandes. All

Car 15 and trolleybus 41 in Rua Manuel Rodrigues on 6 June 1973. *A. G. Murray*

received new, standard Coimbra style bodies, nos.9-12 in the 1950's and no.13 in 1966.

Series 14-15 Steel cars

These two closed cars were also bought from Brill, which received the order (no.22623) on 9 September 1927 and despatched them from Philadelphia on 31 January 1928. They had 79E trucks and steel bodies seating 28: five rows with 2+2 benches and a longitudinal bench for two in each of the corners. They were a success and were the example for the later build "Standard Coimbra" type bodies. Both were in service until the end of the tram operations in 1980. Car no.14 made several trips on 1 June 1981 on track that still existed between Alegria and Portagem, which the SMC wanted to retain as museum line. Alas this track was lifted a few months later by the CMC and the museum line was never realised.

Because of the truck and steel bodies they were considered by some as Birney Safety cars, but these two Coimbra cars were of the classic design with end platforms separated by bulkheads from the saloon and folding gates. Birney cars were designed with the objective of one-man operation, the driver also taking care of the fare collection. They didn't have platforms separated by bulkheads from the saloon. The doors were controlled by the driver but could not be opened when the car was running, also the car could not run with open doors. Birney cars also had a "Deadman" control. These types of features were later also adopted with other modern car designs, but around 1920 they were what distinguished a Birney from a classic tram. The only features that the two Coimbra

cars had in common with the some 6,000 Birney Safety cars in America, Australia and New Zealand was that they had steel bodies with the type of windows normally used on Birneys and the 79E truck which was the most common type of truck used with Birneys.

Series 16-18 Belgas

These three trams were ordered from the Ateliers Familleureux (Belgium) in 1929. The trucks were made in Germany by MAN. They entered service in 1930 and were the small two-axle version of the ten bogie trams ordered by the CCFP (Porto) in 1928. The original German equipment was replaced in 1948-53 by BTH 116 motors and BTH OK33B controllers. They were the only Coimbra cars with Westinghouse air brakes, which were installed in 1955. All three were withdrawn from service in 1973.

Car no.19 Standard Coimbra

The body of this car was in 1934 built by the workshops of the CCFP in Porto on a 79E type truck ordered from Brill. The body was to the same design as those of the cars 14-15, but made of wood instead of steel. The car was withdrawn in the 1970's.

Car no.20 Standard Coimbra

The body of this car was equal to that of no.19, but built by the Coimbra workshops in 1940. It had the Brill 21E truck that was first used with the zorra. The ability to perform this construction was indicative of the high technical capability achieved by the workshops of the SMC. The car was in service until the end of the tram operations in 1980.

No.	Type (original)	In service	New Standard Body	Out of service	Fate (status October 1999)
1	Brill SC	1910	1961	1980	Museum
2	Brill SC	1910	n.a.	1969	Scrapped
3	Brill SC	1910	1959	1980	Museum
4	Brill SC	1910	1969	1978	Museum
5	Brill SC	1910	1967	1980	School in Solum, since 1997 site of the CMC in Eiras (poor condition)
6	Brill SC	1912	1959	1980	Museum
7	Brill SC	1912	1967	1979	School in S.António dos Olivais, later site of the CMC (poor condition)
8	Brill SC	1926	n.a.	1970's	Scrapped
9	Open	1927	1957	1980	Scrapped
10	Open	1927	1956	1970's	Scrapped
11	Open	1928	1950's	1980	Museum
12	Open	1928	1957	1979	Playground Peneda de Saudade, since 1997/8 site of the CMC in Eiras (bad condition)
13	Open	1928	1966	1980	Portugal dos Pequenitos, since 1997 Fundação Biscaia Barreto (poor condition)
14	Steel	1928	n.a.	1980	Museum
15	Steel	1928	n.a.	1980	Guarda-Inglesa, since 1997 Museum
16	Belga	1930	n.a.	1973	Museum
17	Belga	1930	n.a.	1973	Sold to a private person
18	Belga	1930	n.a.	1973	Sold to a private person
19	Standard	1934	n.a.	1970's	Scrapped
20	Standard	1940	n.a.	1980	Guarda-Inglesa, later school in Solum, since 1997 site of the CMC in Eiras (poor condition)

Re-bodied trams series 1, 3-7, 9-13 Standard Coimbra
During the 1950's and 1960's eleven of the oldest trams got new bodies to the same design as the cars 14-15 and 19-20. This time the bodies were composite: steel frames with wooden panels. Sources differ about the exact year for each car that they were re-bodied. Due to the long time-span there were differences between the cars, the most prominent visible being the type of direction indicators. The cars 4, 5, 7 and 13 had the destination blinds enclosed in the front of the roofs, while all the others had the indicators on top of the roofs like all older cars had. On the basis that the four cars which had the destination films enclosed in the roof instead of being mounted on the top and with reference to old photos, and also comparing that with other sources, the most likely years for the construction of the new bodies are given in the table with general information about the individual trams. Most of these cars were in service until the end of the tram operations in 1980.

The foregoing following table shows cars now in the museum, however the museum has been closed for some time and as at January 2021 no date has been given for re-opening.

Trailers
There were two trailers used with the electric fleet. Probably both were bought by the CCFC from the CCFL in Lisboa and later re-gauged by the CMC from 900 mm to 1000 mm. But it is possible that they might have been acquired directly by the CMC from the CCFL as trailer usage in Coimbra started only in 1918, a period when the CCFL abandoned part of its oldest trailer fleet. Trailer no.1 was a closed former mule car built by John Stephenson (New York). Trailer no.2 was originally an open mule tram. In the 1930's, but at an unknown date, it received a closed body being a copy in wood of the trams 14 and 15, but to a reduced size: six windows at each side instead of seven. Probably it was also narrower as a photo suggests it was used together with Brill SC cars and it is good practice that a trailer isn't wider than the car pulling it. Both trailers were withdrawn from service in 1954. Trailer no.1 is part of the museum collection, but needs restoration. Trailer no.2 went to a school in Bencanta, owned by the Fundação Biscaia Barreto.

Freight trams
There were three freight cars, an electric zorra (open freight car) and two wagons. The zorra was built in 1926 on a Brill 21E truck by the Companhia Alliança in Porto, which made similar zorras for the Porto tram company. The car was used to bring coal from the Coimbra-B railway station to the power plant in Rua da Alegria. When the power plant was put on standby in 1933, the car received a water tank for cleaning purposes. In 1940 the truck was used for the construction of tram no.20.

The wagons were probably inherited from the CCFC mule tram company and might have been used for the transport of mail bags between the post office and the Coimbra-B railway station. They had an obscure life but were retained in the fleet until 1971.

Car 1 approaching Santo Antonio from Tovim. This car still has a cream rocker panel. *P. Haseldine*

Car 11 in Rua Ferreira Borges on 8 May 1972. *Michael Russell*

Car 1 at Praca 8 de Maio on 6 June 1973. *Michael Russell*

Metro-Mondego

In 1906 the Ramal da Lousã was opened. This was a broad gauge heavy railway but is only mentioned here because it was the precursor of a failed project for the return of trams in Coimbra. The line linked the Coimbra-A (Estação Nova) station with the towns of Miranda do Corvo and Lousã to end in Serpins, a village part of the municipality of Lousã. The line crossed central Coimbra with a street track along the Mondego river. Old postcards show steam trains on the Av.Navarro. In 1976 a new stop was made near Portagem called Coimbra Parque, a few years later this was moved to a location close to the Alegria tram depot. At this time most trains terminated in Coimbra Parque.

In order to access the workshops there were scheduled trains using the street track to Coimbra-B. These trains ran late at night or early in the morning because during daytime police escort was required, but not during the night.

After many years of launching ideas, proposals and plans to convert the railway into a light-rail line, the Metro-Mondego company was established in 1996 to develop definitive plans and manage the projects to realise them. The project always remained controversial because of the costs. The continuing discussions and different opinions with the consecutive governments meant that the definitive decision to realise the system was only taken in 2006. In January 2010 the Lousã railway was closed, except

Car 13 approaching Praca da República on Rua Lourenco de Almeida Azevedo under a menacing sky on 2 April 1974. *Michael Russell*

Only minutes later, car 9 is approaching Praca da República on Rua Lourenco de Almeida Azevedo with the weather now clearing on 2 April 1974.
Michael Russell.

Car 6 at the Largo da Manutencao Militar junction on 15 October 1976. *Michael Russell*

Car 10 seen between Santo Antonio and Tovim on 15 October 1976. *Michael Russell*

Tramcar 13, built as a crossbench car in 1928 on a Brill 21E truck with General Electric electrical equipment, rebuilt as a semi-convertible in the 1930s and rebodied in 1966, waits at Largo de Portagem while its trolley is turned having arrived from Alegra depot on 20 September 1973. Note the broad gauge railway track to the left. *D. Pearson*

Seen passing Praça da República bound for Largo da Portagem on line 4 is car No. 1 of 1910, rebodied in 1961, on 20 September 1973. *D. Pearson*

Heading towards Praça 8 de Maio on 18 August 1978, tramcar 1 precedes trolleybus 44 in Avenida Sâo da Bandeira. *D. Pearson*

for the section between the Coimbra-A and Coimbra-B stations, which was to be closed for rebuilding only later, and works started.

The foreseen network using 1435 mm gauge was:
• Light-rail line from Coimbra-B until Serpins using the route of the Lousã railway with a length of 37.6 km and a total of 33 stops. In the São João area a diversion from the old railway route was planned to serve the football stadium and a shopping mall. There would have been double track between Combra-B and Alto de São João and single track with passing loops beyond Alto de São João.
• Urban light-rail line from Aeminium / Loja da Cidadão to Hospital Pediatrico. This would be 4 km long with 11 stops. At Aeminium, which was projected on the railway just outside Estação Nova in the direction of Coimbra-B, this line was to be connected with the Lousã line. The route would be Aeminium / Loja da Cidadão - Camara - Mercado - Praça da República - Universidade (bottom of the monumental stairs) - Arcos de Jardim - Sereia - Celas - Universidade Polo III - H.U.C (Hospital Universidade de Coimbra) - Hospital Pediatrico (children's hospital). Between Sereia and Celas was a tunnel projected under the Rua Dr. Augusto Rocha. There were also tunnels projected between the stops Universidade Polo III, H.U.C and Hospital Pediatrico. All stops however would be at surface level.

The economic crisis in November 2011 forced the then new government to decide to stop the project. At that time the old railway tracks had been lifted. The construction or renovation of many parts, including stations, stops and access roads, for the line outside the urban area were close to conclusion except for track laying and installing of the catenary and signalling. In the urban area little was done, except for demolishing most of the buildings in the area between Aeminium and Rua Sofia to give space to the light-rail line.

Protest came from the three municipalities involved, Coimbra, Miranda do Corvo and Lousã, and their inhabitants. These protests continued over the years, the government responded with promises to study the issue further.

In June 2017 another new government proposed a busway with electric buses on the old railway route. This solution is also controversial and decisions still have to be made. Perhaps trams will once again return in Coimbra, but at the moment of writing the future doesn't look positive.

COIMBRA Tramway Chronology 1944 - 1966

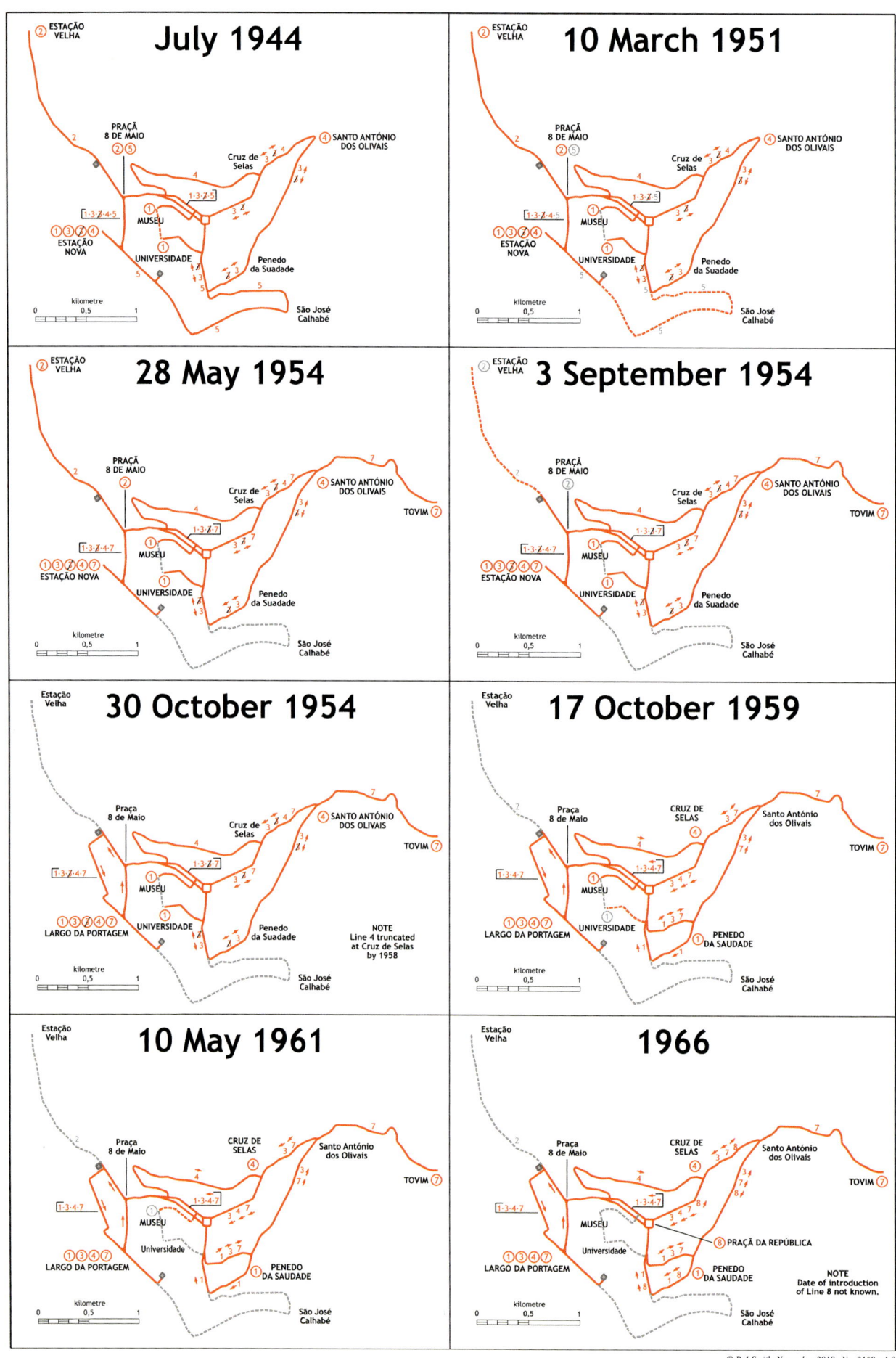

© R.A.Smith, November 2018. No. 2158, v1.0

COIMBRA

Trolleybuses

By Pedro Costa, translation by Andy Steel †,
contributions by Ernst Kers

40 is seen climbing up to Santa Clara on
20 May 1980. *Michael Russell*

Coimbra council had first shown an interest in trolleybuses in 1938, recognising their advantages on steep hills. On 16 February 1947 they opened Portugal's first trolleybus route, from Largo da Portagem across the narrow lattice-girder Mondego Bridge and up the hill beyond to Santa Clara monastery as "single line" throughout. The whole route was installed by Swiss engineers, and was operated by two Swiss trolleybuses, Saurer 3TP/Saurer/Sécheron numbered 21 and 22. The wider new Santa Clara Bridge was inaugurated on 30 October 1954 and trolleybus line 6 was made "double line" to the Pequenitos model village and single beyond, until doubled throughout in the 1960s. From the north bridgehead (Largo da Portagem) the route was extended with the same Baixa loop that was also introduced for the trams on 30 October 1954.

The two Swiss built trolleybuses subsequently received new bodies, 21 was rebodied by UTIC in 1972 and 22 rebodied by Caetano in 1974. 22 is preserved.

Six further trolleybuses, British built Sunbeam MF2Bs with Park Royal bodies numbered 23 to 28, arrived in 1951 to take over the São José tramline 5, which was replaced from 10 March 1951 by a circular trolleybus route 5 (anti- clockwise) and 5 barré (clockwise), traversing a new dual-carriageway road from São José to Arcos de Jardim. From 1959, trolleybus 5 also replaced the Arcos de Jardim to University branch tramway, and on 10 May 1961 route 5 barré was diverted (downhill only) to replace the branch tramway to the Museum. In 1959 a trolleybus branch was opened from São José to Estadio, and the result was a circular route 5 with two branches, two short-working points and about five destinations, which must have confused even the residents at times.

The subsequent introduction of trolleybuses on routes 1, 3 and 4 has been referred to in the tramway chapter, and was followed by the start of an eastern outer circle (7, clockwise; 7T, anticlockwise) on 13 May 1991, which provided a direct link between Tovim and São José. The narrow Rua Ferreira Borges is now pedestrianised and the route via Estação Nova is used in both directions. Trolleybus deliveries subsequent to 21-2 and 23-8 were three British-built BUT LETB1 with Park Royal bodies delivered in 1954 and numbered 29-31. In 1958 four BUT LETB1 with Portuguese UTIC bodies numbered 32-5 were delivered whilst a further batch of six similar vehicles, numbered 36-41 arrived in 1961. More Sunbeams in the form of six MF2NS with UTIC bodies arrived in 1966 and took the fleet numbers 42-47.

The next arrivals came in 1983-88, Caetano-EFACEC 175Tr110 numbered 50-69, and were the first trolleybuses built wholly in Portugal. Trolleybuses 23-31 were rebodied by Caetano in 1974 and 1979, and Braga's trolleybus fleet was acquired in 1980 but not used. In the 1980s, most of the 1958 and 1961-built BUT LETB1 trolleybuses and all of the 1966-built Sunbeam MF2NS trolleybuses received modifications either to the front or rear ends, in some cases both, and some also received modifications to the side windows. A further order for new trolleybuses was cancelled, and motor buses took over certain trolleybus duties.

In late 2003 the two Caetano/EFACEC trolleybuses retained by Porto after the system closure in 1997 were acquired. 167, an articulated 190Tr110, and 74, a two-axle 175Tr110, became 70 and 71 respectively in the Coimbra fleet. The two vehicles,

Saurer 22 is seen in original condition with 21 immediately behind at Rua da Alegria depot on 9 September 1969. *A. G. Murray*

which were the only two trolleybuses in the Coimbra fleet to be equipped with auxiliary diesel engines, usually operated on route 4 but 70 was noted out of service with technical problems from November 2004.

By May 2006 construction had started of a new route to Estádio Cidade de Coimbra. In May 2008 a new route 60 was inaugurated along this extension operating Monday to Friday during university term time only, providing a 30 minute service between 08.00 and 18.00 on an anti-clockwise circular route, as follows: São José, Rua Dom João III, Rua General Humberto Delgado, Rua Miguel Torga, Av. Dias da Silva to Santo António dos Olivais. The service was reduced to a 40 minute frequency later in 2009 operated with one trolleybus.

Also in 2009 an order was placed in March with Solaris for a single Trollino 12AC trolleybus. It was delivered in September and was numbered 75. Operation of route 60 ceased with effect from 1 October 2011, which made about 2.7 km of one-way wiring redundant, including about 1.7 km which been erected specially for the route's opening.

Trolleybus operation was suspended in the spring of 2017 because of major roadworks in the city centre. During the suspension the opportunity was taken to upgrade some overhead and substation equipment, and to refurbish the vehicles. On 31 May 2018 trolleybus operation resumed. The current serviceable fleet of five comprises Solaris Trollino 12 number 75, and four of the EFACEC vehicles, numbers 54, 55, 58 and 63. The remaining members of this batch are withdrawn permanently and are used as a source of spare parts.

Trolleybus operation in Coimbra was suspended at the end of January 2022

COIMBRA TROLLEYBUS FLEET LIST

Fleet number	Chassis	Bodywork	Electrical Equipment	Built New	Acquired	With-drawn	Notes
21-22	Saurer 3TP	Saurer	Sécheron	1947		1984-87	21 rebodied by UTIC in 1972. 22 rebodied by Caetano in 1974. 22 preserved
23-28	Sunbeam MF2B	Park Royal	BTH	1951		1993	23/25/27/28 rebodied by Caetano in 1974, 24/26 rebodied by Caetano in 1979
29-31	BUT LETB1	Park Royal	BTH	1954		1993	29 rebodied by Caetano in 1974. 30/31 rebodied by Caetano in 1979
32-35	BUT LETB1	UTIC	BTH	1958		1986-93	Some received modification to windows in the 1980s, see text.
36-41	BUT LETB1	UTIC	AEI	1961		1989-93	Some received modification to windows in the 1980s, see text.
42-47	Sunbeam MF2NS	UTIC	BTH	1966		1993-94	All received modification to windows in the 1980s, see text.
48-49	Henschel II-6500	CAMO (1976)	Siemens	1951	1980	1980	Ex Braga 7/8, ex Heilbronn 204/205, did not enter service
50-69	Caetano 175TR110	Caetano	EFACEC	1983-88		2000-	
70	Caetano 190TR110	Caetano	EFACEC	1985	2003	2009	Ex Porto 167, returned to Porto
71	Caetano 175TR110	Caetano	EFACEC	1984	2003		Ex Porto 74
75	Solaris Trollino ST12	Solaris	Skoda	2009			

and back by the same route

Sunbeam MF2B 25 is at Rua da Alegria depot on 9 September 1969, note the centre exit. *A. G. Murray*

Saurer 21 after being rebodied, is seen turning at Largo do Arnado on 6 June 1972. *Michael Russell*

No. 23, Sunbeam MF2B / Park Royal / BTH, the initial vehicle of the six-vehicle batch, which was exhibited at the London Commercial Motor Show in 1950. It is seen on 20 September 1973 in front of the Aqueduto before climbing alongside it on Bairro Sousa Pinto to serve the University. *D. Pearson*

No. 47, Sunbeam MF2B / UTIC / BTH built 1965, and numerically the last of the British-built trolleybuses supplied to Coimbra, passes Largo de Portagem en route to Estadio on 20 September 1973. Note the railway track to the left. *D. Pearson*

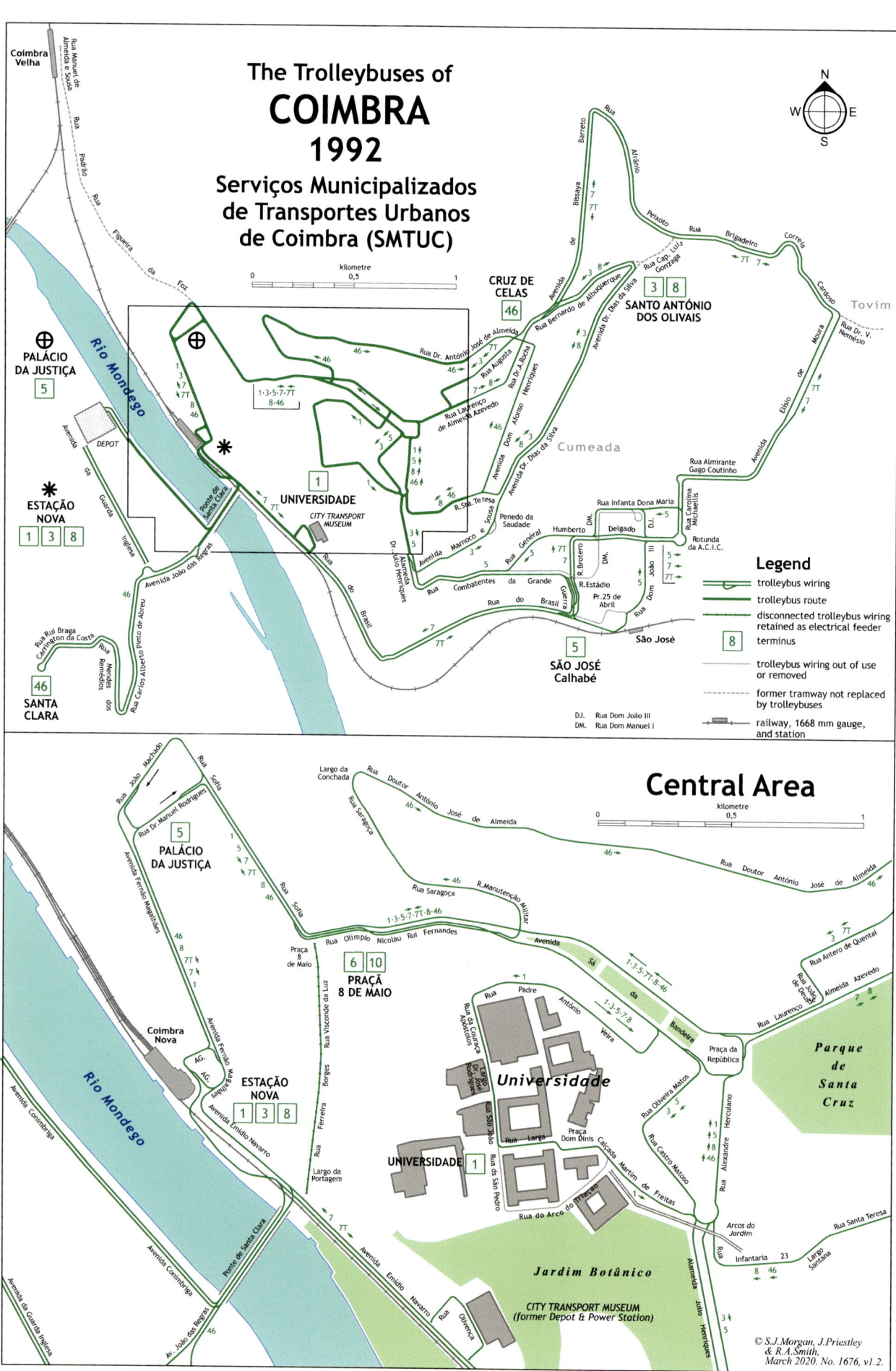

The Trolleybuses of
COIMBRA
1992
Serviços Municipalizados
de Transportes Urbanos
de Coimbra (SMTUC)

Central Area

Legend

trolleybus wiring	
trolleybus route	
disconnected trolleybus wiring retained as electrical feeder	
8	terminus
trolleybus wiring out of use or removed	
former tramway not replaced by trolleybuses	
railway, 1668 mm gauge, and station	

DJ. Rua Dom João III
DM. Rua Dom Manuel I

© S.J.Morgan, J.Priestley & R.A.Smith, March 2020. No. 1676, v1.2.

BUT 35 turning at Alameda Dr Júlio Henriques, Jardim Botânico on 6 June 1973. *A. G. Murray*

A nearside view of an unidentified Sunbeam at Largo da Portagem on 6 June 1973. *A. G. Murray*

Rebodied 25, a 1951 Sunbeam, is passing Estação Nova in 1974. *P. Haseldine*

47, the last UK built trolleybus (apart from the Dennis Doncaster demonstrator) a Sunbeam built 1965 is seen in 1969. *P. Haseldine*

26 with its second body is at Largo da Portagem on 15 October 1976. *Michael Russell*

Saurer 22 with its second body is at the Santa Clara terminus on 15 October 1976. *Michael Russell*

31 with its second body is seen at San Jose with 37 and 28 behind on 20 May 1980. *Michael Russell*

36 with a rebuilt front end is seen at Praça 8 de Maio with 51 behind on 11 July 1986. *Michael Russell*

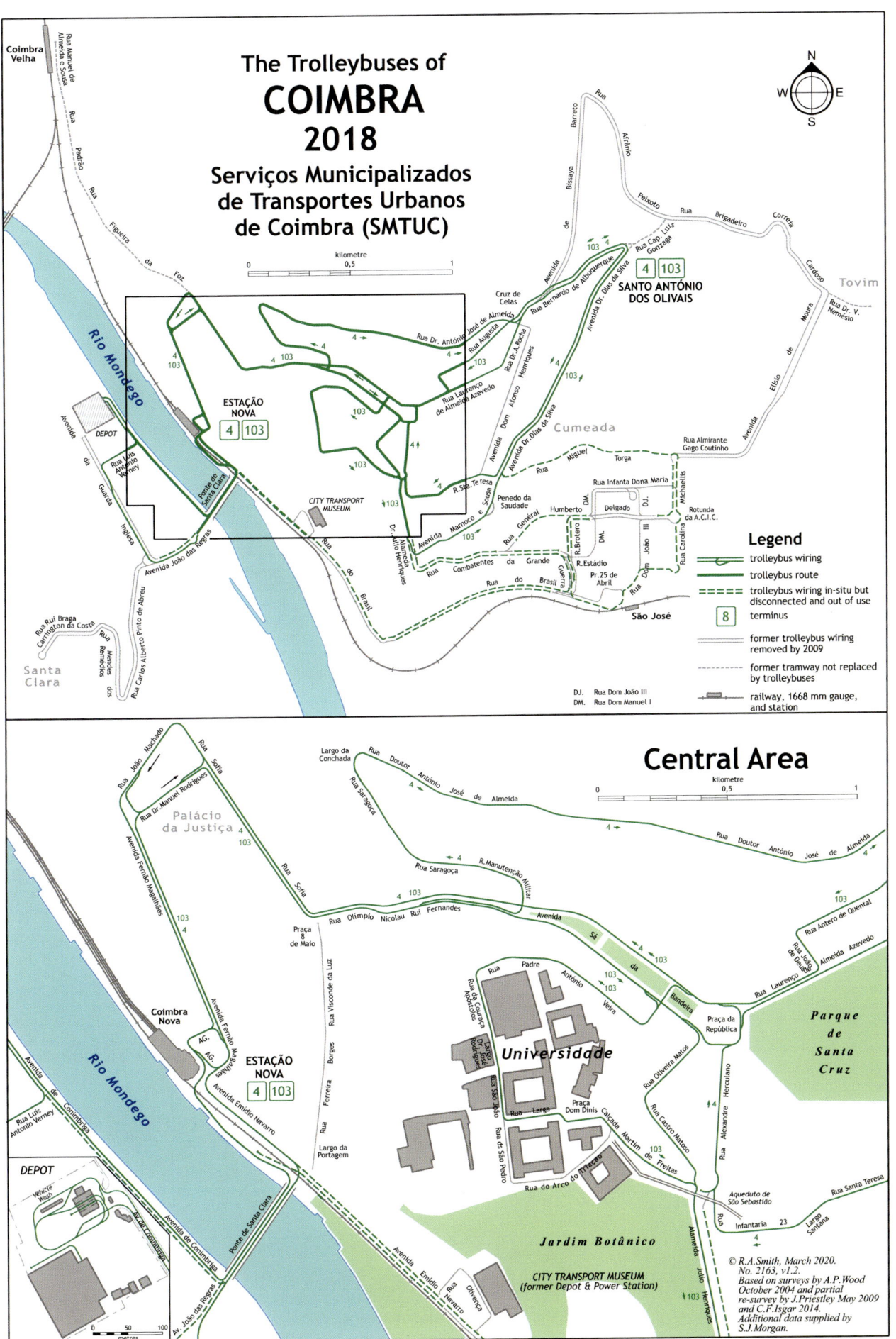

The Trolleybuses of
COIMBRA
2018
Serviços Municipalizados
de Transportes Urbanos
de Coimbra (SMTUC)

kilometre
0   0,5   1

Legend

〰 trolleybus wiring

━━━ trolleybus route

┅┅┅ trolleybus wiring in-situ but
disconnected and out of use

[8] terminus

──── former trolleybus wiring
removed by 2009

------ former tramway not replaced
by trolleybuses

🚂 railway, 1668 mm gauge,
and station

D.J.   Rua Dom João III
DM.   Rua Dom Manuel I

Central Area

kilometre
0   0,5   1

*Parque
de
Santa
Cruz*

Universidade

Jardim Botânico

CITY TRANSPORT MUSEUM
(former Depot & Power Station)

© R.A.Smith, March 2020.
No. 2163, v1.2.
Based on surveys by A.P.Wood
October 2004 and partial
re-survey by J.Priestley May 2009
and C.F.Isgar 2014.
Additional data supplied by
S.J.Morgan.

64 turns at the Largo da Manutencao Militar junction on 12 July 1986. *Michael Russell*

With its rebuilt front end prominent, 45, BUT / UTIC / BTH of 1965 arrives at Santo António dos Olivais, terminus of line 8, on 28 September 1992.
D. Pearson

70, the sole articulated vehicle which was on loan from Porto is seen working a service in Rua João Machado on route 4 on 6 May 2004. *Michael Russell*

Trollino 75 is approaching Cruz de Celas on 16 May 2011. *Michael Russell*

Trollino 75 is just arriving at Santo Antonio on 5 May 2014. *Michael Russell*

Coimbra 21, one of the two Saurer 3TP / UTIC (1972) / Sécheron trolleybuses is seen in Rua Ferreira Borges, Portagem on 6 June 1973. *A.G. Murray*

BRAGA

Trams and Trolleybuses

By Carl Isgar

Car 1 at the São João
da Ponte terminus.
A. P. Tatt/R. Copson – OTA

A mule car no 11 operated by Companhia de Carris e Ascensor do Bom Jesus pulled by three mules seen prior to 1908 in Praca de Liberdade (then Largo da Arcada). *Photographer unknown*

Most visitors to Braga, 55km north of Porto, come for the pilgrimage church of Bom Jesus do Monte, situated about 5km east of the town and commanding a magnificent view. The church is approached by a huge baroque stairway up the hillside, but for those who felt themselves exempt from such a penance a standard-gauge funicular up the hill alongside was opened on 26 March 1882. The funicular has two parallel tracks and was constructed by Niklaus Riggenbach and SLM (Swiss Locomotive and Machine Works) With a length of 270 metres and a maximum gradient of 52 per cent, this was the first funicular in Portugal and the fourth only in antiquity to the two Lausanne funiculars of 1877 and Riggenbach's pioneer line of 1879 at Giessenbach. Like the Giessenbach line, it relied on the weight of the water ballast for motive power, and on a central Riggenbach ladder rack for braking. The funicular still uses the original cars and the water counter-balance system.

Mule Trams

The railway from Porto to Braga had opened on 21 May 1875 with the station at the west of the town. A separate company, the Companhia de Carris de Ferro de Braga, formed in 1876 had obtained a concession for a mule tramway of 900 mm gauge from Braga station through the town and on towards Bom Jesus. The urban section of the line was opened on 20 May 1877, before work began on the funicular.

The route of the tramline was from the station via the Rua do Souto, which is the narrow main street of Braga, to the Largo de São Francisco where the depot of the trams was located. Soon, probably on 2 July 1877, the line was extended to the Portico of Bom Jesus at the bottom of the stairway to the pilgrimage church. The timetable for July 1877 shows ten return trips for the section Largo de São Francisco - Bom Jesus and six return trips for the section Largo de São Francisco - Railway station, the latter apparently in connection with the trains. The trip from Largo de São Francisco to the Portico of Bom Jesus took 40 minutes, in the opposite direction 30 minutes, a difference which without doubt was caused by the gradient over the route. The urban route took ten minutes for both directions.

The tram track was soon re-laid with new Vignoles steel

A view thought to have been taken in the early days of the system of car 8 at Portico Alameda at the foot of the staircase leading to Bom Jesus. *Portimage Foundation*

Car 1 with trailer 4 at Praça in 1959. *Courtesy of the National Tramway Museum, photographer J.H. Price*

rails on oak sleepers to permit the use of steam locomotives, coinciding with the opening of the funicular on 25 March 1882. The tramway and funicular operators joined forces in June 1883 to form the Companhia Carris e Ascensor do Bom Jesus.

In 1886 a second, more northern route was opened between Largo de São Francisco and the railway station. This route was via Rua dos Capelistas, Campo da Vinha (nowadays Praça do Conde de Agrolongo) and Rua dos Biscanhois where it joined the first route again just west of the Arco da Porta Nova. This route might have been made to ban the steam trams from the busy but narrow Rua do Souto.

Steam Trams

The operators had meanwhile obtained two small second-hand steam locomotives. They were short unenclosed 0-6-0 well tanks with Brown's patent boilers, and had been built in 1877 with Winterthur works numbers 112 and 113 for a 900 mm gauge line from Regua (in the Douro valley) to Vila Real de Trás-os-Montes. This line had a short life and was many years later replaced by a metre gauge railway taking a different route. The Winterthur engines had worked as a pair, coupled together by a load-carrying framework. At Braga they ran on the public highway, without any form of side skirting. From 8 July 1891 and decided after several serious accidents, the locomotives were only used between Ponte de Santa Cruz (Peões) and Bom Jesus. Since that date only mules were used to haul the trams in the urban area. The steam engines were used mainly on Sundays and holidays, and postcards show that at less busy times the through service was provided by single cars hauled by mules. A third locomotive of unknown (possibly German) origin was added in 1887.

Electrification

In 1905 the first proposal was made to replace the mule and steam trams by electric trams; however this failed to be realised just as numerous other proposals that followed. Finally in May 1913 it was decided that the municipality would take over and electrify the system. The tramway paid a regular 6% dividend, but the municipality bought out the company, placing a contract in 1913 with AEG and Thomson-Houston Iberica and retaining the 900 mm gauge. A new depôt bearing the initials CMB (Camâra Municipal de Braga) and the date 1914 was erected south of the station. The power station was located on the same site as the new depot. Two boilers, two steam engines and two dynamos could deliver together 600 hp of electricity, which was not only used for the trams but also for electric street lights and pumps to supply water from the Rio Cavado to the town.

In the night of 3 to 4 October 1914, just after midnight the first electric tram started test running. On 14 October eight experienced tram-drivers from Porto went to Braga for training of the local drivers. On 18 October the official inauguration took place and electric services with the new J. G. Brill four-wheel cars began on 19 October 1914. The system was still single-track, with loops spaced to allow a ten-minute service. The three steam locomotives were disposed of, but the trailers were retained for use on Sundays and holidays and whenever pilgrimage traffic was heavy. The trams were unique in Portugal in using bow collectors instead of trolley poles.

The basic time-table of 1914 showed trams every 20 minutes in both directions via Rua do Souto and via Campo da Vinha, resulting in every 10 minutes a tram as far as Peões. The trams continued every 30 minutes from Peões to Bom Jesus. On

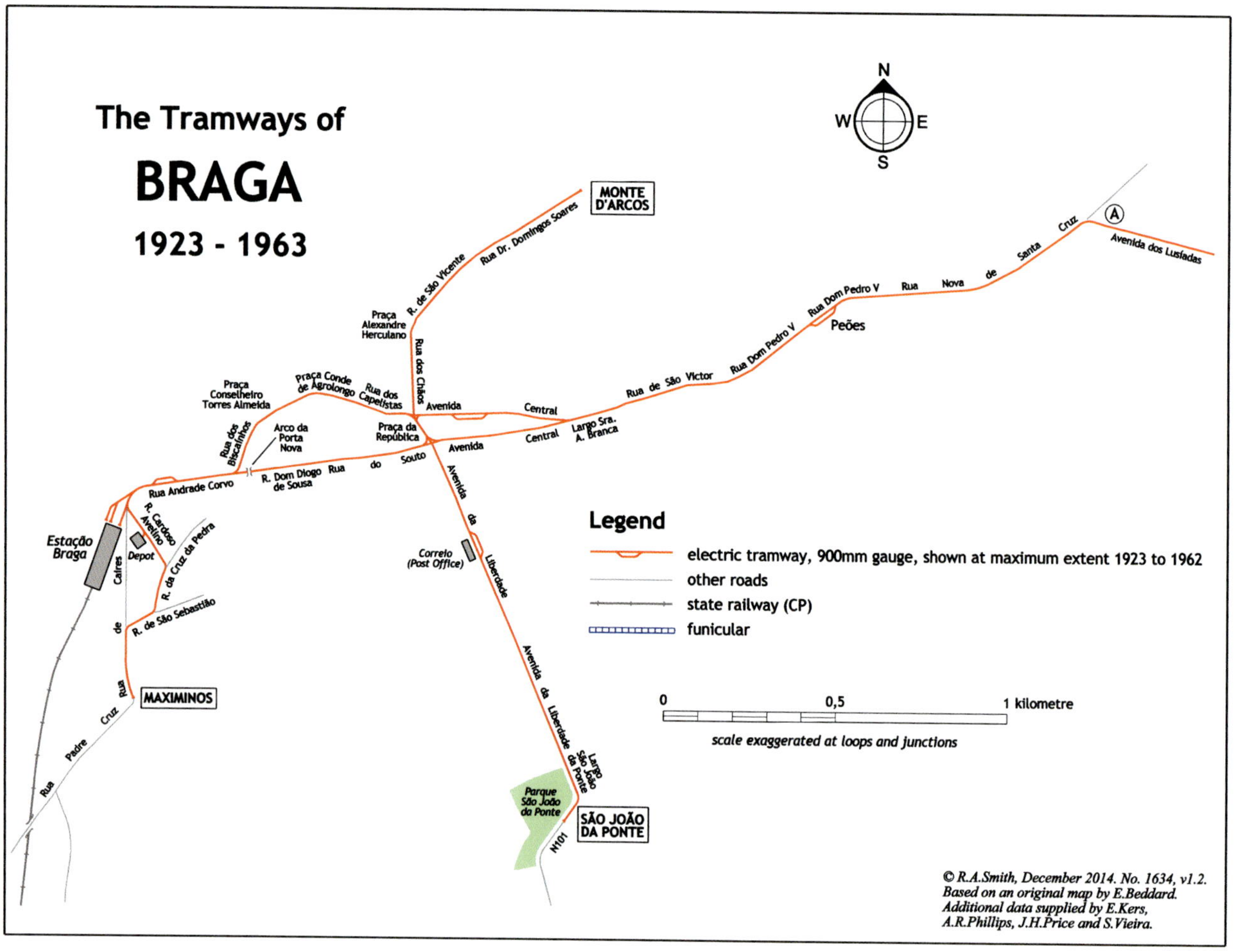

The Tramways of
BRAGA
1923 - 1963

N
W • E
S

MONTE D'ARCOS

R. de São Vicente
Rua Dr. Domingos Soares
Avenida dos Lusíadas
Cruz
A
Santa
Rua Nova
Rua Dom Pedro V
Rua Dom Pedro V
Peões
Praça Alexandre Herculano
Rua de São Victor
Rua dos Chãos
Praça Conselheiro Torres Almeida
Praça Conde de Agrolongo
Rua dos Capelistas
Avenida
Central
Central
Largo Sra. A. Branca
Arco da Porta Nova
Praça da República
Avenida
Souto
Rua dos Biscaínhos
R. Dom Diogo de Sousa
Rua Andrade Corvo
Rua
do
Avenida da
R. Cardoso Avelino
Estação Braga
Depot
R. da Cruz da Pedra
Correlo (Post Office)
Liberdade
R. de São Sebastião
Rua
de
Caires
MAXIMINOS
Rua
Cruz
Rua Padre
Avenida da Liberdade
Largo São João da Ponte
Parque São João da Ponte
SÃO JOÃO DA PONTE
N101

Legend

electric tramway, 900mm gauge, shown at maximum extent 1923 to 1962	
other roads	
state railway (CP)	
funicular	

0   0,5   1 kilometre

scale exaggerated at loops and junctions

© R.A.Smith, December 2014. No. 1634, v1.2.
Based on an original map by E.Beddard.
Additional data supplied by E.Kers,
A.R.Phillips, J.H.Price and S.Vieira.

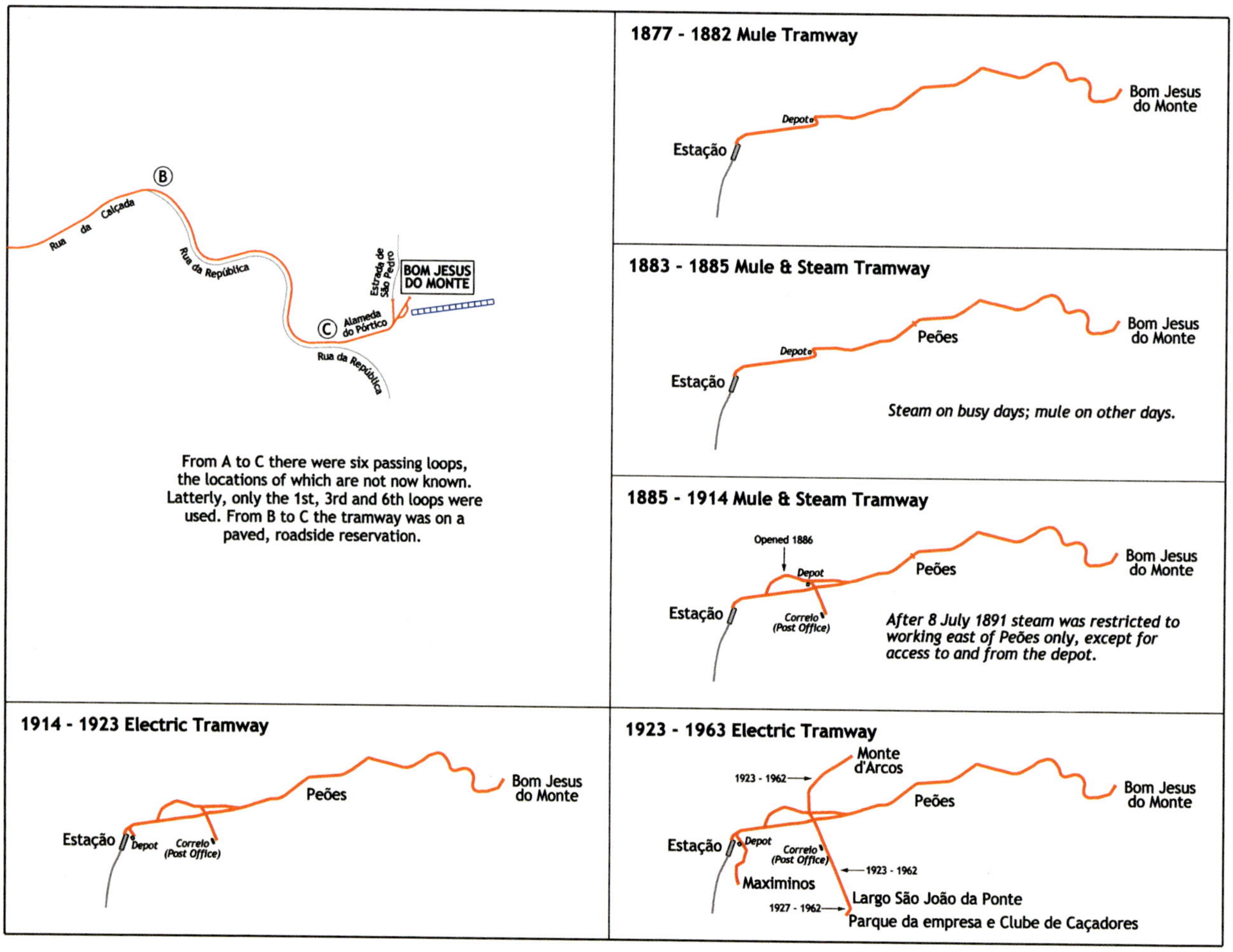

B
Rua da Calçada
Rua da República
C
Alameda do Pórtico
Rua da República
Estrada de São Pedro
BOM JESUS DO MONTE

From A to C there were six passing loops,
the locations of which are not now known.
Latterly, only the 1st, 3rd and 6th loops were
used. From B to C the tramway was on a
paved, roadside reservation.

1877 - 1882 Mule Tramway

Bom Jesus do Monte
Depot
Estação

1883 - 1885 Mule & Steam Tramway

Peões
Bom Jesus do Monte
Depot
Estação

Steam on busy days; mule on other days.

1885 - 1914 Mule & Steam Tramway

Opened 1886
Depot
Peões
Bom Jesus do Monte
Estação
Correio (Post Office)

*After 8 July 1891 steam was restricted to
working east of Peões only, except for
access to and from the depot.*

1914 - 1923 Electric Tramway

Peões
Bom Jesus do Monte
Estação
Depot
Correio (Post Office)

1923 - 1963 Electric Tramway

Monte d'Arcos
1923 - 1962
Peões
Bom Jesus do Monte
Estação
Depot
Correio (Post Office)
Maximinos
1923 - 1962
1927 - 1962
Largo São João da Ponte
Parque da empresa e Clube de Caçadores

Bom Jesus Funicular, Braga looking up from lower station in 1959.

Courtesy of the National Tramway Museum, photographer J.H. Price

A view of one of the funicular cars at the upper terminus of the line.

R. Buckley

Sundays and days of pilgrimage with heavy traffic to the church, there were separate services between the station and the Avenida Central and the Avenida Central and Bom Jesus. The most important extension of the Braga tramways was a line Largo da Ponte - Cemitério de Monte de Arcos, intersecting the main route at Praça de Republica, which was already proposed soon after the inauguration of the electric trams in 1914. The construction of this line started only seven years later in September 1921. Severe problems in acquiring the necessary

materials meant that that the construction took almost two years and the line was only opened on 30 June 1923.

On Sunday 20 May 1923 the most severe accident in the history of the Braga tram occurred, a runaway with disastrous results. Eight people died and dozens were wounded. The open trailer involved was almost completely destroyed.

Another extension was beyond the depôt to the suburb of Maximinos, introducing a service Maximinos – Peões, on 7 July 1923. A tradition continued from steam days was the carriage

Car 6 in the town centre on 7 September 1962. *W C Janssen/Online Transport Archive*

of mail between the railway station and the main post office in the town centre, using two four-wheeled vans attached to the service trams. On 19 June 1927 the Largo da Ponte - Cemitério de Monte de Arcos line was extended at the southern end over the bridge to the entrance of the Parque da empresa e do Clube de Caçadores.

In a report made in 1938 the director of the SMB described the tram system to be in a state of dilapidation. He included in his report the question of whether the trams should be maintained or abandoned. As an option he proposed to replace the trams by trolleybuses. However, it was going to be another 25 years before that happened.

Trolleybuses

By 1950 the track was badly worn and in need of renewal. The Serviços Municipalizados de Braga introduced bus operation, using locally-bodied British vehicles, but the government favoured electric traction from its new hydro-electric schemes and persuaded Braga council to replace trams with trolleybuses. In 1961, the municipality bought (with State aid) the complete trolleybus system of Heilbronn, Germany, which had been installed in 1951 and closed on 30 December 1960. The purchase comprised nine trolleybuses, six by Henschel and three by MAN (all with Siemens Schuckert equipment) together with two elderly Hansa-Lloyd motor tower-wagons. It also comprised the complete overhead, poles, feeder cables, rectifiers, switchgear, tools and spares, which had been supplied originally to Heilbronn by BBC. The assorted material was sent by barge in October 1961 from Heilbronn to Rotterdam and shipped from there to Leixões with final delivery by road. In Heilbronn the trolleybuses bore the numbers 201-6 (Henschel) and 207-9 (MAN); three other MAN vehicles (210-12) had been sold previously to the Bavarian town of Landshut.

Early in 1962 a section of trolleybus wiring was erected from the depôt to Porta Nova for driver training. Since the trams had bow collectors the wiring could not be shared, but at one point the tram wire and the positive trolleybus were crossed at a sharp angle, with only a small air-gap. This gap was fouled by the carbon shoe of trolleybus 9, resulting in a spectacular and unexplained electrical disaster, including a short-circuit. Only then did it emerge that in Braga (due probably to a contractor's error) the tramway overhead polarity was negative and the rails positive.

MAN trolleybus 3 still in its original livery is seen near the depot in August 1967. *Courtesy of the National Tramway Museum, photographer, A. R. Philips*

This had been discovered earlier by a French consultant called in to trace the enormous power leakage, but his recommendations had not been implemented; the undertaking had apparently increased the tramway supply voltage from 600 to 800 volts to "compensate for the leakage" and trolleybus 9 had effectively received 1400 volts, and was thereafter used solely as a source of spares. The French consultant had said that the power loss was equivalent to central heating of the streets. The conversion was delayed while the trolleybus wiring was altered to run above the tram wires. The "learner" vehicles had to coast under the junctions with their poles down until the trams were withdrawn. The last service trams ran on 20 May 1963, and buses provided a service until trolleybus operation commenced on 28 May, on which day tram 2 made a brief farewell run from the railway station to the Porta Nova and back to the depôt.

Change of Ownership

In 1967 the Serviços Municipalizados handed over its transport undertaking to a private company trading as Empresa de Transportes Urbanos but whose legal title is Sociedade de Transportes Urbanos de Braga (SOTUBE). The funicular with its two venerable Swiss cars was handed over to the Confraria do Bom Jesus (a religious organisation) who run it today, and have added a brakeman's cabin to each car.

Final Years and Closure

The final section of trolleybus route from Gualtar to Bom Jesus had only a single pair of wires suspended from the 1914 bowstring bracket at the roadside. In 1967 this section was replaced by motor buses running into town along a new dual-carriageway road, the trolleybus services becoming route 5 (Maximinos – Gualtar) and route 6 (Ponte – Monte d'Arcos). When the Bom Jesus route was truncated at Gualtar, the route terminated in a reversing triangle. Early in 1969 the Maximinos route was extended to a new terminus, and a one-way system was introduced around the block containing the depôt. At some stage between 1972 and 1976, the Gualtar reverser was replaced by a turning circle at the same point. The cross-town trolleybus route 6 ran for the last time in 1969. The remaining trolleybus route (Maximinos – Gualtar) closed on 10 September 1979 and was sold to the Coimbra undertaking (SMC), whose staff dismantled the Braga installation early in 1980. Six trolleybuses

One of the ex-Heilbronn trolleybuses still in its original livery seen at the depot in v1964. *Courtesy of the National Tramway Museum, photographer, A. R. Philips*

7, a Henschel II-65000 / Wankmiller / Siemens is at Praça da Republica on 12 September 1969. *A. G. Murray*

The north-south Monte d'Arcos - Ponte route closed relatively early. This shows 4 northbound in Avenida Gomes da Costa, having just left the Ponte terminus, on 20 September 1969. Note the double tyres on the front wheels. *P. Haseldine*

were delivered to a scrap merchant at Curia, north of Coimbra, and the others were kept by Coimbra until 1985.

Rolling Stock

Mule Trams
Two open and two closed mule trams were ordered in January 1883 from Starbuck. The open cars were of the cross-bench type with five benches and seating for 20. The closed cars had a seating capacity for 12.

Electric Trams
At its maximum extent the Braga tram fleet comprised 11 electric motor cars, 11 (or 13) trailers and two mail vans, with repetition of fleet numbers between these three groups.

In readiness for the start of electric services in October 1914, Braga obtained eight 20-seat semi-convertible four-wheel cars from the J. G. Brill Company of Philadelphia, similar to Coimbra 1-7 except for their 900 mm gauge and their bow collectors and wooden seats. These cars had six windows at each side and benches in rows of 2+1. They were mounted on 1.98 m wheelbase Brill 21E trucks and were numbered from 1 to 8. Motors and controllers were by AEG, supplied by Thomson-Houston Iberica. Slight modifications were made to the platforms and seating of some cars over the years, and some car bodies eventually had to be reinforced with steel flitches. In the 1920's and 1930's three further cars were needed. The first two cars, 9 and 10, were identical with 1-8. The builder is not known, but no. 10 was involved with the disaster of 20 May 1923, so they were built before that time. The last car added to the fleet was built by the local workshops in 1936/1937 with the use of parts supplied by the CCFP in Porto. It was the Braga version of the Porto Brill-28 type with seven windows at each side and seats with rows of 2+2. This car no. 11 was significantly longer and wider than the other ten cars.

The original trailer stock, quoted as 13 cars in pre-1914 steam days, comprised two-axle cross bench cars (1-7) with panelled and glazed bulkheads, to be hauled singly by mules or in trains of three by the steam engines. Four others (8, 10, 12?, 13?) were closed mule cars with six windows per side and turtle-back roofs, closely resembling various cars for other countries featured in the "John Stephenson Album"; they may have been second-hand purchases from Lisboa. Car 11 was a cross-bench car with turtle-back roof of unknown origin, and car 9 was a relatively modern saloon car with a flat roof and with six drop-sash windows per side, resembling a cut-down double-decker. This may have been a rebuild of the trailer car involved in the 1923 disaster. The tram fleet was dismantled in the depôt in 1963, except for motor 7 and trailers 11-13 which were broken up earlier. The tram livery was originally buff, then claret, and finally some cars were deep yellow, with cream as the second colour. The two mail vans were finished in grey.

Trolleybuses
1 - 3 were MAN type MKE 2 with Kässbohrer bodywork built in 1951 for the Heilbronn opening, ex Heilbronn 207 - 209. Numbers 4 - 9 were Henschel type II 6500 built 1951, ex Heilbronn 201 - 206. Heilbronn 201 - 203 (Braga 4 - 6) had Drögmöller bodies and 204 - 206 (Braga 7 - 9) Wankmiller. The main difference was that the Drögmöller bodies had windows with pans with radiused corners, whereas the Wankmiller bodies had squared-off corners. The Henschel II 6500 vehicles were distinctive in having double tyres on the wheels of the front axle and this is apparent in photographs, but after rebodying nos. 7 & 8 had only single tyres on the wheels of the front axles. The trolleybuses entered service in their German cream livery with blue waistband and lettering "Cidade de Braga". By 1969 they had been painted in a new livery of cherry red and cream with grey roofs. Trolleybuses 7 and 8 were re-bodied in the mid-1970s by Carrocerias Modernas (CAMO) and spent the years 1980-1985 in Coimbra's depôt, without being used. When rebodied they retained their SSW booms.

BRAGA TROLLEYBUS FLEET LIST

Fleet number	Chassis	Bodywork	Electrical Equipment	Built New	Acquired	With-drawn	Notes
1-3	MAN MKE 2	Kässbohrer	Siemens	1951	1961	19xx-79	Ex Heilbronn 207-209
4-6	Henschel II-6500	Drögmöller	Siemens	1951	1961	1976-79	Ex Heilbronn 201-203
7-9	Henschel II-6500	Wankmiller	Siemens	1951	1961	1972-79	Ex Heilbronn 204-206. 7/8 rebodied by CAMO in 1976. 7/8 to Coimbra 48/49. 9 never used

The tower wagon seen parked in the depot on 16 October 1976. *Michael Russell*

A rear view of 1 in Largo de Madre de Deus on 16 October 1976. *Michael Russell*

Trolleybuses 3 and 1 are about to pass each other on the single wire section near Maximinos on 16 October 1976, one of them will need to lower its trolley booms. *Michael Russell*

Braga 1, a M.A.N MKE2 / Kässbohrer / SSW, formerly Heilbronn 207, is arriving at Gualtar terminus on 18 September 1973. *D. Pearson*

Braga 4: a Henschel 6500 / Wankmiller /SSW, built 1951, formerly 204 in the Heilbronn fleet, is seen approaching Gualtar terminus on 18 September 1973. *D. Pearson*

Maximinos terminus on 16 August 1978 showing number 8 with its new Portuguese body by CAMO. This vehicle, originally Heilbronn 205, was a 1951 Henschel II 6500 with electrical equipment from Siemens Schuckert Werke, and carried a Wankmiller body until the rebodying, along with sister vehicle number 7, in 1976. *D. Pearson*

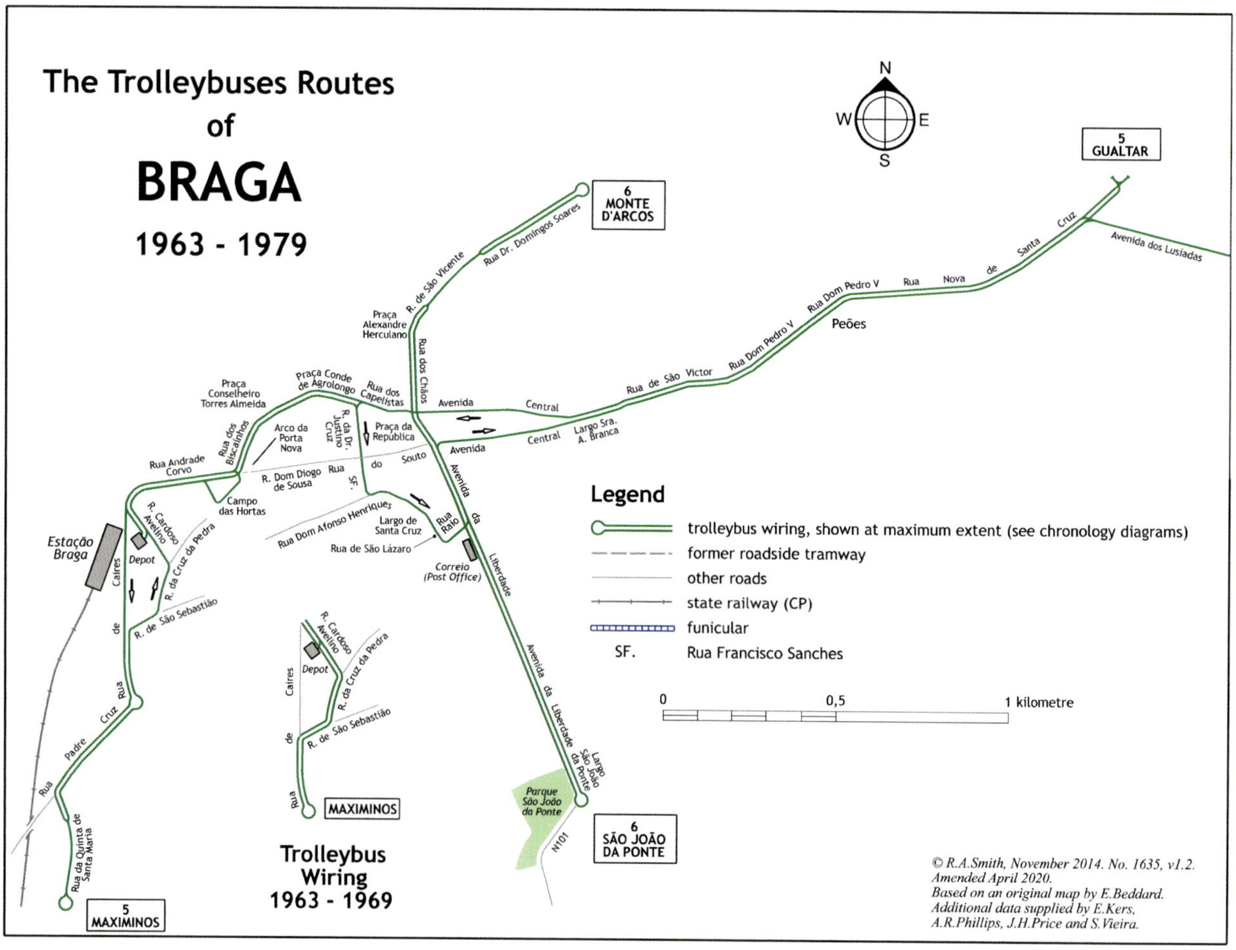

The Trolleybuses Routes
of
BRAGA
1963 - 1979

N
W E
S

5 GUALTAR

6 MONTE D'ARCOS

Rua Dr. Domingos Soares
Rua de São Vicente
Praça Alexandre Herculano
Rua dos Chãos
Rua Dom Pedro V
Rua Nova de Santa Cruz
Avenida dos Lusíadas
Peões
Rua de São Victor
Rua Dom Pedro V
Praça Conde de Agrolongo
Rua dos Capelistas
Praça Conselheiro Torres Almeida
Arco da Porta Nova
R. da Dr. Justino Cruz
Praça da República
Avenida
Central
Central
Largo Sra. A. Branca
Rua dos Biscaínhos
do
Souto
Avenida
Rua Andrade Corvo
R. Dom Diogo de Sousa
SF.
Rua
Rua Dom Afonso Henriques
Largo de Santa Cruz
Campo das Hortas
R. Cardoso Avelino
R. da Cruz da Pedra
Rua de São Lázaro
Rua Raio
Avenida da Liberdade
Estação Braga
Depot
Cairés
Correio (Post Office)
R. de São Sebastião
R. Cardoso Avelino
Cairés
Depot
R. da Cruz da Pedra
R. de São Sebastião
Rua
de
Cruz
Rua Padre
Rua da Quinta de Santa Maria
MAXIMINOS
Trolleybus Wiring 1963 - 1969
Parque São João da Ponte
Largo São João da Ponte
N101
6 SÃO JOÃO DA PONTE
5 MAXIMINOS

Legend

⊸⊸⊸	trolleybus wiring, shown at maximum extent (see chronology diagrams)
– – –	former roadside tramway
——	other roads
✛—✛	state railway (CP)
▭▭▭	funicular
SF.	Rua Francisco Sanches

0   0,5   1 kilometre

© R.A.Smith, November 2014. No. 1635, v1.2.
Amended April 2020.
Based on an original map by E.Beddard.
Additional data supplied by E.Kers,
A.R.Phillips, J.H.Price and S.Vieira.

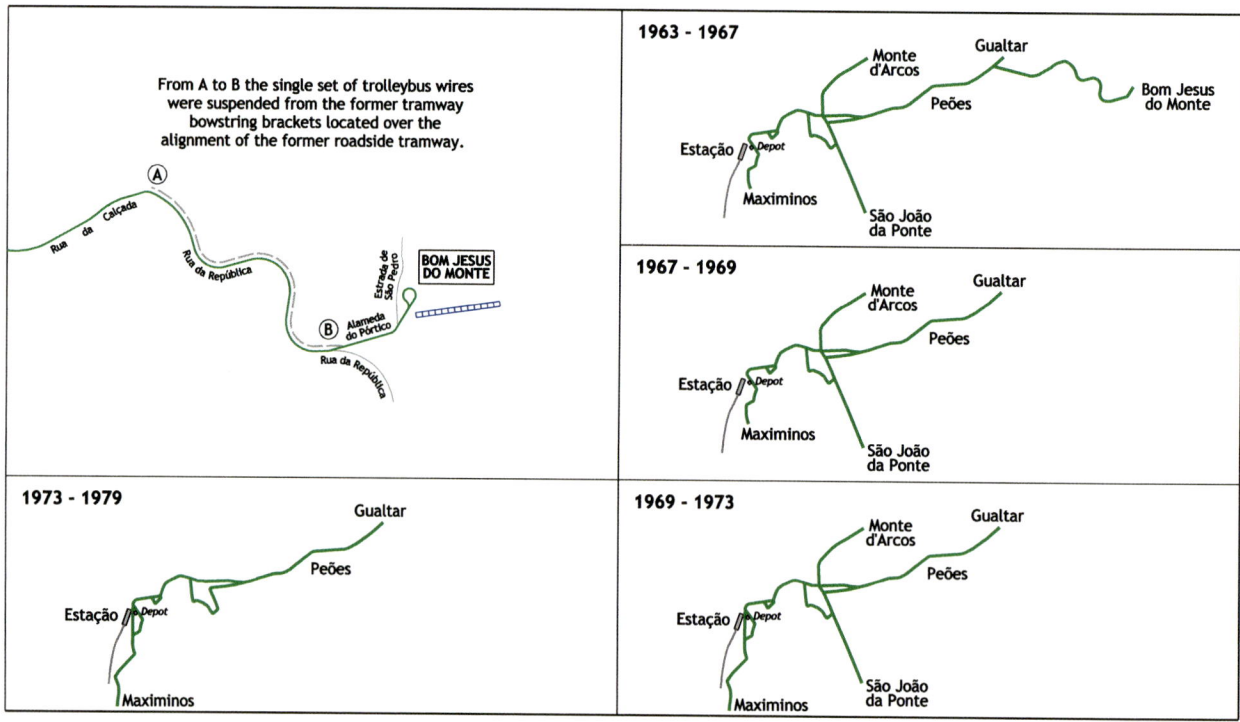

From A to B the single set of trolleybus wires were suspended from the former tramway bowstring brackets located over the alignment of the former roadside tramway.

Rua da Calçada
Rua da República
Estrada de São Pedro
BOM JESUS DO MONTE
Alameda do Pórtico
Rua da República
A / B

1963 - 1967
Monte d'Arcos
Gualtar
Peões
Bom Jesus do Monte
Estação / Depot
Maximinos
São João da Ponte

1967 - 1969
Monte d'Arcos
Gualtar
Peões
Estação / Depot
Maximinos
São João da Ponte

1969 - 1973
Monte d'Arcos
Gualtar
Peões
Estação / Depot
Maximinos
São João da Ponte

1973 - 1979
Gualtar
Peões
Estação / Depot
Maximinos

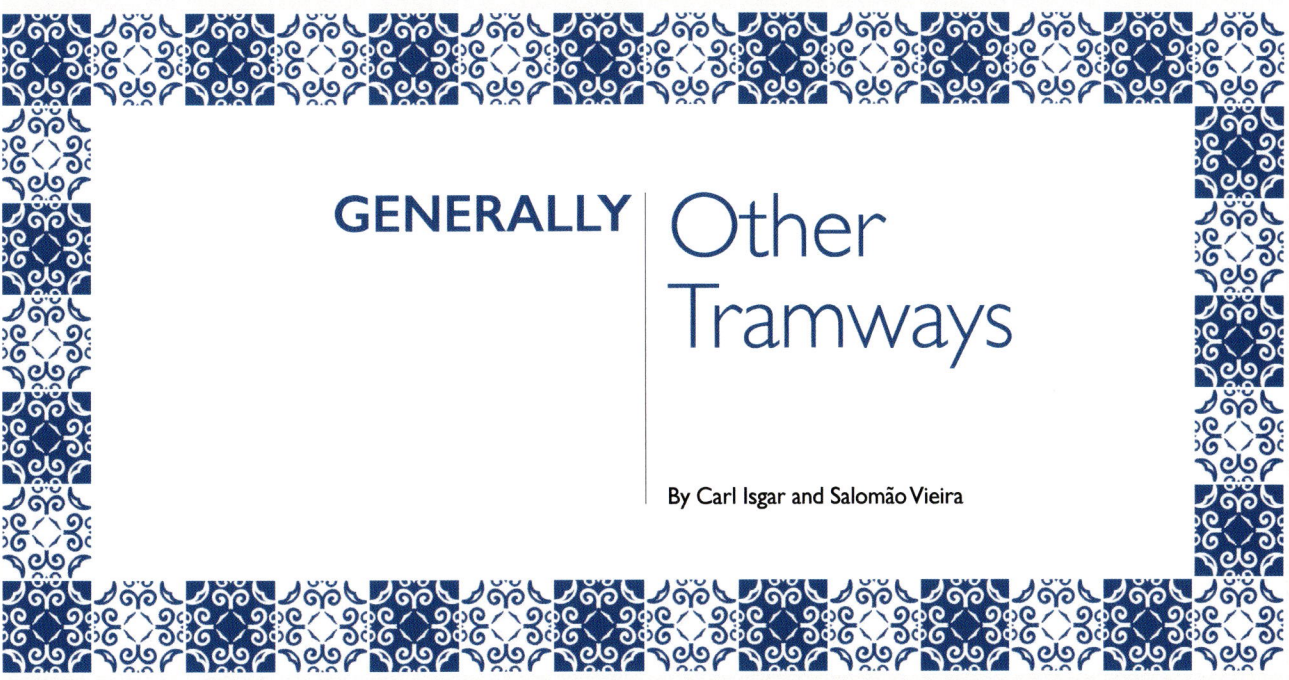

GENERALLY | Other Tramways

By Carl Isgar and Salomão Vieira

Since the third and fourth editions of this book were published, research in Portugal has revealed the former existence of several more animal-powered tramways, some of which also employed steam locomotives for goods and passenger service. They are listed here in approximate geographical order from north to south.

We commence with the 3.67 km 1400-mm gauge mule tramway from Vila do Conde to Povoa de Varzim, on the coast north of Porto. Operation began on 15 October 1874 and by 1882 required six cars and 20 mules, when a new owner, José Galiza and family, took charge of the operation. Some branch lines were added in 1898 to link the railway stations of two towns with the surrounding beaches. Also at this time the fleet was enlarged at least to thirteen vehicles with acquisitions from Companhia Carris de Ferro do Porto and others. In 1926 experiments were carried out to equip of one or two cars with heavy oil "oleos pesados" engines, but an accident put an end to this two years later. This was Portugal's longest-lived mule tramway (1874 to October 1934) and the last to survive.

In 1873 a German wine-exporter, Maximilian Schreck of Companhia Vinicola Portuguesa obtained a concession for a 26.8-km animal tramway from Regua to Vila Real, with freight haulage by oxen on gradients and horse or mule haulage elsewhere and for passengers. Operation began on 26 September 1875 but ceased in July 1876. The company (Companhia Transmontana) sought the help of SLM of Winterthur, who supplied two 900-mm gauge 0-6-0T locomotives (works numbers 112 and 113 of 1877). They were intended to run as a pair with smokeboxes facing, with a permanent coupling framework 8.10m long and 1.80m wide between them, on whose timber platform a load of 15 tonnes could be carried. Further details appear in "The Locomotive, Railway Carriage and Wagon Review" of May 1941. Steam working was tried between July and November 1877, but permission to continue with steam was refused and the line closed in 1880, the locomotives being transferred to the tramway at Braga. The present railway from Regua to Vila Real follows a different course.

Another little-known steam tramway was the Penafiel - Lixa line (Companhia do Caminho de Ferro de Penafiel a Lixa e a Entre-os-

Rios), whose 49km of metre-gauge track ran from Penafiel north to Lixa and south to Entre-os-Rios, with a depot at Novelas. Since the track was laid in the street in Lixa, in Aguas de Sao Vincente and elsewhere, the company used enclosed tramway engines built by Henschel (11339-40 of 1912, 12180-1 of 1913 and 12982.-3 of 1914). The Penafiel-Lixa line was opened in six stages from 10 November 1912 to September 1914, followed by Penafiel Entre-os-Rios in 1915. Financial difficulties led to suspension of services in 1920, but the track remained intact to 1931, and some was still visible in Lixa 30 years later.

Downstream from Coimbra was the horse (or mule) tramway of the Companhia Mineira e Industrial do Cabo Mondego, opened in December 1875 from Cabo Mondego in Buarcos to the harbour of Figueira da Foz, to carry coal, lime, tiles, glass and cement. The government licence of September 1874 imposed an obligation to carry passengers, which began in August 1876 and used four cars built in the company's own workshops, plus 14 wagons. Some cars from Companhia Carris de Ferro de Lisboa were acquired in 1909. In 1888 the line was extended to the railway station of Figueira da Foz and the system totalled 7.8 km, and was quoted as 850-mm gauge. Two steam locomotives were used between 1902 and 1914, when they were taken out of service. Another two steam locomotives were acquired and were in use from 1921 to 1926, and summer only mule trams continued until March 1933, after a 1927 electrification plan was abandoned. The cars were thereafter referred to as "mulectricos".

Midway along the Atlantic coast between the Vouga and Mondego rivers is Praia de Mira, where the sandy soil produces an aquatic plant used as fertilizer. The Empresa dos Americanos de Mira was formed in 1888 to build 13 km of metre-gauge tramway to handle the crop, with a line from Areão to Mira, via Praia de Mira. Two steam locomotives, one certainly a Krauss imported from Germany, and two cars and five wagons built by M. J. Soares of Coimbra formed the rolling stock. Freight service began in January 1889 and passenger service in July. In 1898 the track was lifted for use elsewhere.

The Portuguese parliament first considered tramways in 1858, and in January 1859 sanctioned Portugal's first tramway, from Pinhal de Leiria to the harbour of S Martinhodo do Porto, chiefly to carry timber. It was 36 km long and of standard gauge, and opened on 1 December 1860. The cars were hauled by oxen, usually in trains of four goods wagons and a two-class passenger car, and took seven hours, using both ox-power and gravity. The right-of-way was reused in 1885 for the west coast railway.

Another early ox-powered tramway, without passenger service, was opened on 24 October 1864 from Braçal mine to the river Mau, a tributary of the Vouga River.

Mention should also be made of proposed or short-lived mule tramways at Aveiro, Elvas and S Jacinto, and a military horse tramway at Oeiras.

A metre-gauge-steam tramway opened at Torres Novas, an agricultural and vineyard zone, promoted by "Companhia do Caminho de Ferro de Torres Novas a Alcanena" promoted by Barão de Matosinhos, the same of the Rail Road Conimbricense. The first section, some 7 km long, opened on 18 June 1889. It ran from the railway station of Torres Novas or Riachos, then via the East Line of Companhia Real, over local roads, to the Torres Novas town centre through some narrow streets. Two locomotives 0-6-0T by La Metallurgique (Belgium) and several cars, eight at least, similar to the cars of Carris de Ferro of Lisboa, were also used for passengers service to connect with trains running on the line of Companhia Real. In 1892 five daily trains in each direction were run to transfer to Companhia Real trains.

Work to extend the line from Torres Novas to Alcanena, a distance of 14 km, began swiftly but inauguration of this second section did not take place until 1 February 1893, when the opening took place accompanied by great festivities. However, inexplicably all the line closed four months later, on 30 June of that year. The line and rolling stock were incorporated in Companhia Real and sometime later the locomotives and eight of the cars were sold to Companhia do CF do Porto à Póvoa e Famalicão (PPF).

The Vale do Lima railway
(Contributed by Salomão Vieira)
Following the First World War, Portugal was awarded reparations against Germany in the form of railway rolling stock. Twelve passenger cars and five vans were supplied by MAN in 1925 for the northern metre-gauge lines, and an order was placed in 1924 to equip a proposed electric light railway running inland from Viana da Castelo.

The Vale do Lima light railway was first proposed in 1900 as a steam-worked 18-km line from Viana de Castelo to Ponte da Lima, following the north bank from Viana to Lanheses and crossing there to the south bank, but work did not start. In 1920, a similar concession but with electric traction was awarded to the Sociedade Industrial do Norte, going beyond Ponte da Lima to Ponte de Barca (40 km), but work was delayed by interminable discussions on the route to be followed, e.g. north bank or south bank. Construction of the Viana-Lanheses section began on 28 February 1930, but ceased in 1932 after the introduction of local bus services.

Other elevators (lifts) and funiculars
(Contributed by Salomão Vieira)
Nazaré
Portugal has a number of funiculars in addition to those in Lisboa

and Braga already described. One is at Nazaré, a small fishing town north of Lisboa lying partly below a 90-m cliff. Following the success of the Lavra and Glória funiculars in Lisboa, the same promoters built a steam-worked funicular to link the upper and lower towns at Nazaré, 318 metres long, with the topmost 50 metres in tunnel. The metre-gauge track was single with a central passing loop, and service commenced on 28 July 1889. Owned for many years by the Brotherhood of Our Lady of Nazaré, it was municipalised in 1932. The water for the boilers was carried up in tanks under the cars. Two 62-seat cars dating from the opening were used until February 1963, when the service was suspended as the result of an accident.

The Serviços Municipalizados de Nazaré (SMN) provided a replacement service with two exLisboa single deck buses, but later it was decided to re-open the funicular in a modern form. New track was laid, new trucks and winding gear were supplied by the Swiss firm of Bell, and new all-metal enclosed car bodies were made by Dalfa of Ovar, who also built the bodies of Porto's Lancia trolleybuses. The upper and lower stations were reconstructed with modern ticket offices and Ultimate ticket machines, and the line was re opened on 6 April 1968.

Viana do Castelo
At Viana do Castelo a funicular was built by "Companhia do Elevador do Monte de Santa Luzia" promoted by Bernardo Pinto Abrunhosa, and was inaugurated on 2 June 1923. It links the railway station with the Sanctuary of Santa Luzia, and is 632 metres long, with a single metre-gauge track and a midway passing loop. The running time is seven minutes. The electric winding gear was supplied by Brown Boveri, and the original cars were replaced in 1945 by two new cars built in the railway works at Campanhã. Originally white, these cars are now in the CP railcar livery of red and white with striped ends.

The original company was wound up in the 1930s and after 1945 the funicular was owned by CP Railways. From April 1988 to April 2001 the concession to exploit the funicular was transferred to the firm SOMALIS, Lda. Later the funicular was abandoned, however in January 2005 the city council, Câmara Municipal of Viana do Castelo took ownership of the funicular and repair work soon began. New building stations, new rails and new cars were introduced and the new funicular opened on 5 April 2007, and has been in service ever since. Other tourist facilities were provided at the two funicular stations.

Porto
Elevador Vila Nova de Gaia - Portugal has no mountain rack railways, apart from the former line at Madeira, but did possess a rack line of 1672-mm (5ft 6in) gauge running in a public street. This was the Elevador de Vila Nova de Gaia, opposite Porto, and its main purpose was to bring wagons down from the main line to the quays of the River Douro. It was built as a single track 750 metres long, with a central Riggenbach rack rail let into the road surface.

The original two 0-4-0T rack engines (numbered 1 and 2) were obtained in 1891 from Kessler of Esslingen (Germany) and works numbers 2431-2 but were actually built under licence at Saronno in Italy. At this period the line carried passengers as well as goods, but by 1906 it had been taken over by the Companhia União Fabril (an industrial group with rail-served factories and many private-owner wagons) and passenger service ceased. A new 0-4-0T rack engine, 3 was obtained in 1906 from Borsig of Berlin (works number 5891) and another in 1921 (Borsig 11027), replacing the Kessler engines. Regular operation is thought to have ceased in the late 1930s, but the locomotives remained for many years in the shed at Gaia and may have seen occasional use. The quayside tracks were lifted by 1958 and the rack line about 1960. The principal traffic handled was port wine in cask, Vila Nova de Gaia being the centre of this famous and old-established trade, where visitors are encouraged to tour the wine lodges and sample the product.

Elevador da Lada - This is a vertical elevator (lift) and is located in the Ribeira zone, near the D. Luís Bridge and Cathedral. It was inaugurated on 13 April 1994 but operation did not begin until 28 June 1995. It has vertical height of 28 m, a single cabin and is operated by Câmara Municipal do Porto (Porto city council). Operation has been interrupted on several occasions.

S. Martinho do Porto

Elevador do Outeiro - This is a vertical lift and staircase linking Rua Vasco da Gama (locally known by Rua dos Cafés) and Largo Comendador José Bento da Silva. Inaugurated on 15 October 2008, it has a vertical height of 30 m, a single cabin with a capacity of 13 passengers, and is operated by Câmara Municipal de Alcobaça (Alcobaça City Council). Tourist facilities are incorporated at the upper and lower boarding points.

Miragaia, Porto

According to an official announcement in "Jornal de Notícias" dated 28th August 2018, Pablo Pita, a local firm of Architects, has won a competition to install two rolling escalators along Escadas da Sereia and Escadas do Monte dos Judeus, adjacent to Largo da Alfândega. No timescale has been given for this proposal.

Coimbra

Elevador do Mercado

In the centre of Coimbra there is a vertical twin car lift with staircase linking the central Market with Rua Padre António Vieira, in the high residential zone near the university. It was inaugurated on 23 November 2001. It has a total vertical height of 51 m including an external tower 20 m high, two separate shafts each with a single cabin and is operated by SMTUC (Serviços Municipalizados de Transportes Urbanos de Coimbra).

Almada

Elevador da Boca do Vento - This is a vertical lift with staircase located close to the south bank of the river Tejo. It was inaugurated on 1 July 2000. It has a vertical height of 50 m, a single cabin and is operated by Câmara Municipal de Almada (Almada City Council)

Viseu

Funicular de Viseu - This is a funicular located in the town centre linking Feira de S. Mateus along Calçada de Viriato to Adro da Sé (Cathedral). It was inaugurated on 25 September 2009 and is a single track line, metre gauge, with a length of 400 m on a 16% gradient and has a midway passing loop, in Rua da Ponte de Pau. There are two cars, each with a capacity of 50 passengers, built by Ingenieria y Servícios de Montaña of Zaragoza. The line was promoted by Sociedade Polis Urbana de Viseu and constructed by Construtora Abrantina. Running time is 5 minutes; it is

operated by Câmara Municipal de Viseu (Viseu City Council) and no fares are charged.

The Funicular de Viseu suspended its activity in October 2015, waiting a general mobility plan for the town. Since then there has been limited operation in the tourist season, three summer months and at Christmas and Easter.

Covilhã

Funicular de Santo André - Situated in the centre of Covilhã, this line links Rua Marquês d'Avila e Bolama to Rua António Augusto Aguiar. It was inaugurated on 15 March 2009 and has a length of 90 m and a gradient of 29%. It is not on rails but has rubber tyred wheels and guide rail. There is a single car with a glass body, constructed by Liftech - Tecnologia para Elevadores Lda, Maia with a capacity of 11 passengers. It has two stops, runs on request and is operated by Câmara Municipal da Covilhã (Covilhã City Council).

Funicular de S. João de Malta - Similar to the Santo André installation but with flanged wheels rather than rubber tyres and located in the centre of Covilhã; it links Rua Mateus Fernandes to Largo de S. João da Malta and Avenida 25 de Abril. The line, constructed by EFACEC and LIFTECH and inaugurated on 1 September 2013 has a length of 200 m and a vertical height of 62 m. There is a single unmanned car constructed by Ingenieria y Servícios de Montaña' of Zaragoza. There are three stops and the line is operated by Câmara Municipal da Covilhã (Covilhã City Council).

Elevador da Goldra - Similar to the Santo André installation but with an unmanned car more like the one on the João de Malta line and located in the centre of Covilhã, it links Rua Marquês d'Àvila e Bolama (top) to Parque da Goldra. The line was inaugurated on 1 September 2013 and has a length of 25 m and a gradient of 32%. There are two stops and the line is operated by Câmara Municipal da Covilhã (Covilhã City Council).

Elevador Jardim - This two car installation opened on 25 April 2015. Other installations similar to the Funicular da Goldra are in study or being discussed.

Oeiras

SATU (Serviço Automático de Transportes Urbanos) - This line linked Paço de Arcos railway station, on the Cascais line, to Oeiras Parque, with an midway station on Tapada, that included a passing loop. Inaugurated on 7 June 2004 it comprised an elevated metre gauge 1150 m long single track line on a viaduct with two unmanned cars each with 8 seats and space for 71 standing passengers. It was operated jointly by Câmara Municipal de Oeiras (Oeiras City Council) and the construction company Teixeira Duarte. Plans to extend the system to Lagoas Parque (second phase) and to Tagus Park (third phase) did not materialise and the line closed on 31 May 2015 due to financial difficulties.

Lisboa

Castelo de S. Jorge - In March 2013 a plan was published to provide three vertical lifts to the castle from Rua dos Fanqueiros and Rua da Madalena which would be partially inside an existing building. They were inaugurated on 31 August 2013. A pedestrian route links Mercado do Chão do Loureiro (Chão do Loureiro Market) with two existing lifts, which also connect with the castle.

A major scheme of accessible routes from 'Baixa' to 'Castle' has already resulted in the opening of a new elevator: Elevador de

Santa Luzia, linking Rua Norberto Araujo to Miradouro de Santa Luzia (Portas do Sol), which was inaugurated on 10 June 2015.

Largo do Martim Moniz to the Castle - This is a group of two escalators linking Largo do Martim Moniz to the Castle following the steep steps of Escadinhas da Saúde. It was installed by Liftech and opened to the public on 13th October 2018.

Other plans include:
• An elevator from Campo das Cebolas to Largo da Sé; anticipated opening in 2017.
• A funicular between Graça and Alta Mouraria.

Panoramic Elevator, adjacent to the Ponte 25 de Abril

Recently a panoramic transparent elevator with a viewing platform, all in glass, was constructed in a tower along the Pilar 7 (north side) of the Ponte 25 de Abril. The tower is 80 metres high, and from the cabin and the platform a superb view of Lisbon, the River Tejo and other sights can be obtained. It is operated by 'Associação Turismo de Lisboa' and opened on 27 September 2017.

Other proposals and new installations

In Vila Nova de Gaia a proposal for a funicular linking the bank of the Douro River to the summit of mount Serra do Pilar was changed to a cableway ('teleférico'), this was inaugurated on 1 April 2011. In Santarém, a funicular was proposed in 2004 which would have linked Ribeira de Santarém, near the railway station, to the town centre. However there have been no further developments on this proposal. In Ourém, near Fátima, proposals for either an elevator or a teleférico linking the village to the castle were published in 2008. In 2016 it was reported that the municipality had submitted the proposal to the EU for funding.

At Montemor-o-Velho, near Coimbra a system of three open air escalators was installed and commenced operation in July 2014.

Nazaré Aerial Ropeway

A cableway (teleférico) was proposed to link the town with Pederneira. It was to be a bi-cable installation, shuttle type, with two cabins (one in each direction), to carry 250 people per hour. Liftech was to be responsible for overall contract work, construction, all electromechanical systems, cabins, all the automation systems, communication, video, fire, etc. Unfortunately this project has been abandoned.

Four public cableways (teleféricos) are now in service in mainland Portugal:
• Guimarães - Teleférico da Penha. Opened on 4 March 1985. Operated by Turipenha.
• Jardim Zoológico (Zoological Garden) de Lisboa. Opened in 1994.
• Parque das Nações – Lisboa. Opened during the "Expo 98" (1998).
• Vila Nova de Gaia – Teleférico da Serra do Pilar. Opened on 1 April 2011.

Madeira

The Atlantic island of Madeira, a Portuguese possession lying to the west of Africa, is regarded as forming part of Portugal and Europe. A brief note is therefore included on Madeira's rail transport.

A metre-gauge rack railway, the Elevador do Monte, was opened on 16 July 1893 from Pombal station in Funchal to Santa Luzia, and continued to Monte station on 5 August 1894. A further extension to Terreiro da Luta (876 m above sea level) was opened on 24 June 1912. The line was owned by the Mount Railway Co, Ltd of 24-25 Fenchurch Street, London EC3, who used the local title Carminho de Ferro do Monte.

The railway was 4 km long, with seven trains a day (eight on Sundays) and a running time of 20 minutes. Most passengers were shore excursionists from cruise ships or from Union Castle or Royal Mail Line vessels. Inclusive tickets were issued on board ship covering travel by oxcart from the quay to the station, the rack train up the mountain, and an exciting descent by toboggans, known locally as 'tea-trays". The company built a large restaurant at the summit station, and in 1926 opened the Grand-Hotel Belmont with its own halt near Monte.

The line was worked by 0-4-0 rack tank locomotives built by Maschinenfabrik Esslingen.

One engine was built in each of the years 1893, 1894, 1903 and 1912, with fleet numbers 1-4, and works numbers 2568, 2654, 3524 and 3668 respectively. There were four passenger cars and each engine propelled a one-car train On 10 September 1919 one of the locomotives suffered a boiler explosion, causing four deaths. A further locomotive (5) arrived in 1925; this was an 0-4-2 rack tank by SLM of Winterthur (works number 3120). The damaged locomotive and one of the original pair were then dismantled and parts of each were apparently used to build a new locomotive (6).

In Funchal town, a narrow-gauge mule tramway was opened on 3 June 1896 from the harbour area to the lower station of the rack railway, using cross-bench cars lettered in English ("Tram to Mount Railway Station. Fare 3d"), English money being accepted. Despite being subsidised by the town council and taken over by the Mount Railway in 1902, the tramway remained unprofitable, and the last day of operation was 28 January 1916. Proposals of 1913 and 1917 for an electric tramway did not materialise.

From the 1930s, the rack railway faced competition from road vehicles on a newly-built motor road, and operation ceased in 1939, the track being lifted in 1942. The company still existed twenty years later, with an office and a manager to administer the company's properties.

In September 1998 a vertical lift was inaugurated at Fajã dos Padres, 7 km W of Funchal. Known as the Elevador da Fajã dos Padres, it has a vertical height of 250 m, a single cabin and is operated by Hidroenergie. In April 2016 a second link was opened in the form of the Teleférico da Fajã dos Padres.

On 15 September 2000 a cableway (teleférico) was inaugurated from the town to Monte, which reinstates the long lost link provided by the Elevador do Monte. It is operated by Teleféricos da Madeira.

In recent years several cableways ("teleféricos') have been constructed or planned. Those constructed are:

• Monte: Funchal centre to Monte (already mentioned).
• Jardim Botânico: Monte to Jardim Botânico. Opened in September 2005.
• Rocha do Navio (Santana, on north coast): Opened on January 1999.
• Rancho: Fajãs do Cabo Girão ao Rancho: (Câmara de Lobos). Private. Opened on 4 August 2003.
• Achadas da Cruz (Porto Moniz): Opened before 2004.

Lisboa 810 is seen at Largo de Dona Estefânia on 7 June 1973. *A.G.Murray*

• Garajau (Santa Cruz): Opened on 22 April 2007.

Mirandela Metro

28 July 1995 saw the opening of Portugal's first surface "metro", 4.1 km in length from Mirandela town to the regional college at Carvalhais with a half-hourly service of ex-Yugoslav diesel railcars rebodied by the coachbuilder, Salvador Caetano.

Tua-Mirandela-Bragança at 134 km was the longest of the four metre-gauge branches running northwards from the Douro valley.

The Mirandela-Bragança section closed definitively on 14 October 1992, but the section from Mirandela to Tua operated until 2008, when several accidents and the construction of a barrage in the River Tua closed the branch between Tua and Cachão.

A new company Sociedade Metropolitano Ligeiro de Mirandela, was set up to reactivate the section from Mirandela to Carvalhais and add three Intermediate stations two of which (acknowledging EU financial help) are named after Jaques Delors and Jean Monnet. The railcars come in two versions known as "CP" with twin seats for 48, and "Metro" with single seats for 24 and space for 90 to stand. Livery is green with white skirting. The company currently also operates the branch Mirandela-Cachão.

A tourist circuit linking Tua, on the Douro railway line, with Mirandela, incorporating a boat trip, an elevator and a trip on the railway with a typical train is planned but at the time of writing (2021) has not been implemented.

Above: The controls of a Lisboa Remodelado car showing the large Kiepe Controller and modern switch panel. *Carris*

Right: In a well-established Lisboa practice, a driver adjusts the destination display by using a mirror specifically provided for the purpose. *K Lomas King*

LISBOA

Fleet List

"Remodelado" cars 542
and 560 ascending Calçada
de Sao Francisco in
September 2014. *L Vieira*

First generation electric trams

Series: 203-282

Type: 2-axle open cross-bench	Truck: Brill 21E
Body: J.G. Brill Company	Wheel base: 1.98 m
Length: 7.95 m	Wheel diameter: 0.85 m
Width: 2.32 m	Motors: 2x GE 59 - 25 hp
Weight: 9.1 t; 203-232 with air	Controllers: GE K10D
brakes 9.4 t	Wheel brakes: manual; 203-232 by
Seats: 32 on 8 cross-benches	1930 also air
Standing: 6	Track brakes: manual
In service: 1901	
Withdrawn: 1932-1937	

Series: 400-474

Type: 2-axle closed cars with drop	Truck: Brill 21E
windows	Wheel base: 1.98 m
Body: St.Louis Car Company	Wheel diameter: 0.85 m
Overall length: 8.12 m	Motors: 2x GE 59 - 25 hp
Saloon length: 5.48 m	Controllers: GE K10D; 400-414
Width: 2.25 m	since 1911: EE DB1/K3
Floor height: 0.89 m	Wheel brakes: manual
Platform height: 0.74 m	Track brakes: manual; 400-414
Weight: 9.4 t	since 1911 also electro magnetic
Seats: 20, 24 after the 1927/31	In service: 1901
rebuilding	Withdrawn: see text.
Standing: 14	

Series: 473-474 rebuilt in 1918

Body: Carris	Platform height: 0.70 m
Overall length: 8.56 m	Weight: 9.8 t
Floor height: 0.86 m	In service: 1918

Series: 283-322

Type: 4-axle open cross-bench	Trucks: Brill 22E
Body: J.G. Brill Company	Pivot distance: 7.68 m
Length: 11.33 m	Wheel base trucks: 1.22 m
Width: 2.29 m	Wheel base overall: 7.99 m
Floor height: 0.92 m	Wt ratio main/pony wheels: 3.0
Platform height: 0.92 m	Diameter main wheels: 0.85 m
Weight: 11.1 t	Diameter pony wheels: 0.52 m
Seats: 48 on 12 cross-benches	Motors: 2x GE 59 - 25 hp
Standing: 6	Controllers: GE K10D
In service: 1902	Wheel brakes: manual
Withdrawn: 1952-1955, 283	Track brakes: n.a.
instruction car 1955-1961	

A rare late view of one of the two-axle cross-bench cars, believed to be 236, taken in the early nineteen-thirties at Corpo Santo. These cars were all withdrawn soon after. Bogie car 356 is waiting to follow. *Col. Estúdio Horácio Novais / FCG – Biblioteca de Arte e Arquivos*

St Louis car (400-474) in original condition seen in Rua Rodrigo da Fonseca. *Photographer unknown*

Preserved cars 444 and 508 are seen at Belem during the celebrations to commemorate the centenary of the electric tramway on 31 August 2001. *Michael Russell*

Works car 397 was originally Brill crossbench motor 247. It is seen in 1967 in the Permanent Way yard at Santo Amaro. A new Daimler "Fleetline" chassis seen in the left foreground awaits bodywork. This car was later sold to a customer in the USA and subsequently operated on the Detroit tramway. *J. Jordan*

Cross bench car 313 is seen in Avenida da Liberdade, opposite the Lis Hotel, on 13 June 1954. *W. C. Janssen/Online Transport Archive*

St Louis car 420 at Santa Apolónia on 3 May 1964. *L. Folkard*

St Louis tourist car 1, formerly 437, is seen at Belem in 1967 on the occasion of a visit by four representatives of the Crich Tramway Museum. The two tourist cars, 1 and 2, were painted red as it was believed that this was the original Lisboa livery. No trace of red has been found on any of the museum restoration projects although it has been suggested that the St Louis cars were delivered in this colour as it was that company's default colour where the livery was not specified by the customer. *J. Jordan*

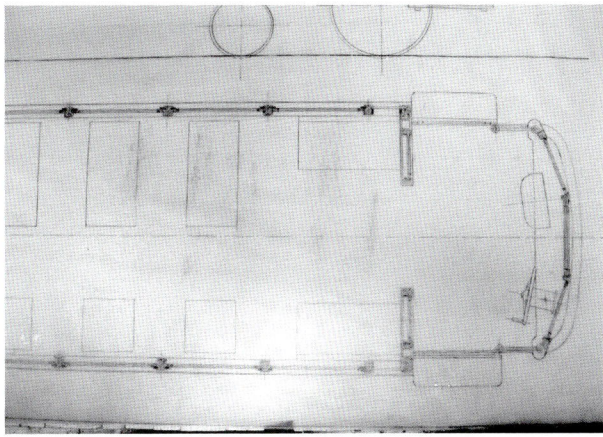

The CCFL archive contains an undated "concept drawing", the salient features of which are shown in these two views. No further information was found, but the fact that the drawing was prepared suggests that at some point consideration was being given to the possibility of converting cars 283-304 and 306-322 to saloons without the expense of building new "standard" 40-seat bodies, as had been done earlier with 305 following an accident. This would have provided 29-seats at a low cost, but with poor passenger accessibility due to the very high platform steps. It seems that the use of the electrical equipment to power 39 "Caixotes" towing trailers was considered the most cost-effective option. *CCFL*

Series: 323-342

Type: 4-axle semi-convertible	Trucks: Brill 22E
Body: J.G. Brill Company	Pivot distance: 5.90 m
Overall length: 11.78 m	Wheel base trucks: 1.22 m
Saloon length: 8.53	Wheel base overall: 6.21 m
Width: 2.36 m	Wt ratio main/pony wheels: 3.0
Floor height: 0.94 m	Diameter main wheels: 0.85 m
Platform height: 0.74 m	Diameter pony wheels: 0.52 m
Weight: 12.5 t	Motors: 2x GE 59 - 25 hp
Seats: 40	Controllers: GE B 18C
Standing: 21	Wheel brakes: manual, rheostatic,
In service: 1906	since 1931/2 air
Withdrawn: by 1996 (see text)	Track brakes: n.a.

Series: 343-362

Type: 4-axle semi-convertible	Trucks: Brill 27GE1
Body: John Stephenson & Co	Pivot distance: 5.49 m
Overall length: 12.09 m	Wheel base trucks: 1.37 m
Saloon length 8.53 m	Wheel base overall: 6.86 m
Width: 2.36 m	Wheel diameter: 0.85 m
Floor height: 0.97 m	Motors: 4x GE 59 - 25 hp
Platform height: 0.74 m	Controllers: GE K 28B
Weight: 18.5 t	Wheel brakes: manual, air
Seats: 40	Track brakes: manual but
Standing: 21	removed.
In service: 1907	
Withdrawn: by 1996 (see text)	

Series: 363-367

Type: 4-axle (original open) with	Trucks: Brill 22E
longitudinal benches	Pivot distance: 5.94 m
Body: J.G. Brill Company	Wheel base trucks: 1.22 m
Overall length: 12.00 m	Wheel base overall: 6.26 m
Saloon length: 7.62 m	Wt ratio main/pony wheels: 3.0
Width: 2.29 m	Diameter main wheels: 0.85 m
Floor height: 0.99 m	Diameter pony wheels: 0.52 m
Platform height: 0.99 m	Motors: 2x GE 59 - 25 hp
Weight: 13.5 t	Controllers: B 18C
Seats: 36, later reduced to 31	Wheel brakes: manual, rheostatic
Standing: 31	Track brakes: n.a.
In service: 1907	
Withdrawn: 1948	

Series: 475

Type: 2-axle semi-convertible	Truck: Brill 21E
Body: Carris	Wheel base: 1.98 m
Length: 8.36 m	Wheel diameter: 0.85 m
Width: 2.36 m	Motors: 2x GE 59 - 25 hp
Floor height: 0.92 m	Controllers: GE K10D
Platform height: 0.74 m	Wheel brakes: manual
Weight: 10.2 t	Track brakes: manual
Seats: 24	In service: 1909
Standing: 14	Withdrawn: 1952

332 is seen at Praça da Figueira on 12 July 1982. *L Folkard*

Car 337 at Belem on 4 April 1974. This car retains the original bodywork but has been rebuilt with an arched roof. *Michael Russell*

324 at Santa Apolónia on 3 September 1962. *W C Janssen/Online Transport Archive*

346 and 349 are seen at Praça do Comércio on 1 September 1962; 346 retains its original body whereas 349 has been extensively rebuilt with an arched roof. *W C Janssen/Online Transport Archive*

356, originally a Brill semi-convertible and rebodied in 1932, is seen turning out of Praça do Comércio on 12 July 1982. *L. Folkard*

A 1930 view of the entrance to Santo Amaro depot showing two of the workman cars with "Providence" lifeguards. Photographs of these cars in service are extremely rare but after nearly century two have emerged together. *Photographer unknown*

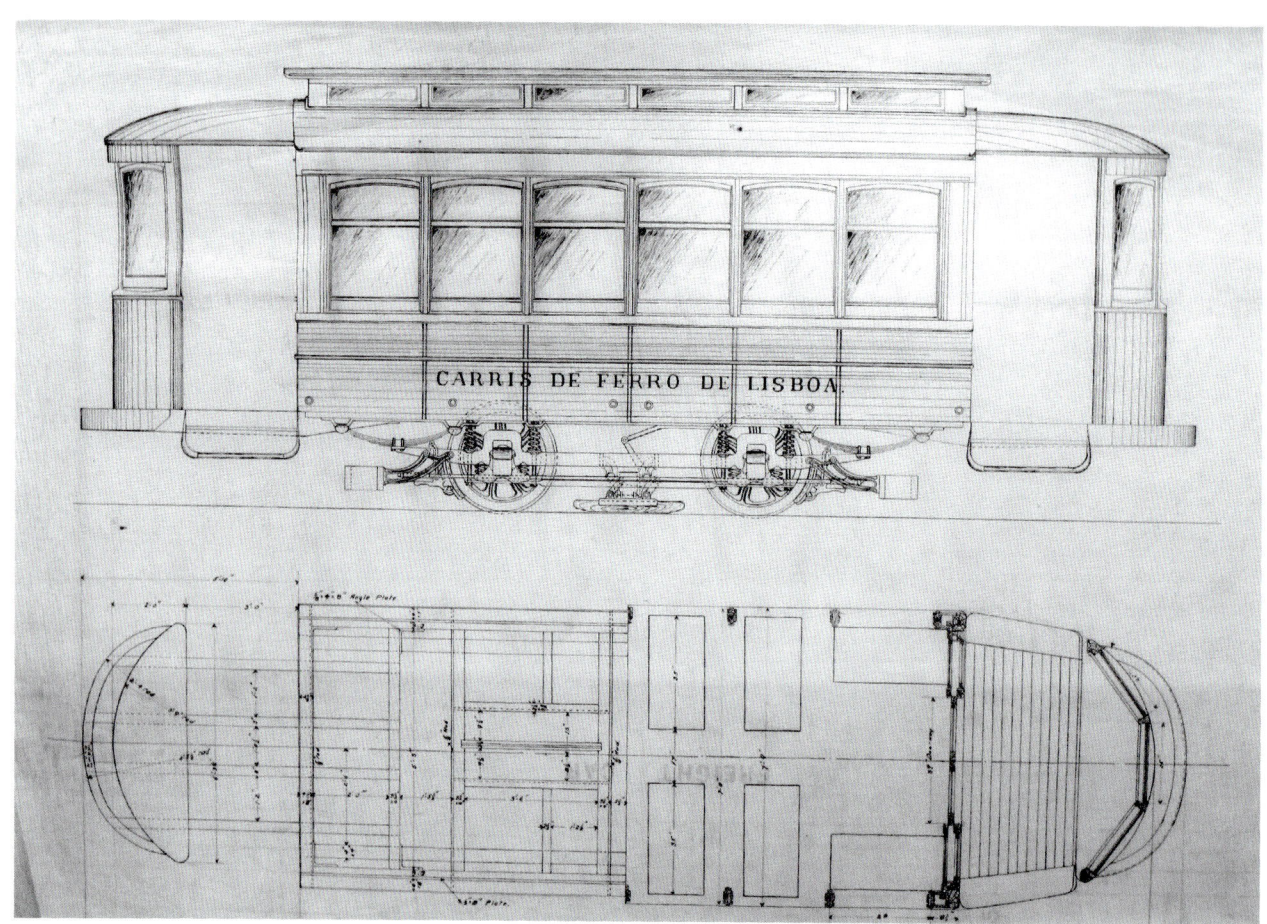

A CCFL scale drawing of car 475. No photographs of this car in its original form have been found, but fortunately this drawing survives in the CCFL archive. *CCFL*

505, Brill 1914, one of the few not rebodied seen in 1969. *P. Haseldine*

501 is seen at Santa Apolónia on 3 September 1962. Note the lady with a basket on her head. *W C Janssen/Online Transport Archive*

Series: 476-499, 503-507

Type: 2-axle closed cars with drop windows Body: J.G. Brill Company Overall length: 8.56 m Saloon length: 5.18 m Width: 2.29 m Seats: 20 Standing: 14	Truck: Brill 21E Wheel base: 1.98 m Wheel diameter: 0.85 m Motors: 2x GE 59 - 25 hp

Series: 476-484

Floor height: 0.87 m Platform height: 0.72 m Weight: 10.6 t In service: 1912; 483-484: 1914 Withdrawn: 1952-1953	Controllers: GE K10D Wheel brakes: manual Track brakes: manual

Series: 485-499

Floor height: 0.86 m Platform height: 0.71 m Weight: 10.6 t In service: 1914 Withdrawn: 1961-1963	Controllers: EE DB1K3 Wheel brakes: manual Track brakes: manual, electro magnetic

Series: 503-507

Floor height: 0.89 m Platform height: 0.73 m Weight: 10.2 t In service: 1914 Withdrawn: by 1971	Controllers: GE K10D Wheel brakes: manual Track brakes: manual

Series: 500-502

Type: 2-axle semi-convertible Body: J.G. Brill Company Overall length: 10.17 m Saloon length: 6.93 m Width: 2.28 m Floor height: 0.86 m Platform height: 0.70 m Weight: 12.3 t Seats: 26 Standing: 14	Truck: Brill Radiax E1 Wheel base: 3.05 m Wheel diameter: 0.85 m Motors: 2x GE 59 - 25 hp Controllers: GE K10D Wheel brakes: manual Track brakes: manual In service: 1914 Withdrawn: 1968

Second generation electric trams

Series: 508-531

Type: 2-axle semi-convertible Body: Carris Length: 8.78 m Width: 2.36 m Floor height: 0.90 m Platform height: 0.75 m Weight: 11.2 t Seats: 24 Standing: 21 In service 508-519: 1924/5; 520-531: 1926 Withdrawn: by 1981	Truck: Brill type 21E swing-link built by Maley & Taunton (508-519) and Brill (520-531) Wheel base: 2.13 m Wheel diameter: 0.85 m Motors: 2x GE 59 - 25 hp Controllers: GE K10D Wheel brakes: manual, air, Track brakes: manual

516 seen at Praça do Comércio on 2 September 1962 on route 17. *W C Janssen/Online Transport Archive*

508 is seen near Campolide on 9 June 1973. *Michael Russell*

One of the 1927 additions to the Lisboa fleet, Estrela car 607, is seen at Santo Amaro depot. *Maley and Taunton*

Series: 601-612

Type: 2 axle closed with sliding windows	Truck: Maley & Taunton
Body: J.G. Brill Company	Wheel base: 2.44 m
Length: 9.14 m	Wheel diameter: 0.74 m
Width: 2.33 m	Motors: 2x Metrovick - 45 hp
Floor height: 0.82 m	Controllers: EE DB1X33
Platform height: 0.82 m	Wheel brakes: manual
Weight: 10.0 t	Track brakes: manual, air, electro magnetic
Seats: 28	In service: 1927
Standing: 21	Withdrawn: 1972

Series: 532-551

Type: 2 axle closed with lifting windows	Truck: Maley & Taunton built type 21E
Body: Carris	Wheel base: 1.98 m
Length: 8.38 m	Wheel diameter: 0.787 m
Width: 2.38 m	Motors: 2x Metrovick - 45 hp
Floor height: 0.84 m	In service: 1927-1928
Platform height: 0.84 m	Withdrawn: 1980-1988
Weight: 9.7 t	
Seats: 24	
Standing: 21	

Series: 532-536	Series: 537-551
Controllers: EE DB1K3	Controllers: GE K10D
Wheel brakes: manual, air, rheostatic	Wheel brakes: manual, air
Track brakes: manual	Track brakes: manual

Above: Lisboa in 1949: Estrela class 608 and St Luis-built 411 in Restauradores at the foot of the Avenida da Liberdade. Note the awkward sliding widows, which because of the configuration of the body only half of which could be opened at any time. *Courtesy of the National Tramway Museum, photographer N.N Forbes*

612, an Estrela long wheelbase car, this car was rebuilt with conventional glazing at an unknown date, at Praça Duque de Terceira, Cais do Sodré, on route 20 on 8 September 1969. *A. G. Murray*

Martim Moniz in 1967, with car 544 and an unidentified trailer leaving Avenida Almirante Reis. Note the high level of the platform on the earlier standards. *J. Jordan*

305, originally a cross bench car, rebodied as a standard saloon in 1936 is in Rua Dr Nicolau Bettencourt at São Sebastião on route 13 on 9 June 1973. *A. G. Murray*

550 a Brill 21E CCFL standard and modern trailer 142, are at the Metro station loop, Sete Rios on route 1, 7.June 1973. *A. G. Murray*

328, rebodied in 1932, is seen at Alcantara on 17 July 1982. *L. Folkard*

Series: 305, 327-329, 338, 340, 345, 353, 355, 356

Type: 4 axle closed with lifting windows Body: Carris Length: 11.57 m Width: 2.37 m Seats: 40 Standing: 21	In service: 328/9, 340/5, 355/6: 1932; 353: 1933; 305: 1936; 327: 1937 Withdrawn: by 1996 (see text)

Series: 305, 327-329, 338, 340

Floor height: 0.92 m Platform height: 0.75 m Weight: 13.3 t; 338: 12.8 t	Truck: Brill 22E Motors: 2x GE 59 25 hp Controllers: B18C; 305, 338: DB1K3 Wheel brakes: manual, air, rheostatic Track brakes: n.a.

Series: 345, 353, 355, 356

Floor height: 0.96 m Platform height: 0.80 m Weight: 17.6 t	Truck: Brill 27GE1 Motors: 4x GE59 25 hp Controllers: B18C Wheel brakes: manual, air Track brakes: air

Series: 552-571

Type: 2 axle closed with lifting windows Body: Carris Length: 8.38 m Width: 2.38 m Floor height: 0.91 m Platform height: 0.80 m Weight: 9.7 t Seats: 24 Standing: 21	Truck: Maley & Taunton built type 21E Wheel base: 1.98 m Wheel diameter: 0.85 m Motors: 2x Metrovick - 45 hp Controllers: Dick Kerr DB1K3 Wheel brakes: manual, air, rheostatic Track brakes: manual In service: 1931 Withdrawn: by 1986 (see text)

566, a Brill 21E CCFL standard, in Rua Aliança Operário, Boa Hora on route 18 on 7.June 1973. *A. G. Murray*

225 is seen at Estrela with 719 and 723 on 3 April 1974. *Michael Russell*

Left: 711 squeezes through the narrow gap at Sao Tome on 3 April 1974. *Michael Russell*

Series: 203-282, 415,455,467,468,483

Type: 2 axle closed with lifting windows	Truck: Brill type 21E
Body: Carris	Wheel base: 1.98 m
Length: 8.38 m	Wheel diameter: 0.85 m
Width: 2.38 m	Motors: 2x GE59 - 25 hp
Floor height: 0.88 m	Controllers: EE DB1K3 or BTH B510
Platform height: 0.73 m	Wheel brakes: manual, air
Weight: 10.3 t	Track brakes: manual
Seats: 24	In service: 1932-1937, 1940
Standing: 21	Withdrawn: 1974-1995 (see text)

Series: 613-617, 701-735

Type: 2 axle closed with lifting windows	Truck: Maley & Taunton built type 21E swing-link
Body: Carris	Wheel base: 1.98 m
Length: 8.38 m	Wheel diameter: 0.85 m
Width: 2.38 m	Motors: 2x Metrovick - 45 hp
Weight: 10.7 t	Controllers: DB1K33
Seats: 24	Wheel brakes: manual, air
Standing: 21	Track brakes: manual, electro-pneumatic
In service: 1935-1936 (613-617 + 701-725), 1938 (726-730), 1940 (731-735)	
Withdrawn: see text	

Series: 801-810

Type: 4 axle closed with large drop windows	Length: 11.58 m
Body: Carris	Width: 2.38 m
	Seats: 36
	Standing: 23

Series 801-805	Series 806-810
Weight: 15.9 t	Weight: 14.4 t
In service: 1939	In service: 1943
Withdrawn: by 1991 (see text)	Withdrawn: by 1994 (see text)
Truck: Maley & Taunton hornless equal-wheel	Truck: Brill 22E
Wheel base: 1.372 m	Wheel base: 1.219 m
Pivot distance: 5.42 m	Pivot distance: 5.34 m
Wheel diameter: 0.74 m	Wheel diameter: 0.85 m / 0.52 m
Motors: 4x Metrovick MV109 25 hp	Motors: 2x GE 59 25 hp
Controllers: DB1K33	Controllers: GE K10
Wheel brakes: manual, air, electro pneumatic,	Wheel brakes: manual, air, rheostatic
Track brakes: electro magnetic.	Track brakes: n.a.

727 passes over the right angle crossing at Sao Bento on 3 April 1974.
Michael Russell

802 is seen on the Santa Amaro depot forecourt on 18 May 2008 during a special trip. Note the streamlined livery and absence of destination boxes over the cab compared with 801. *O. Brison*

![801 is seen when new in 1939]

801 is seen when new in 1939. *B. King collection*

![Interior of 801]

Interior of 801 showing the rattan seating, the open doors from the saloon to the platform and the closed doors to the cab. *B. King collection*

Third generation electric trams

Series: 736-745

Type: 2 axle closed with drop windows	Truck: Maley & Taunton built type 21E swing-link
Body: Carris	Wheel base: 1.98 m
Length: 8.47 m	Wheel diameter: 0.85 m
Width: 2.38 m	Motors: 2x Metrovick - 45 hp
Weight: 10.8 t	Controllers: DB1K33
Seats: 24	Wheel brakes: manual, air
Standing: 16	Track brakes: manual, electro-pneumatic, electro-magnetic
In service: 1947	
Withdrawn: 1985/6	

Series: 901-910

Type: 4 axle closed with drop windows	Truck: Maley & Taunton hornless equal-wheel
Body: Carris	Wheel base: 1.372 m
Length: 11.69 m	Pivot distance: 5.49 m
Width: 2.38 m	Wheel diameter: 0.74 m
Weight: 15.8 t	Motors: 4x Metrovick MV109
Seats: 40	25 hp
Standing: 16	Controllers: DB1K33
In service: 1948	Wheel brakes: manual, air, electro pneumatic
Withdrawn: until 1989	Track brakes: air.

Series: 283-304, 306-322, 403-470 (21 cars), 472-482, 484-499, 506, 609, 610

Type: 2 axle single-end closed with half-drop bus type windows	Truck: Brill type 21E, partly with swing-link
Body: Carris	Wheel base: 1.98 m
Length: 8.47 m	Wheel diameter: 0.85 m
Width: 2.38 m	Motors: 2x 25 hp
Weight: 9.6 t	Controllers: EE DB1K3
Seats: 26	Wheel brakes: manual, air
Standing: 18	Track brakes: manual, electro-magnetic

Lisboa 736 in Praça do Comércio on 4 May 1964. *L. Folkard*

"Caixote" 290 with an unidentified matching trailer enters Praça do Comércio in the early nineteen eighties. The letter A in the windscreen indicates that the motor car is running without a conductor. Passengers not holding passes were required to travel in the trailer; this was an interim step towards full one-man operation. *O. Brison*

Above: 906 departing from Cais do Sodré on 4 April 1974. *Michael Russell*

Right: "Caixote" cars 403, a rebodied St Louis car and 499, a rebodied Brill car, are on routes 5 and 24 at Largo do Rato, 9 June 1973. *A. G. Murray*

An interior view of "Caixote" 298 taken after the car was withdrawn from service but showing the functional design of open plan driving compartment. *O. Brison*

771 showing the offside without doors as a result of the car's conversion to single-ended configuration, seen on route 25 at the summit of Rua de São Domingos in 1994. *B. King*

Single ended cars with updated braking 771-785

Series 771-785	
Type: 2 axle closed with lifting windows Body: Carris Length: 8.38 m Width: 2.38 m Weight: 10.7 t Seats: 24 Standing: 21 In service: 1935-1936 (613-617 + 701-725), 1938 (726-730), 1940 (731-735) Withdrawn: see text	Truck: Maley & Taunton built type 21E swing-link Wheel base: 1.98 m Wheel diameter: 0.85 m Motors: 2x Metrovick - 45 hp Controllers: DB1K33 Wheel brakes: manual, air Track brakes: manual, electro-pneumatic

New fleet no	Former fleet no
771	557
772	554
773	561
774	563
775	553
776	562
777	566
778	571
779	555
780	565
781	556
782	560
783	564
784	559
785	558

Standard cars created in the 1980s using parts of other cars.

Series 737–745, 761-763	
Type: 2 axle closed with lifting windows Body: Carris Length: 8.38 m Width: 2.38 m Weight: 10.7 t Seats: 24 Standing: 21	Truck: Maley & Taunton built type 21E swing-link Wheel base: 1.98 m Wheel diameter: 0.85 m Motors: 2x Metrovick - 45 hp Controllers: DB1K33 Wheel brakes: manual, air Track brakes: manual, electro-pneumatic

Double ended cars 737-745

Fleet no	Truck from	Body from	Notes
737	737	268	
738	738	232	
741	741	233	Temporary fleet number 746 circa 1994
742	742	?	Bodies from cars 231, 234, 259 but not known which
743	743	?	
744	744	?	
745	745	250	

Single ended cars 761-763

Fleet no	Truck from	Body from	Notes
761	736	276	
762	739	226	
763	740	568	

CCFL-built 524 of 1925 seen at Belém with a Brill 8-bench trailer in 1949. *Courtesy of the National Tramway Museum, photographer N.N Forbes*

Brill-built 480 with 9-bench trailer 159 at Praça do Comércio in 1949. *Courtesy of the National Tramway Museum, photographer N.N Forbes*

Trailers

Series: 153-203	
Type: open cross-bench trailers	Wheel brakes: manual, air
Withdrawn: until 1953 (see text)	Track brakes: n.a.
Series: 188-202	
Body: The J.G. Brill Company	Truck: Brill
Weight: 3.3 t	Wheel base: 1.83 m
Seats: 32	Wheel diameter: 0.79 m
Standing: 18	
In service: 1898 (as mule trams)	
Series: 165-187	
Body: Carris	Truck: Carris
Weight: 3.8 t	Wheel base: 1.98 m
Seats: 36	Wheel diameter: 0.79 m
Standing: 18	
In service: 1909 (179-187), 1917 (177/8); 1923 (165-176)	
Series: 153-164	
Body: Carris	Truck: Maley & Taunton
Weight: 4.0 t	Wheel base: 2.44 m
Seats: 36	Wheel diameter: 0.61 m
Standing: 18	
In service: 1927	

A pair of trailer running units in the yard at Santo Amaro; just discernible on the top unit is the M&T cast above the axle box, whereas the lower unit has no markings, circa 1990. *B. King*

Closed Trailers

Series: 101-200	
Type: closed single-end trailers	Truck: Maley & Taunton and Carris
Body: Carris	Wheel base: 2.44 m
Length: 8.44 m	Wheel diameter: 0.61 m
Width: 2.38 m	Wheel brakes: manual, air
Weight: 4.5 t	Track brakes: n.a.
Seats: 28	In service: 1950 101-108, 1951
Standing: 28	109-126, 1952 127-163, 1953
	164-189, 1954 190-194 and 1955
	195-200
	Withdrawn: 1975-1988 (see text)

535, trailer 101 and 247 parked in Arco do Cego depot on 10 October 1978. *Michael Russell*

Low floor car 503 at Cais do Sodré gives a welcome touch of modernity to coastal route 15. *P. Haseldine*

The current fleet
501-510 Low Floor Articulated Cars
541-585 Remodelados
713-745 Standard cars

Cars 501-510
No of cars ordered 10
Plus option for additional 20
No of body sections 3
No of articulations 2
Length: overall (m) 24.02
   end modules (m) 10.08
   centre module (m) 3.86
"Axle" configuration . . . . Bo'2Bo'
Width (m) 2.40
Height (m) 3.215
Doorways and width (m) . . . . 4 x 1.30
Minimum gangway clearance (mm) 520
Floor height, high (mm) 700
Floor height. low (mm) 350
Internal steps (mm) . . . . 2 x 175
Nominal % low floor 62
Weight, empty (t) (estimated) 30
Maximum speed (km/h) 70
Commercial speed (km/h) 20
Acceleration (m/s2) 0.47
Contact wire voltage (V) 600
Motor rating (kW) . . . . 4 x 103
Passenger capacity:
Seats 65

standing (4/m2) 90
crush standing (6.67/m2) 145
Wheel diameter (all wheels)
   maximum (mm) 590
   minimum (mm) 510
Bogie wheelbase:
   power bogie (mm) 1800
   trailing bogie 1700
Track gauge (mm) 900

Cars 541-585
Length over dashes (m) 8 385
Width over panels (m) 2.378
Height over railhead (m) 3.190
Weight, unladen (t) 10:73
truck and motor s (t) 4 55
Maximum speed (km/h) 50
Acceleration (loaded), (m/s2) 1 2
Braking (m/s2) 1.2
Contact wire voltage (V) 600
Motor rating (kW) . . . . 2 x 50
Passenger capacity:
   seats 20
   standing . . . . 38 + 1
Wheel diameter:
   maximum (mm) 750
   minimum (mm) 670
Wheelbase (mm) 1981
Track gauge (mm) 900

Renumbering list (of donor bodies)

541 (ex 270)	550 (ex 278)	559 (ex 222)	568 (ex 254)	577 (ex 252)
542 (ex 241)	551 (ex 235)	560 (ex 238)	569 (ex 257)	578 (ex 225)
543 (ex 230)	552 (ex 242)	561 (ex 483)	570 (ex 244)	579 (ex 702)
544 (ex 236)	553 (ex 237)	562 (ex 275)	571 (ex 223)	580 (ex 262)
545 (ex 277)	554 (ex 264)	563 (ex 261)	572 (ex 255)	581 (ex 724)
546 (ex 227)	555 (ex 246)	564 (ex 266)	573 (ex 272)	582 (ex 701)
547 (ex 273)	556 (ex 256)	565 (ex 245)	574 (ex 247)	583 (ex 714)
548 (ex 228)	557 (ex 240)	566 (ex 265)	575 (ex 267)	584 (ex 707)
549 (ex 221)	558 (ex 243)	567 (ex 281)	576 (ex 415)	585 (ex 719)

Cars 713-745

Part of the 47 cars of the series 613-617, 701-735, 737/8 and 741-745 were retained with the modernisation of the mid 1990's as reserve and for special services. In 2015 several were painted green and allocated to a new regular tourist service.

The current (2018) status:
713 green tourist service
717 (renumbered from 617) green tourist service
722 green tourist service (was used for many years as Christmas tram)
723 was red tourist tram 4. Now in storage
726 was red tourist tram 3. Now in storage
732 storage
733 storage
735 green tourist service
741 green tourist service
744 green tourist service
745 green tourist service (was used for many years as Christmas tram)

Sightseeing cars

On 18 April 1965, Carris introduced a sightseeing tour of Lisboa by restored tram. The cars used were St Louis 1901-built saloons 437 and 435 (now 1 and 2) in an ornate red livery. In the 1990's the tourist service was allocated more trams, first two trams were series 700 and later Remodelados. They all received a new number and were painted red. Bogie tram 355 received the same livery and was for many years used as information office on Praça do Comércio, but is now stored in Santo Amaro.

1 (ex 437 São Luís)	5 (ex 583 Remodelado)	9 (ex 570 Remodelado)
2 (ex 435 São Luís)	6 (ex 585 Remodelado)	10 (ex 355 information office)
3 (ex 726 Standard)	7 (ex 546 Remodelado)	11 (ex 569 Remodelado)
4 (ex 723 Standard)	8 (ex 584 Remodelado)	12 (ex 553 Remodelado)

Works cars, motor

The present fleet numbering dates from 1988

Fleet no	Former fleet no	Notes
Z01	389	Railgrinder/water tank
Z02	390	Works shunter
Z12	394	)
Z13	395	General purpose motor car with crane
Z14	398	)
Z15	399	)

Works cars, trailers

Fleet no	Former fleet no	Notes
Z03	62	)
Z04	63	)
Z05	64	)
Z06	65	)
Z07	68	Rail and sleeper wagons
Z08	69	)
Z09	71	)
Z10	73	)
Z11	74	)
Z16	70	)

The Carris Museum

Museum collection

100 replica mule car built circa 2000 based on a replica circa 1950 made for a publicity film
101 the first of 100 post-war trailers
260 a 1930s "Standard" body with truck and equipment from a 1901 two-axle crossbench car
283 a 1901 crossbench car on maximum traction bogies
329 a 1930s "Standard" rebodying of a 1906 maximum traction bogie car
330 the last of the class to retain a wooden underframe and clerestory roof
348 the least altered of the 1906 equal wheel four motor bogie cars
444 restored to original 1901 condition for the centenary of electric traction
506 post-war lightweight "Caixote" body on 1914 truck
508 first tram to be built rather than "assembled" in Lisboa
535 early "Standard" car with high platforms built 1929
549 early "Standard" car with high platforms built 1929
741 1947 body retained when truck was used to update an older car
777 1980s upgrade with improved braking of an earlier "Standard" car
802 1939 four-motor bogie car
910 1947 four motor bogie car

Tourist car 6 (ex-585) seen in September 2014 on a tourist service passing the cathedral. *L Vieira*

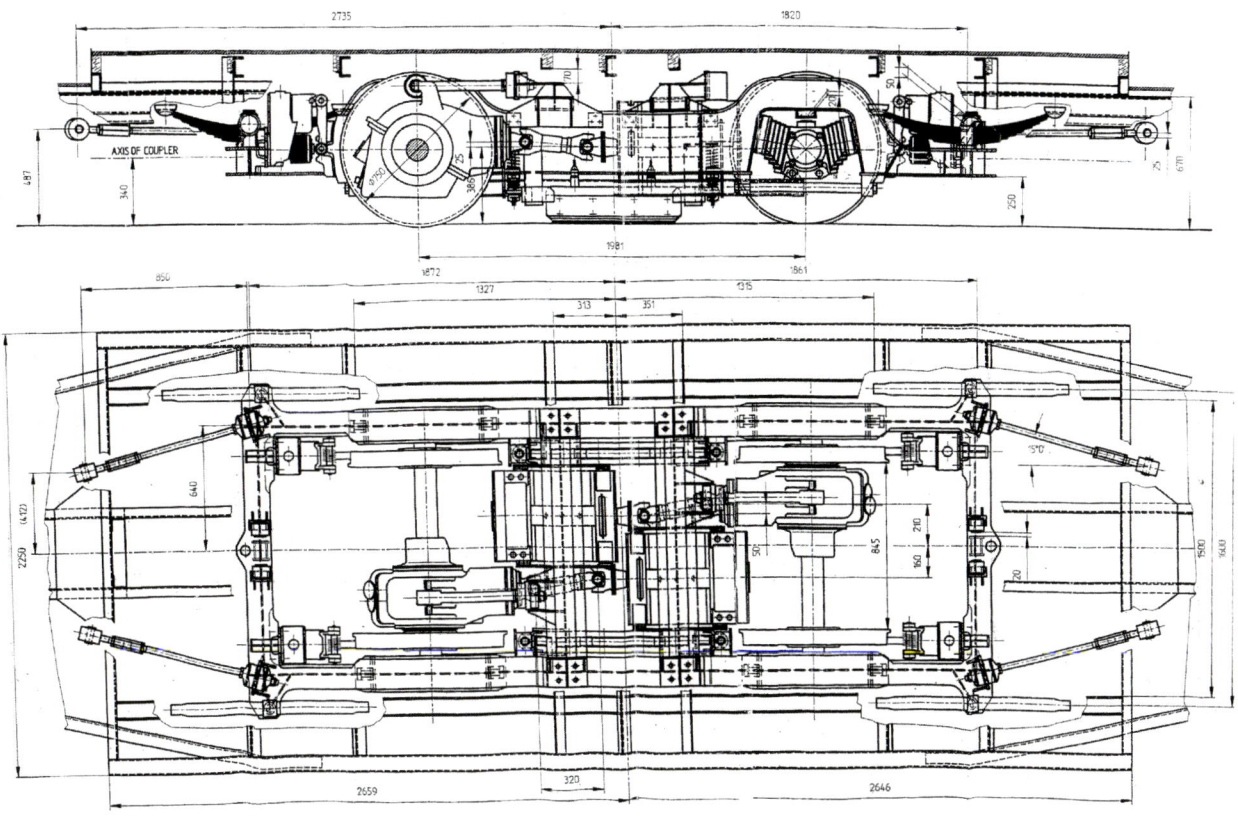

Scale drawing of a Lisboa Remodelado truck

Water car 389 at the Ajuda/S Amaro junction near Largo Calvario on 4 April 1974. *Michael Russell*

390, a works car - the shunter, seen at Santo Amaro depot on 8. September 1969. *A. G. Murray*

Works cars 398 and 399 in Santa Amaro depot yard in the shadow of the 24 de Abril bridge on 15 September 1995. *Michael Russell*

PORTO

Fleet List

271 is passing Massarelos on 16 May 1964 with windows raised and the sun blinds lowered. *L Folkard*

STCP Fleet List

Former CCFP Motor Cars

Carros Antigos 101-104

101 7-bay "transformado" 102-103 5-bay Constructora as rebuilt to 6-bay 104 8-bay "transformado"	All with Brill trucks and Siemens motors and controllers All withdrawn by 1959

Brill-bogie of 1904 (CCFP 90, 190, 191, 226, STCP 249)

Type: 4-axle semi-convertible Body: J. G. Brill Company Length: 10.82 m Width: 2.20 m Weight: 16,230 kg Seats: 32 In service: 1904	Truck: Brill 27G, since 1965 ACF Motors: 4x Siemens, since 1924 2x GE 270, since 1967 2x CGE CT139B Controllers: Siemens, since 1924 GE B54E Wheel brakes: manual, later also air Withdrawn: 1972

A Constructora cars with 6 windows (STCP 105-111)

Type: 2-axle semi-convertible Body: Constructora Length: 8.60 m Width: 2.10 m Weight: 10,225 kg Seats: 22	Truck: Brill 21E Wheel base: 1.98 m Motors: 2x Siemens, later GE 270 Controllers: later GE B54E Wheel brakes: manual, later also air In service: 1904-1906 Withdrawn: see text

A Constructora cars with 7 windows (STCP 112-114)

Type: 2-axle semi-convertible Body: Constructora Length: 8.70 m Width: 2.20 m Weight: 10,225 kg Seats: 23	Truck: Brill 21E Wheel base: 2.15 m Motors: 2x Siemens, later GE 270 Controllers: later GE B54E Wheel brakes: manual, later also air In service: 1907, 1909 Withdrawn: see text

Carros Ingleses (STCP 115-119)

Type: 2-axle semi-convertible Body: UEC Length: 9.00 m Width: 2.20 m Weight: 11,085 kg Seats: 23 In service: 1910	Truck: Brill 21E Wheel base: 2.26 m Motors: 2x Siemens, later GE 270 Controllers: later GE B54E Wheel brakes: manual, later also air Withdrawn: see text

Car 101 at Praça de Liberdade in 1949. *Courtesy of the National Tramway Museum, N.N. Forbes Collection*

Car 103 seen at Lordelo in 1949, car 102 was similar. *Courtesy of the National Tramway Museum, N.N. Forbes Collection*

This tram was the only former mule tram of this type that received an STCP number. It was in use until the late 1950's, the last years as training car. This photo of STCP 104 was taken at Boavista around 1947.

Photographer Unknown

An aquatint showing a Constructora car, either 33 or 34, with the original five windows in Praça da Batalha. The building on the extreme left of the picture was the office of A Constructora, the builders of many early electric cars. A cinema now occupies the site. *Commercial postcard.*

The Brill-bogie car acquired in 1904. Apparently a photo taken to demonstrate the semi-convertible nature of the car as half of the windows are fully open and the other half closed. The tram has open platforms and the equal-wheel bogies are clearly visible. Probably taken in 1904, source unknown but could be CCFP. It is likely that this is the photo sent in 1904 to Brill to show the assembled and painted car. The original might be in the archives in Philadelphia. *Photographer unknown*

Car 249 seen some time after 1967 on maximum traction bogies at São Bento station. *Commercial Postcard*

Tramcar 163 was built in 1905 by A Constructora. It was one of a series of 24 supplied in 1904-1906. The photo was taken between 1907 and 1911. This photo was used to aid the restoration of car 107 which is now museum car 163, the original number of car 107 is lost in the mists of time. *Scan from book*

STCP 107 in front of the Boavista depot, 14 August 1972. *www.tramway.com*

Tramcar 199 was one of two built by A Constructora in 1907. At the same time the CCFP workshops reconstructed another tram similar to these. The photo was taken on Praça da Batalha between 1912 and 1924. This is the best photo known of this type in its almost original state. Café Java still exists a century later. *Photographer unknown*

STCP 113 is one of the four cars with seven windows at each side built by A Constructora. At one time it received closed platforms. Later these were extended and it was fitted with retractable steps. Tim Boric took this photo in September 1981, when the tram had just turned on the shady Rotunda da Boavista, so that it could set off on the next trip on route 19 to Matosinhos. *Tim Boric*

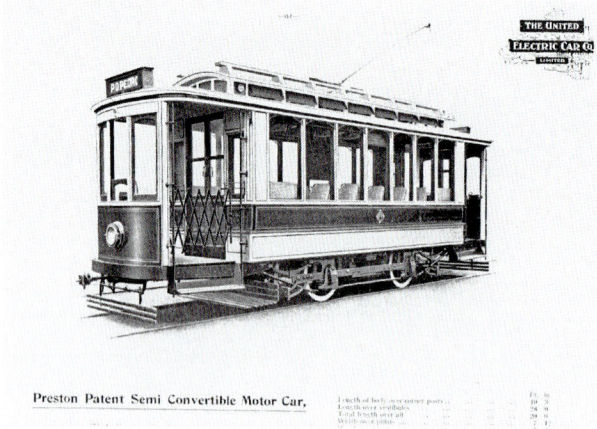

An image of a Porto car from the United Electric Car Co Limited catalogue. *Courtesy of the National Tramway Museum*

STCP 119 entering Rua 31 de Janeiro, August 1965. This was the most extensively rebuilt of the UEC cars. *Ron Jedlicka*

Brill-23 (STCP 120-146)

Type: 2-axle semi-convertible	Truck: Brill 21E
Body: J. G. Brill Company	Wheel base: 2.15 m
Length: 8.70 m - 9.10 m	Motors: 2x Siemens or GE80, later GE 270
Width: 2.20 m	
Floor height: 0.90 m	Controllers: later GE B54E
Platform height: 0.72 m	Wheel brakes: manual, later also air
Weight: 10,550 kg	
Seats: 23	In service: 1910-1914
	Withdrawn: see text

Brill-28 (STCP 150-199)

Type: 2-axle semi-convertible	Truck: Brill 21E
Body: CCFP	Wheel base: 2.15 m
Length: 8.70-9.10 m	Motors: 2x GE 270
Width: 2.40 m	Controllers: GE B54E
Floor height: 0.97 m	Wheel brakes: manual, later also air
Platform height: 0.77 m	
Weight: 11,500 kg	In service: 1926-1937
Seats: 28, later reduced to 23	Withdrawn: see text

Brill-28 Plataforma Salão (STCP 200-223)

Type: 2-axle semi-convertible	Truck: Brill 21E
Body: CCFP	Wheel base: 2.26 m
Length: 9.60 m	Motors: 2x GE 270
Width: 2.40 m	Controllers: GE B54E
Floor height: 0.97 m	Wheel brakes: manual, air
Platform height: 0.77 m	Track brakes:
Weight: 12,600 kg	In service: 1938-1946
Seats: 28, later reduced to 23	Withdrawn: see text

Brill-32 (CCFP 250-253 + 256-259 + 284-287, STCP 250-261)

Type: 2-axle semi-convertible	Truck: Brill 21E
Body: CCFP	Wheel base: 2.36 m
Length: 9.60 m	Motors: 2x GE 270, later 2x CGE CT 139B
Width: 2.40 m	
Floor height: 0.97 m	Controllers: GE B54
Platform height: 0.77 m	Wheel brakes: manual, later also air
Weight: 12,600 kg	
Seats: 32, later reduced to 26	In service: 1926-1928
	Withdrawn: 1967-1981

Brill-bogie (CCFP 268-273 + 200-201, STCP 270-277)

Type: 4-axle semi-convertible	Truck: 2 x Brill 39E, manufactured under licence by Bergische Stahl-Industrie of Remscheid, Germany
Body: CCFP	
Floor height: 1.02 m	Pivot base: 4.82 m
Platform height: 0.78 m	Wheel base: 1.51 m
Seats: 40, later reduced to 32	Motors: 2x GE 270, later 2x BTH 114DR
Wheel brakes: manual, air	
Withdrawn: see text	Controllers: GE B54, later Siemens K7731-1 (270 + 277 at one end BTH OK33B)

Brill-bogie (CCFP 200-201, STCP 276-277)

Length: 11,43 m	Weight: 17,200 kg
Width: 2.35 m	In service: 1926

Car 150, originally a narrow bodied car that was widened by CCFP approaches Ermesinde on the terminal loop in 1967. *J. Jordan*

Brill-23 STCP 124 not only received extended platforms and retractable steps, but also, uniquely, a roof without a clerestory. It is seen here at Praça de Liberdade. *Slide*

Car 144 is at Carmo in 1973. This was the last surviving unmodified narrow Brill car in service. *Michael Russell*

Brill-28 Plataforma Salão 160 on Avenida da Boavista. This was originally a narrow bodied car that was widened by CCFP. *Slide*

Brill-28 Plataforma Salão no.169 on the Esplanada do Castelo, February 1992. This is one of the narrow Brill cars that was widened in the 1920s as evidenced by the narrow corner pillars. *Ernst Kers*

Brill-28 Plataforma Salão 244 is apparently just finished by the workshops and has already has the STCP crest but still a CCFP number. The STCP renumbered it shortly after to 201. *STCP*

Car 206 is seen departing from Praça de Liberdade in the late seventies on route 18 but the destination is not shown. Note the matchboarded sides; this car was one of the last cars with this feature to survive. *B. King*

257 is seen at Rua Sa Da Bandeira on 16 May 1964. Similar car 250 is in the museum collection. *L. Folkard*

Brill-bogie (CCFP 268-273, STCP 270-275)

Length: 11.56 m	Weight: 17,400 kg
Width: 2.43 m	In service: 1928

Belga (CCFP 274-283, STCP 280-289)

Type: 4-axle closed	Truck: 2x ACF
Body: Familleureux	Pivot base: 5.50 m
Length: 11,65 m	Wheel base: 1.50 m
Width: 2.45 m	Motors: 2x GE 270, later 2x CGE
Floor height: 0.95 m	CT 139B, later 2x BTH 114DR
Platform height: 0.79 m	Controllers: GE B54, later
Weight: 16,000 kg	Siemens K7731-1 (281 + 288
Seats: 40, later reduced to 32	Kiepe NF51)
In service: 1928-1929	Wheel brakes: manual, air
	Withdrawn: see text

Fumista (CCFP 296-311, STCP 300-315)

Type: 2-axle convertible, later closed	Truck: Brill 21E
Body: CCFP	Wheel base: 2.15 m
Length: 9.50 m	Motors: 2x GE 270
Width: 2.50 m	Controllers: GE B54
Floor height: 0.94 m	Wheel brakes: manual, later also
Platform height: 0.77 m	air
Weight: 12,000 kg	In service: 1929-1930
Seats: 28, later reduced to 22	Withdrawn: 1967-1984

Fumista-bogie (CCFP 312-315, STCP 266-269)

Type: 4-axle convertible, later closed	Truck: 2 x Brill 39E type, understood to be manufactured
Body: CCFP	locally
Length: 11,53 m	Pivot base: 5.25 m
Width: 2.49 m	Wheel base: 1.51 m
Floor height: 1.06 m	Motors: 2x GE 270, later 2x CGE
Platform height: 0.83 m	CT 139B, later 2x BTH 114DR
Weight: 16,400 kg	Controllers: GE B54, later Kiepe
Seats: 40, later reduced to 31	NF51 (266, 268, 269 at one end
In service: 1931	BTH OK33B)
	Wheel brakes: manual, air
	Withdrawn: 1980, 1990, 1981, 1991

Car 270 in Praça de Liberdade in 1962. Note that this car still retains its trailer hose connections. *W C Janssen/Online Transport Archive*

Brill-bogie no.271 at the Passeio Alegre 1978. *Tim Boric*

Belga 279 still has its original number, Providence lifeguard and trailer brake connections on Praça da Liberdade, 22 September 1945. *Photo-print*

Car 288 receiving attention in Massarelos depot in 2009. *Michael Russell*

Fumista STCP 301 at Boavista around 1948. The tram still has two trolley-poles. *MCE*

Car 267 at Foz seen during a run for the purpose of taking official pictures of the car in 1995. *B. King*

Interior of car 267 May 2014. *B. King*

Cars built by STCP

Motor Cars

Pipis (STCP 351-373) Pipizinhos (STCP 400-405)	
Type: 2-axle closed Body: STCP Floor height: 0.98 m	Truck: Brill 21E Motors: 2x GE 270 Controllers: GE B54 Wheel brakes: air, manual
Pipis (STCP 351-373)	
Length: 9.80 m Width: 2.40 m Weight: 11,535 kg Floor height: 1.03 m	Seats: 28, later reduced to 21 Wheel base: 2.36 m In service: 1947-1953 Withdrawn: 1967-1974
Pipizinhos (STCP 400-405)	
Length: 8.25 m Width: 2.35 m Weight: 10,385 kg Floor height: 0.98 m	Seats: 20, later reduced to 15 Wheel base: 2.26 m In service: 1947-1949 Withdrawn: 1955, 1967

STCP 500	
Type: 2-axle closed Body: STCP Length: 9.69 m Width: 2.27 m Floor height: 0.82 m Weight: 12,250 kg Seats: 22 Standing: 22	Truck: Brill 79E Wheel base: 2.36 m Wheel diameter: 0.84 m Motors: 2x CGE CT139B Controllers: GE B54 Wheel brakes: air, manual In service: 1951 Withdrawn: 1974

Trailers

Exact details are unclear; as far as can be ascertained from photographs and archive records the maximum extent of the STCP trailer fleet was as follows.

Bogie trailers (CCFP 1-7, STCP 1-7)	
Type: 4-axle closed Body: Constructora Length: 10.83 m Width: 2.30 m Floor height: 0.91 m Platform height: 0.74 m Weight: 10,140 kg Seats: 32	Truck: 2x Brill 23D Pivot base: 4.04 m Wheel base: 1.25 m Wheel diameter: 0.68 m Wheel brakes: manual, air In service: 1910-1912 Withdrawn: 1966

Sundry ex-mule cars and early ex-motor cars

8-10 early Starbuck saloons, ex-CCAPFM
11-12 seven window saloons
13 four window saloon
14, 17 withdrawn 1946, nothing else known
15-16 crossbench, nothing else known, withdrawn 1946

"Pipi" 373 stands at Pereiro terminus in 1967. *J. Jordan*

Car 400 with 183 behind at Castelo do Queijo in 1962. *W C Janssen/Online Transport Archive*

CE 500 was the prototype for a fleet of modern trams which were never constructed. *Photographer Unknown*

Fumista trailers (CCFP 18-20, STCP 18-20)	
Type: 2-axle convertible, later closed Body: CCFP Length: 7.70 m Width: 2.43 m Floor height: 0.85 m Platform height: 0.80 m Weight: 7350 kg	Seats: 20 Truck: Brill 21E Wheel base: 2.15 m Wheel brakes: manual, air In service: 1934 Withdrawn: 1966

Trailer 7 and car 199 at Castelo do Queijo in 1949. Note the brakesman on the front platform of the trailer and that windscreens have been fitted by this date. *Courtesy of the National Tramway Museum - N.N. Forbes Collection*

Right: Starbuck trailer 9 of 1873-4 on workmen's service in 1949 believed to be at Alto da Maia. It ran for 85 years and is now exhibited at Crich. *Courtesy of the National Tramway Museum, photographer N.N Forbes*

Right, bottom: The three Starbuck trailers, 8, 9 and 10, probably new to CCAPFM, at Massarelos depot in 1959. *J. H. Price*

21, 22, 24 seven window saloons, ex motors (22 re-motored for the museum)
23, believed similar, no known photographs
25-27 believed similar, but not put into service; no known photographs
28, new trailer built by STCP and then motored to become car 405.

Surviving works cars

Fleet no	Type
48	Rail grinder, body ex-passenger car 111 circa 1960
49	Tower wagon
58	Zorra with cabs
66	Zorra with open platforms
76	Breakdown van
77	Zora with cabs
80	Open freight trailer

Also two Hansa-Lloyd motor tower wagons circa 1930.

Trailer 11 around 1950 in front of the Massarelos power plant and depot. This former mule tram is thought to have originated from the CCFP. The platforms were probably extended a little in the first decade of the 20th century after mule tram operations ended.

Photographer Unknown

Rail grinder 48 and car 213 in Rua da Restauraçao in 2002. *B. King*

An unidentified two-axle car with trailer 13 seen at Ermesinde/ Aguas Santas in 1949. *Courtesy of the National Tramway Museum, photographer N.N Forbes*

Overhead wire repair tram no.49, breakdown assistance tram 76 and rail grinder 48 on a temporary, one hour exhibition at Infante, 18 May 2013. *Ernst Kers*

An unidentified bogie car with trailer 19 at Castelo do Queijo in 1949.

Courtesy of the National Tramway Museum, photographer N.N Forbes

Rail grinder 48 with a Transformado type body. This body was replaced by the body of Constructora tram 111 in about 1960. *STCP*

Earlier Electric freight and works cars

No.	Elec. eq.	Body	Capacity	In service	Type
220	?	antique	5.5 t		In 1913/4 this was called a zorra, in 1915/6 and 1919 a wagon
221-222	GE80	Companhia Alliança	6 t	1910	wagon
223-224	?	antique	6 t		wagon
225-226	?	antique	5.5 t / 10 t		zorra. Until 1915 mentioned as 5.5 t, in 1919 as 10 t
227-230	GE80	Constructora	6 t	1910	zorra
231-236	Siemens	Companhia Alliança	6 t	1910	zorra
237-248	GE80	Companhia Alliança	6 t	1913	zorra

Mule- and trailer zorras

Number	Type
60	zorra atrelada 4000 kg
61-62	wagon atrelado 4500 kg *(van)*
63	zorra via e obras (line and works)
64-65	carro de Socorro via e obras (line and works support)
66-79	zorra atrelada 5500 kg antigo
80-82, 84-85, 88-96	zorra atrelada 4000 kg
97	zorra via e obras (line and works)
98	tanque irrigação (water car)

Zorra 60 at the quarry spur and ash dump at Alto da Serra, near São Pedro da Cova in the late 1950s/early 1960s. *P Milheiro*

Car 58 with a small trailer carrying what appears to be a rotary converter, seen at Foz in 1987. *Michael Russell*

Vagão de socorro 76 is probably a former closed freight tram. *STCP*

"Zorra" 77 at the Castelo do Queijo permanent way yard in 1978.
Tim Boric

Works car 66 propelling a length of rail on two "Joaninhas" at Foz with a member of staff riding at the front sounding a gong to warn oncoming traffic and pedestrians. *Photographer unknown*

Car 112 with former sardine trailer 83 unloading scrap at Boavista foundry. *G B Claydon*

Museum Collection

An overview of the museum collection, although not all cars are on display:

1 - Trailer on Brill 23D bogies built in 1909 by the local company A Constructora as the first of a series of seven, the other six being built in 1911.

8 - Americano or mule-tram built by Starbuck, probably in 1872/3 for CCAPFM. Probably the car was reconstructed receiving a reinforced underframe and extended platforms in the first decade of the 20th century.

18 - Trailer Fumista built in 1934 by the workshops of CCFP as one of a series of three trailers of this type.

22 - 4-wheel motorcar of the first type probably built in 1900 by the local company A Constructor" but representing the first type of Transformado electric tramcars in Porto converted out of existing mule cars in 1895.

25 - Trailer Pipizinho built by the workshops of STCP in 1946 as one of a series of six, although only one car was temporary used as trailer, all others were used as motorcars.

49 - Overhead repair motorcar built by the workshops of CCFP in 1932.

58 - Electric freight tramcar of unknown builder and year of construction.

66 – Cabless electric freight tramcar of unknown builder and year of construction.

80 - Open freight trailer of unknown builder and year of construction.

100 - Open cross-bench motorcar, replica built by the workshops of STCP in 1995 using car 129 as the basis, representing a type of open mule-car Transformado converted in 1904 by the local company A Constructora to electric tram. In fact this replica is significantly larger than the original ex-mule tramcars.

104 - 4-wheel replica of a Transformado motorcar converted from mule-cars, built using car 144 by the workshops of STCP in 1997 representing a type of mule tramcar built in 1871/2 in England for CCAPFM and converted by CCFP to electric tram around 1899-1902.

163 - Six window 4-wheel motorcar built in 1904/6 by the local company A Constructora as one of a series of twenty-four, latterly numbered 107.

247 - 4-wheel motorcar Carro Inglês built by UEC of Preston as one of a series of five, reconstructed by the workshops of CCFP probably in the 1930's with changing the platforms into the Brill type latterly numbered 118.

250 - 4-wheel 8-window motorcar Brill-32 or Italiano built in 1926 by the workshops of CCFP as one of a series of twelve.

267 - Bogie motorcar Fumista built in 1930/1 by the workshops of CCFP as one of a series of four.

269 - Bogie motorcar Fumista built in 1930/1 by the workshops of CCFP as one of a series of four.

Car 100 in Rua de Monchique during the annual parade in May 2014. *Michael Russell*

274 - Bogie motorcar Brill built in 1928 by the workshops of CCFP as one of a series of eight.

288 - Bogie motorcar Belga built in 1928 by the Belgian company Familleureux as one of a series of ten, replacements for cars lost in the 1928 Boavista fire.

315 - 4-wheel motorcar Fumista built in 1929 by the workshops of CCFP as one of a series of sixteen.

373 - 4-wheel motorcar Pipi built by the workshops of STCP in 1952 as the last of a series of 24 cars of this type.

500 - 4-wheel motorcar Carro 500 built by the workshops of STCP in 1951 as the prototype of a proposed new generation of tramcars.

In later years the museum became responsible for renting of trams. For this several other trams are available:

143 - Brill-23 built in 1910 or 1912 modernised and equipped with extended platforms in the 1950's

191 - Brill-28 built new by the workshops in 1929.

203 - Brill-28 Plataforma Salão built in 1942 by the workshops probably reconstructed from an original Brill tram.

275 - Bogie motorcar Brill built in 1928 by the workshops of CCFP as one of a series of eight.

143, 191 and 203 have rail-brakes. Also replica trams 100 and 104 of the collection are used for rental services. Although they have never been on display in the museum, the works-cars 48 (rail grinder, ex passenger car 111), 76 (rescue and shunting) and 77 (zorra) are now considered to belong to the museum collection. Also two petrol engined 1920's Hansa-Lloyd tower wagons.

In storage (and hopefully to be restored) are:

113 - 7-window Constructora car modernised and equipped with extended platforms in the 1950's.

134 - Brill-23 of the 1909 order modernised and equipped with extended platforms in the 1950's.

169 (currently numbered 217) - Brill-28 Plataforma Salão converted out of a Brill tram of the 1909 order by widening the bodywork.

177 - Brill-28 built by the workshops in 1934.

201 - Brill-28 Plataforma Salão probably reconstructed in 1946 by the workshops from an original Brill tram.

222 - Brill-28 Plataforma Salão probably reconstructed in 1939 by the workshops from an original Brill tram.

270, 271, 272, 276 and 277 - Brill-bogies built by the workshops in 1928 (270-272) and 1926 (276-277)

280, 283, 284 and 285 - Belgas built by Familleureux in 1928.

Currently operational are:
131, 205, 213, 216, 220, 287

Gallery 5

Porto Museum Fleet

Car 104 and trailer 8 in Rua de Monchique during the annual parade in May 2014. *Michael Russell*

Cars 22, 104 and 163 seen at Foz during the annual procession in May 2000. *B. King*

Cross bench replica car 100 at Massarelos depot during the annual procession in May 2014. *B. King*

Cars 247 and 288 in Rua de Monchique during the annual parade in May 2014. *Michael Russell*

Cars 269 and 315 at Infante during the annual procession in May 2014. *B. King*

Interior of car 250 in 1994, passengers entered at the rear, left at the front, and the last passenger off closed the bulkhead doors. Rattan seat covering and elaborately carved woodwork gave an exotic semi-tropical feel. *B. King*

Car 269 with trailer 1 at Foz during the annual parade in May 1993. *Michael Russell*

Car 315 with trailer 18 at Castello do Queijo during the annual procession in 1995. *B. King*

Car 191 at Viriato during the annual procession in 2002. *B. King*

Museum car 373 with matching trailer 25 in Avenida da Boavista in 1995. *B. King*

The austere interior of 373. *B. King*

Car 500 at Boavista depot in 1999, this was a prototype for a fleet of modern trams which were never constructed. *B. King*

"Zorra" 66 without an enclosed cab is seen in the STCP Museum. *B. King*

Museum cars 267 and 315 both in the green colour scheme are seen side by side at Massarelos during the annual parade in May 2022. *K. Lomas King*

Museum car 250 is on display at Massarelos in the yellow colour scheme at the annual parade in May 2014. *C.F.Isgar*

SINTRA | Tram Fleet List

Details of the original fleet are given in the text, the current fleet is a follows:

Motor Cars

Brill 1, 6 and 7

Type 2-axle cross bench, 8 benches	Truck: Brill 21E
Body: J. G. Brill Company, 5.02m between bulkheads	Wheel base: 1.98m
	Motors: Westinghouse (France)
Length: 7.56m	2x 50hp
Width: 1.93m	Controllers: BTH B18
Floor height:	Wheel brakes: manual
Platform height:	Also equipped with
Weight: 8.5 tonnes	Westinghouse-Newell
Seats: 32	electromechanical track brakes
	In service: 1903

Cars 2, 4 and 5

Type 2-axle saloon, 6 window	Truck: Brill 21E
Body: J. G. Brill Company, but rebodied locally 1940s	Wheel base: 1.98m
	Motors: Westinghouse (France)
Length:	2x 50hp
Width:	Controllers: BTH B18
Floor height:	Wheel brakes: manual
Platform height:	Also equipped with
Weight:	Westinghouse-Newell
Seats: 24	electromechanical track brakes
	In service: 1903

Brill 3

Type 2-axle saloon, 5 window	Wheel base: 1.98m
Body: J. G. Brill Company, 4.47m between bulkheads	Motors: Westinghouse (France)
	2x 50hp
Length: 7.46m	Controllers: BTH B18
Width: 1.95m	Wheel brakes: manual
Floor height:	Also equipped with
Platform height:	Westinghouse-Newell
Weight:	electromechanical track brakes
Seats: 17 Truck: Brill 21E	In service: 1903

In recent years cross bench cars 1, 6 and 7 have been repainted in historic liveries and are seen here at Ribeira depot. *Photographer unknown.*

Cars 1, 3 and 10 are seen at Banzao on 18 May 1980 with in the foreground the freight van which served as a ticket office. *Michael Russell*

Trailers

Brill 8, 10

Type 2-axle saloon, 5 window Body: J. G. Brill Company, 4.47 between bulkheads Length: 7.46m Width: Floor height: Platform height: Weight: Seats: car 8 with 20 longitudinal, car 10 with 17 transverse	Truck: Trunnion Wheel base: 1.98m Wheel brakes: manual In service: 1903

Brill 9, 11, 12, 13, 14

Type 2-axle cross bench, 8 benches Body: J. G. Brill Company, 5.02m between bulkheads Length: 7.56 Width: 1.93 Floor height: Platform height: Weight: Seats: 32 Truck: Trunnion	Wheel base: 1.98m Wheel brakes: manual In service: 1903

Brill 15

Type: saloon (body only) 5 widow Body: J. G. Brill Company, see text for further details Length: 7.46 Width: 1.95 Floor height: Platform height: Weight: Seats: 20 longitudinal	

Ex-Lisboa Cars

Cars 703, 709. See Lisboa fleet list for further details. Cars in store at depot pending further use.

Brill saloon motor car 3 at Casal da Nora. *B. Lennox collection*

Car 3 on depot track, and car 6 on the main line in 2009. Note the manager's house in the background. *P. Haseldine*

Car 7 on works duties at Ribeira in 1959. G. Krambles *– B. Lennox collection*

Car 4 is seen between Monte Santo and Ribeira on 5 May 2005 with the El Castelo dos Mouros and Pena Palace on the hill in the background and the Palácio Nacional de Sintra further down the hill. *Michael Russell*

Car 6 with trailers 9 and 10 at the depot in 2009. *P. Haseldine*

Cars 6 and 7 are seen descending the slope at Monte Santo on 1 May 2005. *Michael Russell*

Trailer 13 being unloaded into the rebuilt Ribeira depot, note that the dash panel has been rubbed down to reveal the previous fleet number.
B. Lennox collection

Above: Trailer 14 in yellow livery at the depot. A rare colour photograph of this livery. *W C Janssen/Online Transport Archive*

Above, right: The body of trailer car 15 in 1967, resting on trestles as it has done since the 1950's and still does in 2020. *J. Jordan*

Right: A weeding train at Casal da Nora. *B. Lennox collection*

Below: Deutz "dresine" and tower car erecting overhead wiring at the junction with Ribeira upper level. *B. Lennox collection*

Tram
Fleet List

By Ernst Kers, Owen Brison & Pedro Mendes

Car is 13 seen working a service on route 1 on 4 September 1962. This is the original crossbench body rebuilt earlier as a convertible car with an 8-window saloon and deep removable windows reaching down to seat level, and later rebuilt as pictured in this view. *W C Janssen/Online Transport Archive*

Brill 1-5

Type 2-axle semi-convertible 6 window Body: J. G. Brill Company, 4.887m between bulkheads Length: Width: Floor height: Platform height: Weight: Seats: 20	Truck: Brill 21E Wheel base: 1.98m Motors: 2x GE 58, 25hp Controllers: GE B54E Wheel brakes: manual These cars at least originally had track brakes In service: 1910 – 1978/80

Cars 1, 3-5 Rebodied

Body SMC 7 window Seats: 28 1, rebodied 1961	3, rebodied 1959 4, rebodied 1969 5, rebodied 1967 1, 3 and 4 in Museum collection

Brill 6-7

Type 2-axle semi-convertible 6 window Body: J. G. Brill Company, 4.877m between bulkheads Length: Width: Floor height: Platform height: Weight: Seats: 20 Truck: Brill 21E	Wheel base: 1.98m Motors: 2x GE 58, 25hp Controllers: DE B54E Wheel brakes: manual These cars at least originally had track brakes In service: 1912 – 1979

Cars 6-7 Rebodied

Body SMC 7 window Seats: 28 6, rebodied 1959	7, rebodied 1967 6 in Museum collection

Brill 8

Type 2-axle semi-convertible 6 window Body: J. G. Brill Company, 4.877m between bulkheads Length: Width: Floor height: Platform height: Weight: Seats: 20 Truck: Brill 21E	Wheel base: 1.98m Motors: 2x GE 58, 25hp ? Controllers: GE B54 ? Wheel brakes: manual In service: 1926 – 1970s

Brill 9-13

Type 2-axle cross bench Body: 9 J. G. Brill Company, 10 bench Body 10-13, SMC, 10 bench Length: Width: Floor height: Platform height: Weight: Seats: 40 Truck: Brill 21E	Wheel base: 1.98m Motors: 2x GE 58, 25hp ? Controllers: GE B54E ? Wheel brakes: manual Rebuilt as convertible cars in 1930s Further modified 1941 In service: 1927/28 – 1979

Cars 9-13 Rebodied

Body SMC, 7-window Seats: 26 9, rebodied 1957	10, rebodied 1956 11, rebodied 1950s 12, rebodied 1957 13, rebodied 1966 11 in Museum collection

Brill 14-15

Type 2-axle dome roof saloon 7 window Body: J. G. Brill Company, steel, 5.61m between bulkheads Length: Width: Floor height: Platform height: Weight: Seats: 28 Truck: Brill 79E	Wheel base: 1.98m ? Motors: 2x GE 58, 25hp ? Controllers: GE B54E ? Wheel brakes: manual In service: 1928 – 1980 14 in Museum collection

Belga 16-18

Type 2-axle dome roof saloon 4 indow Body: Familleureux steel framed Length: Width: Floor height: Platform height: Weight: Seats: Truck: MAN	Wheel base: 1.98m ? Motors: German, re-equipped in 1948-53 BTH 116 Controllers: German, re-equipped in 1948-53 BTH OK 33 Wheel brakes: manual, fitted with Westinghouse air in 1955. In service: 1930 – 1973 16 in Museum collection

Car 19

Type 2-axle dome roof saloon 7 window Body: CCFP *(Porto)* Length: Width: Floor height: Platform height: Weight: Seats: 28 Truck: Brill 79E	Wheel base: 1.98m ? Motors: 2x GE 58, 25hp ? Controllers: GE B54E ? Wheel brakes: manual In service: 1934 – 1970s

Car 20	
Type 2-axle dome roof saloon 7 windows Body: SMC Length: Width: Floor height: Platform height: Weight: Seats: 28 Truck: Brill 21E	Wheel base: 1.98m Motors: 2x GE 58, 25hp ? Controllers: GE B54E ? Wheel brakes: manual In service: 1940 – 1980

Car 1 at Largo de Celas on 9 September 1969. *A. G. Murray*

Two trailers were operated, numbered 1-2, thought to have been obtained from Lisboa. No 1 is an enclosed former mule car built by John Stephenson in New York, it still exists and is unrestored in the Museum collection. No 2, originally an open mule car, was rebodied in the 1930s in the style of motor cars 14 and 15 but constructed of timber and only six windows. Trailer operation in Coimbra commenced in 1919 and ceased in 1954.

Car 3 leaves Santo António for Largo da Portagem via Rua Lourenço de Almeida Azevedo on 18 August 1978. *D. Pearson*

Car 15 heading for Santo Antonio on route 3 in 1967. *J. Jordan*

Belgian car 16 in service in Avenida Sâo da Bandeira. *P. Haseldine*

Belgian-built car 18 rests on the traverser at the former depot in Rua Sofia in 19674. *J. Jordan*

Car 19 en route to Cruz de Celas (Rua Dr. Antonio Jose de Almeida) on 8 May 1972. *Michael Russell*

Trailers 2 and 1 in Rua Sofia depot on 13 October 1959. *Courtesy of the National Tramway Museum, photographer J.H. Price*

BRAGA

By Ernst Kers, Owen Brison & Pedro Mendes

Tram
Fleet List

Car 10 in the town centre in June
1962. *W R Stillman - OTA*

Brill 1-8

Type 2-axle semi-convertible 6 windows Body: J. G. Brill Company, 4.87 between bulkheads Length: Width: Floor height: Platform height: Weight: Seats: 20	Truck: Brill 21E Wheel base: 1.98m Motors: AEG Controllers: AEG Wheel brakes: manual In service: 1914 – 1963

Cars 9-10

Type 2-axle semi-convertible 6 window Body: Unknown Length: Width: Floor height: Platform height: Weight: Seats: 20	Truck: Brill 21E Wheel base: 1.98m Motors: Controllers: Wheel brakes: manual In service by 1923 - 1963

Car 11

Type 2-axle semi-convertible 7 window Body: CCFP (Porto) assembled locally Length: Width: Floor height: Platform height: Weight: Seats: 28	Truck: Brill 21E Wheel base: 1.98m Motors: Controllers: Wheel brakes: manual In service: 1937 – 1963

Trailers 1-7

2–axle cross bench	

Trailers 8-10, 12? 13?

2–axle enclosed saloon	

Trailer 9

2–axle saloon Probably a new body following the 1923 accident	

Trailer 11

2–axle cross bench, with turtle roof	

Car 11 on a private hire on 6 September 1962. W C Janssen - *OTA*

Trailer 6 at Bom Jesus at an unknown date. *Brian G Dutton/Online Transport Archive*

Trailer 9 with three recently delivered ex-Heilbronn trolleybuses in the depot in the early 1960s. This car is believed to be the one involved in the 1923 accident and subsequently rebuilt. *Brian G Dutton/Online Transport Archive*

Trailer 11 at the depot on 8 September 1962. The origin of this trailer is unknown. *W C Janssen/Online Transport Archive*

Trailer 12 with trailer 9 behind and another unidentified trailer at the rear, at the depot. *W C Janssen/Online Transport Archive*

Mail van 1, one of two in the fleet, seen at Estação in 1949. *Courtesy of the National Tramway Museum, photographer N.N Forbes*

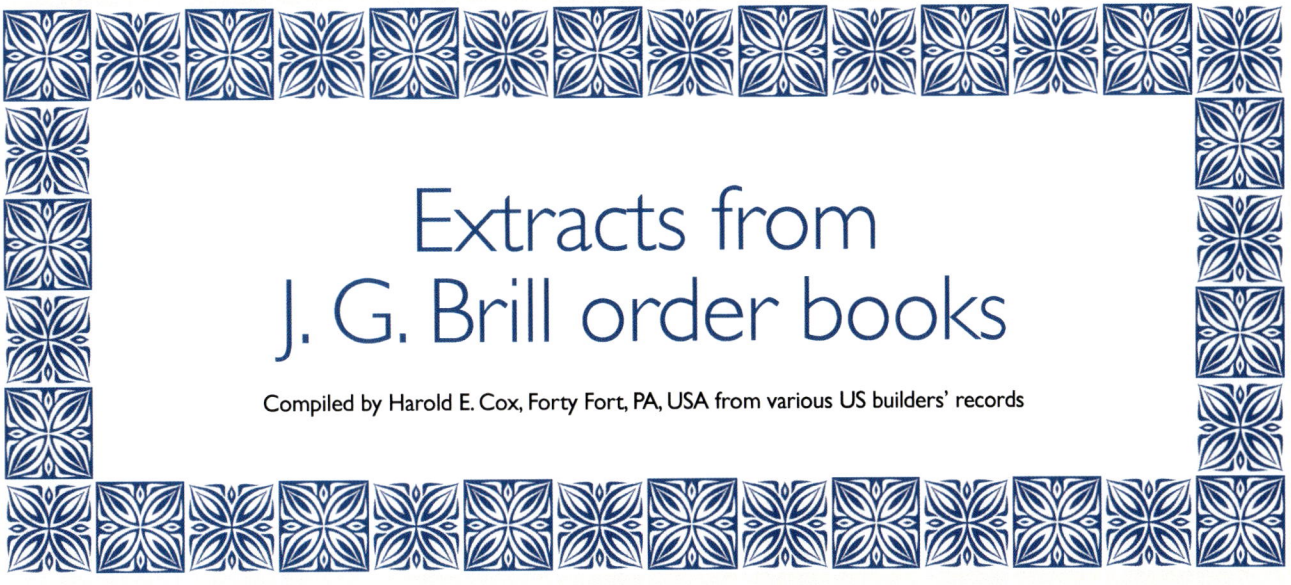

Extracts from
J. G. Brill order books

Compiled by Harold E. Cox, Forty Fort, PA, USA from various US builders' records

Brill Order	Date of Order	Number and detail of cars				Finish Date	Builder photo	Probable Fleet Nos	Notes
		No	Type	Truck(s)	Description				
Lisboa									
7888	07.07.97	20	2xT	pedestal	8-bench open	1898-9	193	183-202	A
9679	23.09.99	80	2xM	21E	8-bench open	1901	214	203-282	B
10320	10 05.00	2	2xM	No 7	sprinkler car	1901		not known	C
SL84	16.11.99	75	2xM		5468-mm enclosed	1901		400-474	L
10623	24-09-00	75	(2x)	21E	trucks only			400-474	L
11096	04.05.01	40	4xMmt	22E	12-bench open	1902	284	283-322	D
14449	31 05.05	20	4xMmt	22E	8534-mm SC	1906	323	323-342	D
14614	29.08.05	1	2xM	21E	7620-mm crane	1906	343		C
JS1150	07.09.06	20	4xM		8534-mmSC	1907		343-362	M
15479	07.09.06	20	(4x)	27GE1	trucks only	1907		343-362	M
15561	26.10.06	5	4xMmt	22E	2nd-cl	1907	365	363-367	D
16026	24.05.07	1	2xM	21E	freight car	1907	501		C
18038	02.11.11	7	2xM	21E	5334-mm enclosed	1912	476	476-482	
18090	26.12.11	3	(2x)	21E	trucks only	1912			
18168	03.02.12	2	(4x)	22E	trucks only	1912			
18938	14.04.13	12	2xM	21E	5334-mm enclosed	1913-4	489	483-494	
18940	14.04.13	3	2xM	Radiax	6947-mm enclosed	1913	501	500-502	
19055	14.06.13	5	2xM	21E	5334-mm enclosed	1914		495-499	
19057	14.06.13	5	2xM	21E	5334-mm enclosed	1914	503	503-507	
20772	20.06.19	12	2xM	21E sl	trucks only	1919			
22287	17.08.25	12	2xM	21E sl	trucks only	1925		520-531	
22435	19.06.26	12	2xM	21E	Estrela enclosed	1926	602	601-612	

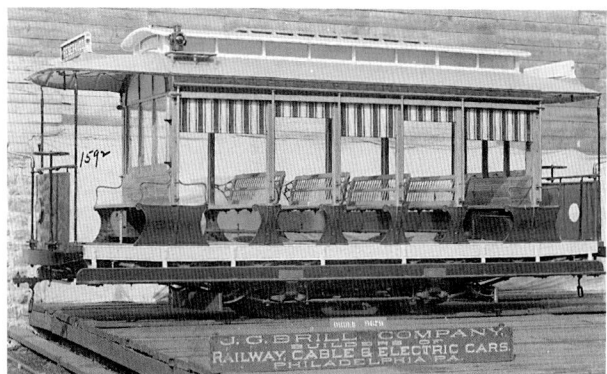

An official J G Brill photograph taken at their factory of car 214. Although the cars were shipped "knocked down and in the white" a sample was completed to check assembly and obtain a photograph. *J. G. Brill – Historical Society of Pennsylvania*

The sparse interior of workman car 365. *J. G. Brill – Historical Society of Pennsylvania*

"Radiax" Car 501. *J. G. Brill – Historical Society of Pennsylvania*

Car 602, body only, *J. G. Brill – Historical Society of Pennsylvania*

Car 503. *J. G. Brill – Historical Society of Pennsylvania*

Works car 343 with two cranes installed, a similar car was supplied with no cranes. *J. G. Brill – Historical Society of Pennsylvania*

Car 365 as built, a workman car without glazing but fitted with canvas blinds. *J. G. Brill – Historical Society of Pennsylvania*

Brill Order	Date of Order	Number and detail of cars				Finish Date	Builder photo	Probable Fleet Nos	Notes
		No	Type	Truck(s)	Description				
Porto									
8687	22.08.98	6	(2x)	21E	trucks only	1898		various	Q, Y
9342	15.05.99	5	(2x)	21E	trucks only	1899		various	Q, V
11611	27.12.01	5	(2x)	21E	trucks only	1902		various	F, Q
12511	27.01.03	4	(2x)	21E	trucks only	1904?		various	X
13052	29.08.03	12	(2x)	21E	trucks only	1903		various	S, X
13592	03.05.04	1	4xMmt	22E	12-bench open	cancelled			H
13594	03.05.04	3	(2x)	21E	trucks only	1904		various	S, X
13661	07.06.04	1	4xM	27G1	7772-mm SC	1904		90	G, H, S
13986	02.11.04	3	(2x)	21E	trucks only	1905		various	S, X
14087	11.01.05	6	(2x)	21E	trucks only	1905		various	S, X
14725	20.10.05	6	(2x)	21E	trucks only	1906		various	S, X
14906	06.01.06	2	(2x)	21E	trucks only	1906		various	S, X
16028	24.05.07	2	(2x)	21E	trucks only	1907		199-200	N, S
16511	08.10.08	6	(2x)	21E	trucks only	1909			S, W
16717	12.04.09	20	2xM	21E	5563-mm SC	1909		251-270	G, R
16824	07.07.09	6	(2x)	21E	trucks only	1909		various	R, U
17076	31.12.09	2	(4x)	23D	trucks only	1910		1	I, N
17092	17.01.10	8	(2x)	21E	trucks only	1910		various	S, T
17162	23.02.10	6	(2x)	21E	trucks only	1910		231-236	P, S
17302	11.05.10	25	2xM	21E	5893-mm SC	1910		271-295	G, S
17568	01.12.10	12	(4x)	23D	trucks only	1911		2-7	I, N
18184	15.02.12	12	(2x)	21E	trucks only	1912		237-248	P, R
18349	21.05.12	20	2xM	21E	5969-mm SC	1912		171-190	G, S
21180	21.07.20	30	2xM	21E	5969-mm SC	cancelled			

An official J G Brill photograph taken at their factory of one of the first batch of twenty Brill 23 type cars supplied to Porto in 1909. Although the cars were shipped in "knocked down and in the white" a sample was completed to check assembly and obtain a photograph. *J. G. Brill – Historical Society of Pennsylvania*

An interior view of one of the first batch of twenty Brill 23 type cars supplied in 1909. *J. G. Brill – Historical Society of Pennsylvania*

Brill Order	Date of Order	Number and detail of cars				Finish Date	Builder photo	Probable Fleet Nos	Notes
		No	Type	Truck(s)	Description				
Braga									
19136	04.10.13	8	2xM	21E	4887-mm SC	1914		1.8	

Brill Order	Date of Order	Number and detail of cars				Finish Date	Builder photo	Probable Fleet Nos	Notes
		No	Type	Truck(s)	Description				
Coimbra									
16980	16.10.09	5	2xM	21E	4877-mm SC	1910	1	1-5	
18033	27.10.11	2	2xM	21E	4877-mm SC	1912	6	6-7	
22372	03.03.26	1	2xM	21E	4877-mm SC	1926	8	8	
22571	30.04.27	5	2XM	21E	10-bench open	1927		9-13	G
22572	30.04.27	2	2xM	21E	4877-mm SC	cancelled			
22623	09.09.27	2	2xM	79E1X	5613-mm enclosed	1928		14-15	G
	?.?.34	1	2xM	79E1X	truck only	1934		19	

An official J G Brill photograph of Coimbra car 1 taken at the factory.

J. G. Brill – Historical Society of Pennsylvania

An official J G Brill photograph of one of the Sintra saloon motor cars. The car appears to have a large oil headlamp. *J. G. Brill – Historical Society of Pennsylvania*

An official J G Brill photograph of one of the Coimbra series 9-13 seen in white primer at the factory. *J. G. Brill – Historical Society of Pennsylvania*

An official J G Brill photograph of the interior of one of the Sintra saloon motor cars. *J. G. Brill – Historical Society of Pennsylvania*

An official J G Brill photograph of one of the Coimbra series 14-15 seen in white primer at the factory. *J. G. Brill – Historical Society of Pennsylvania*

An official J G Brill photograph of two Sintra crossbench cars, a trailer on the left and motor car on the right. *J. G. Brill – Historical Society of Pennsylvania*

Brill Order	Date of Order	Number and detail of cars				Finish Date	Builder photo	Probable Fleet Nos	Notes
		No	Type	Truck(s)	Description				
Sintra									
12642	05.03.03	4	2xM	21E	8-bench open	1903	B13		K
12644	05.03.03	4	2xT	pedestal	8-bench open	1903	D19		K
12646	05.03.03	3	2xM	21E	4470-mm enclosed	1903			K
12648	05.03.03	2	2xT	pedestal	4470-mm enclosed	1903			K

NOTES to Table:

A = Delivered as 900-mm mule cars to implement change of gauge; these cars ran as electric trailers from 1901.

B = The initial order for 120 cars was reduced to 80. The remainder was built as bogie cars; see order 11096.

C = Renumbered into 389-99 series in or post 1907.

D = Under "car type", mt = maximum-traction, other four-axle cars had equal wheels.

F = Trucks for ex-mule cars.

G = Photographed and shipped un-numbered.

H = Order 13592 was cancelled and replaced by order 13661.

I = 14 individual trailer bogies for seven closed trailers.

K = Original numbers not known, later numbered 1-7 (motors) and 8-13 (trailers).

L = Bodies built by the St.Louis Car Company, trucks by Brill.

M = Bodies built by J.Stephenson, trucks by Brill.

N = Bodies built by A Constructora (local Porto company)

O = Open freight trams, bodies built by A Constructora

P = Open freight trams, bodies built by Aliança (local Porto company)

Q = Motors from Walker

R = Motors from General Electric (GE80)

S = Motors from Siemens-Schuckert

T = One enclosed built by A Constructora no.250; one closed freight tram built by Aliança (unknown number); six trucks with unknown use.

U = Two closed freight trams built by Aliança nos.221-222; four open freight trams built by A Constructora nos.227-230.

V = Two trucks were fitted with 2x50 hp, and three trucks with 2x25 hp Walker motors. The 2x50 hp trucks were likely used for the "rebocadores" for hauling freight trailers. For the other three, see Note Y. Order sent by Carris on 4-5-1899.

W = Likely replacement of older non-Brill trucks of older tramcars.

X = Of the 36 in the period 1903-1906 delivered 21E trucks, likely 26 (38-39, 43-54, 57-66, 68-69) were for new built cars, 4 (83-86) for new freight cars, 4 (40-42, 67) for electrifying mule cars and 2 (55-56) as replacement of older, non-Brill trucks. It's not possible to combine exactly the orders with the car numbers.

Y = Former mule cars and new built trams with the bodies to the same design as mule cars. Exact use not known.

ABBREVIATIONS: JS = J. Stephenson; SL = St. Louis; SC = semi-convertible; M = motor cars; T = trailers; x = axles; sl = swing link

Lisboa maximum traction bogie car 323. *J.G.Brill – Historical Society of Pennsylvania*

Interior of maximum traction bogie car 323. *J.G.Brill – Historical Society of Pennsylvania*

1904 original semi-convertible bogie car supplied to Porto. A pictured of the assembled and painted car appears on page 283. *J.G.Brill – Historical Society of Pennsylvania*

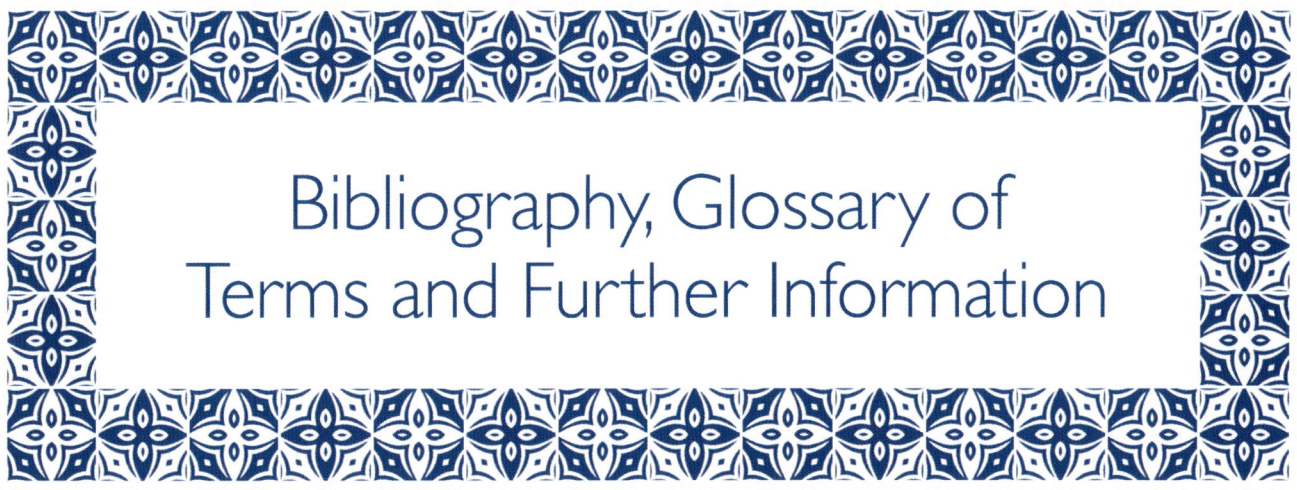

Bibliography, Glossary of Terms and Further Information

BIBLIOGRAPHY

Lisboa

Electricos de Lisboa, Aventuros sabre Carris, by Cristina Ferreira Gomes and Neni Glock (Gradiva, Lisboa 1994) This book contains a complete list of Portuguese-language sources on Lisboa tramways.

Lisboa, a cidade dos Elevadores, (CCFL, Lisboa, 1986) (Portuguese/English text)

As Rodas da Capital, by Vasco Callixto (Junta Distrital de Lisboa, 1967)

Tram Tours of Lisbon, by Joseph Abdo (Ediciôes Especializidos, Lisboa 1992)

Lisbon Area Tickets, (Transport Ticket Society Journal, November 1968)

Lisbon Electric Tramways Ltd, Reports and Accounts 1966 to 1975

Records of English Electric Co, Maley & Taunton and EMB, held by the National Tramway Museum, Crich

Situação da Frota de Electricos (tram fleet list), compiled for CCFL internal use. Various issues, 1981-1995

Transportes da Regias de Lisboa – Bases Gerais do Programa de Acção, (Ministero dos Transportes & Comunicac;oes), Lisboa April 1981

Mountain Climbing sans Rack, by J. H. Price (in Modern Tramway, March to November 1979)

Porto

Porto Origem, Evolução e Transportes by Guido de Monterey 2.A edição published by Guido de Monterey in 1972.

Os Velhos Eléctricos do Porto by Manuel Castro Pereira published in 1995 by José Carvalho Branco and Soc. Editorial da Beira Douro.

Material Circulante sabre Carris, (STCP, various issues)

Photos from the following sources:
CPF = Centro Português de Fotografia - Porto
MCE = Museu do Carro Eléctrico - Porto
SCA = Siemens Corporate Archives - Munich
Postcards and other photos are from the collection of the authors.

Coimbra

Servir Coimbre, (special number of SMC staff magazine, August 1982)

Various

The Tramways of Portugal by John Price and co-author Brian King for the 3rd and 4th editions and published by the LRTL/LRTA; 1st ed. of 1964, 2nd ed. of 1972, 3rd ed. of 1983 and 4th ed. of 1995.

Modern Tramway, various issues (especially January 1950, October 1957, October and November 1980)

Brill Magazine, 1907-1912, (Available in reprint form)

Railway Holiday in Portugal, by D. W. Winkworth (David and Charles, 1968)

The Bus Fleets of Lisbon and Oporto, (PSV Circle, 1972) (Second Edition in preparation)

Schienenseilbahnen in aller Welt, by Walter Hefti (Birkhauser Verlag, 1975)

Unkonventionaller Bergbahnen, by Walter Hefti (Birkhauser Verlag, 1978)

The Continental Steam Tram, by G. E. Baddeley (LRTL, 1980)

Zahnradbahnen in aller Welt, by Wolfgang Messerschmidt (Franck'sche Verlag, 1972)

Portugal Antigo e Moderno, by A. Barbosa de Pinho Leal (pages 479-485, Linhas ferreas americanas)

Gazete dos Caminhos de Ferro

Continental Railway Journal, various issues

50 Anos Sorefame, (special number of APAC Journal Bastao Pilato, July 1993)

Trolleybus Magazine, issues 62, 168, 169 and 189 (1972-91)

United Electric Car Company, works manager's monthly reports, 1908-14

Reports of Ministerio das Obras Publicas

GLOSSARY OF TERMS

These are mostly tramway and railway terms that are frequently encountered.

Aberto	open
Agulha	points
Agulha a abrir	facing points
Agulha a fechar	trailing points
Agulheiro	pointsman
Aluguer	on hire
Apeadeiro	halt
Areia	sand
Ascensor	funicular
Atenção	warning
Atrelado	trailer
Autocarro	motor bus
Automotora	railcar
Aviso	notice
Bifurcação	junction
Bilhete	ticket
Bilheteira	ticket office
Cais	platform, quay
Caminho de ferro	railway
Carreira	route, service
Carril	rail
Carro electrico	tramcar
Carruagem	railway coach
Chegada	arrival
Circulação	circular route
Cobrador	conductor
Combinador	controller
Combóio	train
Companhia	company
Completo	full
Correspondencia	Metro interchange
Desdobramento	duplicate
Eixo	axle
Elevador	funicular, lift
Entrada	entrance
Escada mecânica	escalator
Estação	station
Estação de recolha	depot
Estrângula	interlaced track
E proibido fumar	no smoking
Fechado	closed
Fiscal	inspector
Galeria	tunnel
Guardo-freio	motorman
Guia Oficial	timetable (booklet)
Horario	timetable (poster)
Informações	information
Janela	window
Lanternim	clerestory
Linha	line, track
Lugar	seat
Mapa da rede	system map
Máquina	locomotive
Metropolitano	underground railway
Monitor	instructor
Motorista	driver
Oficina	workshop
Operário	workman
Paragem	stopping place
Paragem zona	fare stage
Partida	departure
Passageiro	passenger
Porta	door
Preço do bilhete	fare
Rede	network, system
Reservado	reserved
Retrete	toilet
Ramal	branch line
Roda	wheel
Saída	exit
Salva-vidas	lifeguards
Tabela de preços	fare list
Tensão	voltage
Tracção a vapor	steam traction
Tracção electrica	electric traction
Transferência	transfer ticket
Tranvia	suburban train
Travões	brakes
Trolleicarro	trolleybus
Velocidade	speed
Via dupla	double track
Via estreita	narrow gauge
Viagem	journey
Via larga	broad gauge
Via simples	single track
Zorra	all-purpose freight vehicle

FURTHER INFORMATION

Websites

Many of the tramway and light rail operators in Portugal have dedicated websites to provide information for those wishing to visit and travel around by local public transport.

Listed below are the websites for the main urban transport operators.

City/Town	Operator	Website	Note
Lisboa	Carris	www.carris.pt	A
Almada	Metro Transportes Sul do Tejo (MTS)	www.mts.pt	
Porto	STCP	www.stcp.pt	A
Sintra	Sintra Atlántico	www.cm-sintra.pt	
Coimbra	SMTUC	www.smtuc.pt	
Braga	Transportes Urbanos de Braga	www.tub.pt	

Note A. These websites include a link to the local tramway museum.

All of the above listed websites provide information on tickets and tariffs, including day tickets, and have some maps that may be downloaded.

Social Media

Social media platforms such as Facebook provide a wealth of information about trams and light rail systems and Portugal is no exception. Currently there are Facebook groups for Lisboa, Sintra, Coimbra, Porto and Braga as well as more general groups covering Spain and Portugal and tramways and light rail around the world.

NOTES

NOTES

NOTES

The Light Rail Transit Association

Advocating modern tramways and light rail systems

LRTA
Since 1937

The LRTA is an international organisation dedicated to campaigning for better fixed-track public transport, in particular tramways and light rail. The Association celebrated its 80th anniversary on 30 June 2017.

Membership of the LRTA is open equally to professional organisations, transport planners and individuals with a particular interest in the subject. Members receive free of charge by post *Tramways & Urban Transit*, the Association's all-colour monthly magazine, as part of their subscription. With tramway and light rail systems being adopted not only in Europe but world-wide, this high-quality journal features topical articles and extensive in-depth news coverage as well as trade news and readers' letters. Details of local meetings in the British Isles are also included.

The LRTA also publishes *Tramway Review* – a quarterly journal devoted to historical material.

Officers of the Association – many with transport industry experience – form part of an extensive network of light rail and tramway information sources, which includes the comprehensive LRTA library.

For more information visit our website: **lrta.org**

To become a member of the LRTA go to: **lrta.info/shop** or e-mail **membership@lrta.org**
Postal address: **LRTA Membership, 38 Wolseley Road, SALE, M33 7AU**

For general enquiries contact: **secretary@lrta.org**
Postal address: **LRTA Secretary, 8 Berwick Pace, Welwyn Garden City, Al7 4TU**

To order copies of our wide range of books go to: **lrta.info/shop**
Orders may be sent by post to: **LRTA Publications, 38 Wolseley Road, SALE, M33 7AU**

Books due to be published by the LRTA over the next eighteen months include:

Upper Silesia Tramway Guide

Bluebird Reborn – LCC No 1 Restored

Traditional Tramway Architecture

Modern Trams - Volume 1

Rail Based Public Transport in Canada

In addition further tramway colour albums based on the successful Vienna and Milan albums are planned on the following:

Tramways in Colour in Northern Italy   Tramways in Colour in Southern Italy

Tramways in Colour in the former DDR (3 volumes)

Tramways in Colour in Spain (3 volumes)

Potential authors of books on subjects relevant to the Association's interests are invited to contact the Publications Director of the LRTA at:
24 Heath Farm Road, FERNDOWN, BH22 8JW